An Introduction to Biological Rhythms

An Introduction to Biological Rhythms

by JOHN D. PALMER
Zoology Department
University of Massachusetts
Amherst, Massachusetts

With Contributions by

FRANK A. BROWN, JR.
Department of Biological Sciences
Northwestern University
Evanston, Illinois

and

LELAND N. EDMUNDS, JR.
Division of Biological Sciences
State University of New York at Stony Brook
Stony Brook, New York

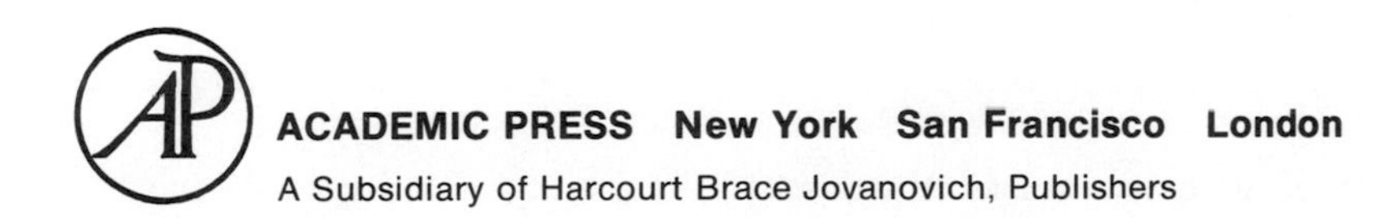
ACADEMIC PRESS New York San Francisco London
A Subsidiary of Harcourt Brace Jovanovich, Publishers

ACADEMIC PRESS, INC.
111 Fifth Avenue, New York, New York 10003

United Kingdom Edition published by
ACADEMIC PRESS, INC. (LONDON) LTD.
24/28 Oval Road, London NW1

Library of Congress Cataloging in Publication Data

Palmer, John D (Date)
An introduction to biological rhythms.

Includes bibliographies and index.
1. Biological rhythms. I. Brown, Frank Arthur, (Date) joint author. II. Edmunds, Leland N., joint author. III. Title.
QH527.P34 574.1 75-36653
ISBN 0-12-544450-8

PRINTED IN THE UNITED STATES OF AMERICA

to students entering the field: may one of you find the clock

and

to my parents, who introduced me to
nature, never tried to point me
in any direction, and watched
quietly while I made my first
crude estimates of which end
was really up: thanks

Contents

Preface xi

Biographical Sketch of Frank A. Brown, Jr. xv

Biographical Sketch of Leland N. Edmunds, Jr. xvii

1 INTRODUCTION TO BIOLOGICAL RHYTHMS, THEIR PROPERTIES, AND CLOCK CONTROL

Tidal, Daily, and Annual Rhythms 1
Rudimentary Terminology 6
Properties of Clock-Controlled Rhythms 7
The Coupling Unit 26
Master or Multiple Clocks 27
Summary and Conclusions 28
Selected Readings 30

2 A SURVEY OF RHYTHMS IN PLANTS AND ANIMALS

Plants 31
Animals 60
Summary and Conclusions 90
Selected Readings 92

3 TIDAL (BIMODAL LUNAR-DAY) RHYTHMS

A Survey of Tidal Rhythms 93
Determination of Phase 107
Summary and Conclusions 118
Selected Readings 119

4 HUMAN RHYTHMS

Rhythms in Alcohol Metabolism 120
Rhythms in Pain Tolerance 122
Man as an Experimental Subject 123
Sleep-Wakefulness Rhythms 125
Temperature and Performance Rhythms 131
Rhythmic Moods 143
Cell Division Rhythms 143
Heart-Rate Rhythms 145
Rhythms in Renal Processes 149
Rhythms in Birth and Death Rates 153
Rhythmic Sensitivity to Allergens and Drugs 155
The Effect of a 10 Hz Electric Field on Human Rhythms 157
Longitudinal Travel and Biological Rhythms 160
Shift Work 169
Monthly Rhythms 169
Summary and Conclusion 173
Selected Readings 174

5 CLOCK COMPENSATED ANIMAL ORIENTATION

Sun-Compass Orientation 175
Star-Compass Orientation 187
Moon-Compass Orientation 187
Summary and Conclusions 190
Selected Readings 191

6 THE CLOCK CONTROL OF PLANT AND ANIMAL PHOTOPERIODISM

Plants 192
Animals 201
Summary and Conclusions 207
Selected Readings 208

7 EVIDENCE FOR EXTERNAL TIMING OF BIOLOGICAL CLOCKS

FRANK A. BROWN, JR.

Comparison of the Internal and External Hypotheses 209
Rhythms in the Fiddler Crab 212

Translocation Experiment with Oysters 215
Statistical Analysis of Time Series 217
On the Absence of Constant Conditions 219
Responsiveness to Very Weak Magnetic Fields 222
Responsiveness to Very Weak Electric Fields 232
Responsiveness to Background Radiation 233
Geoelectromagnetic Fields and "Clocks" 236
Lability of the Rhythmic Cycles 239
Phase Maps and Phase Dissociation 241
Solar-Day and Lunar-Day Rhythms 244
Autophasing 247
Geophysically Dependent Rhythms 255
Precision of Solar-Day Cycles 261
Propensity for Lunar Periodisms 264
Annual Rhythms 270
External Timing as a Scientific Hypothesis 270
General Conclusions 273
Summary 276
Selected Readings 278

8 MODELS AND MECHANISMS FOR ENDOGENOUS TIMEKEEPING

LELAND N. EDMUNDS, JR.

Introduction 280
Formal Models for an Autonomous Clock 284
Cellular and Biochemical Clock Mechanisms 303
The Intrinsic versus Extrinsic Timing Problem 356
Summary and Conclusions 357
Selected Readings 359

Glossary 363

Index 367

Preface

When I first became interested in the subject of biological rhythms, I was still a graduate student. In discussing my new-found interest with other students and members of the faculty, I found that our conversations centered mainly on whether or not rhythms even existed. Presenting a convincing argument at that time was very difficult for three reasons: (1) I had to assert that Claude Bernard's almost sacrosanct concept of a virtual straight-line homeostatic constancy was in fact not accurate but had to be revised to a rhythmic stasis; (2) I had to confront them with a paradox—biorhythms such as activity cycles persisted in the laboratory in the *absence* of day–night cycles in light and temperature; and (3) I had not yet become a "card-carrying" Ph.D. and so had not received my assigned credibility. But all that was a decade and a half ago, and the skeptics have now been tamed and the discipline accepted as a bona fide byway of biology. The major questions now are: Where in the cell are the so-called clockworks that govern the rhythms? How do they work?

Along with the establishment of a new field come the symposia, and eight major ones, along with more than twice that many satellite meetings, have been held—in fact, in traditional biological fashion, symposium overkill is becoming a problem. Along with the meetings has come an extensive and complex literature. The field boasts two of its own journals, twenty-three major review articles, and about twenty books. All but four of the latter are technical, and serve, mainly, only the people in the discipline. The nontechnical ones are out of print. Being a fascinating subject, mention of biorhythms is made in most of the newer introductory biology books. Even courses on the subject are being offered.

That the subject is becoming a part of the life science curriculum is good, because the omnipresence of rhythms throughout the plant and animal kingdoms suggests that they represent a fundamental property of life itself. As such, the elucidation of the clockworks is a very important and basic task of biology. I would proffer that ten to twenty years after this task is completed, the discovery may well be dignified with a Nobel Prize. I mention this as an indication of how strongly I feel about the importance of the subject, and this brings us to the reason for the existence of this volume.

This is an introductory book, pitched at a level between the handful of popular books on the subject and the technical treatises. It is an attempt to distill a great deal of complex and sometimes confused literature and to present the most important, lucid, and interesting information. I have tried to communicate the distillate in a straightforward uncomplicated way, and in doing so have occasionally had to simplify to a degree that may evoke some raised eyebrows from the initiated. But this is not intended to be a book for the expert. It is for the inquiring student who wants an introduction to the subject; it is for busy biologists in other fields who want to get a "feel" for the subject; and I especially hope that it will serve as a basic textbook for the existing biorhythms courses and act as a seed for the inauguration of new courses.

So that the book does not represent only my bias of what should be known about the subject and as a means of introducing the presently popular hypotheses of how the clock mechanism is thought to work, I asked two eminent scholars to write the concluding chapters. Frank A. Brown, Jr., Morrison Professor of Biology at Northwestern University, describes in Chapter 7 the timing of biological rhythms by periodic geophysical forces. And in Chapter 8 Leland N. Edmunds, Jr., Professor and Head of the Division of Biological Sciences at the Stony Brook campus of the State University of New York, outlines the evidence supporting the existence of an environment-independent, autonomous living clock. He also exposes the reader to most of the jargon and shorthand notations used in the field (these and all other rhythm terminology are defined in the Glossary at the end of the book). To further introduce these two men, I have included biographical sketches of them which follow.

I hope that students of all ages will find this a useful book and that they will be as enthusiastic about the subject as we are. I hope that some, especially those with biochemical and geophysical orientations, will be stimulated to enter the field. I also hope that among you will be the ones to unlock, finally, the secrets of the thus far enigmatic living clock.

I wish to thank Elizabeth Brooks and Judith L. L. Goodenough for their invaluable help and suggestions during the writing and production of this book.

John D. Palmer

Biographical Sketch of Frank A. Brown, Jr.

Dr. Frank A. Brown, Jr., the author of Chapter 7, is Morrison Professor of Biology at Northwestern University. He received his doctorate at Harvard in 1934, taught at the University of Illinois for three years, and then transferred to Northwestern where he has remained ever since. He served as Chairman between 1949 and 1957. He is a member of numerous, learned scientific societies. He has served as Treasurer and Vice President of the American Society of Zoologists, Vice President of the American Society of Naturalists, and President of the Society of General Physiology. He has also been a member of the Editorial Boards of four major biological journals.

For many years he has spent his summers at the prestigous Marine Biological Laboratory at Woods Hole, Massachusetts, where a great deal of the work reported in Chapter 7 was done. From 1945 through 1949 he was the head of the laboratory's invertebrate zoology course. He served as a trustee for twenty-five years and as a member of the executive committee for eleven years. He has also worked at the Bermuda Biological Station and the Mt. Desert Laboratory in Maine.

Professor Brown's early research endeavors focused on the endocrine systems of invertebrates. After fifteen years of pioneering investigations in this field, his studies on color change in the fiddler crab led him into the field of biological rhythms. The latter subject has occupied a major share of his time for the past twenty-five years, although for the last fifteen years he has added to his interests extensive work on animal orientation and biogeophysics. His vigor, perspicacity, and success in all these endeavors are reflected in his publication record and the number of doctorate students he has produced. He has directed forty doctoral projects, and is the author, coauthor, or editor of four books and over three hundred research papers and other scholarly writings. For the record, Chapter 7 is his three hundred and thirty-sixth contribution to science.

Biographical Sketch of Leland N. Edmunds, Jr.

Dr. Leland N. Edmunds, Jr., the author of Chapter 8, is Professor and Head of the Division of Biological Sciences at the Stony Brook campus of the State University of New York. He received both his Master's and Doctorate degrees at Princeton (the latter in 1964), and then moved as a National Science Foundation Research Fellow to the Organization for Tropical Studies at the University of Costa Rica. The following year he joined the SUNY network at Stony Brook where he has since remained. During the 1972–1973 academic year, he held a Visiting Investigatorship at the Carlsberg Foundation in Copenhagen.

Professor Edmunds has devoted over a decade of work to the study of rhythms in single-celled organisms, especially the cell-division cycle in *Euglena*. As a biochemist, he has been mainly interested in events occurring at the levels of cell organelles, membranes, and enzyme systems. As a working hypothesis, he has assumed that an autonomous, internal clock governs overt rhythms.

His laboratory is a very active one, producing, on the average, more than five major papers per year. Since 1969 Professor Edmunds has been an invited speaker at twelve international symposia. In the little free time he has available, he is writing a multivolume treatise on all aspects of biological rhythms, a magnum opus which is certain to become the definitive work on the subject. A preview of what it will resemble is found in Chapter 8.

1

Introduction to Biological Rhythms, Their Properties, and Clock Control

TIDAL, DAILY, AND ANNUAL RHYTHMS

As I sit here in my office writing, in the adjacent laboratory the spontaneous running activity of 20 fiddler crabs—very common, non-edible crustaceans living along our shoreline—is being automatically measured in the dark interior of an incubator. In their natural habitat on the shore, the crabs live in self-constructed burrows during high-tide inundations, but when the tide recedes, they emerge and scurry around the beaches to feed, mate, and threaten each other with their giant fiddlelike claws. With the return of the flood tide, they again retreat underground and wait complacently for the next low tide.

Conditions in the incubator are much different from the shoreline environment: there are no high or low tides; one cannot distinguish between day and night because the chamber is dark and the door is opened only once every 10 days. The temperature is held constant at 20°C. Consequently, the crabs get no inklings from these monotonous surroundings as to the state of the tides outside their prison. Yet, with amazing precision, the animals run around in their containers during the times of each low tide (Figure 1-1)! So accurate are their responses that the students working in the lab use the crab behav-

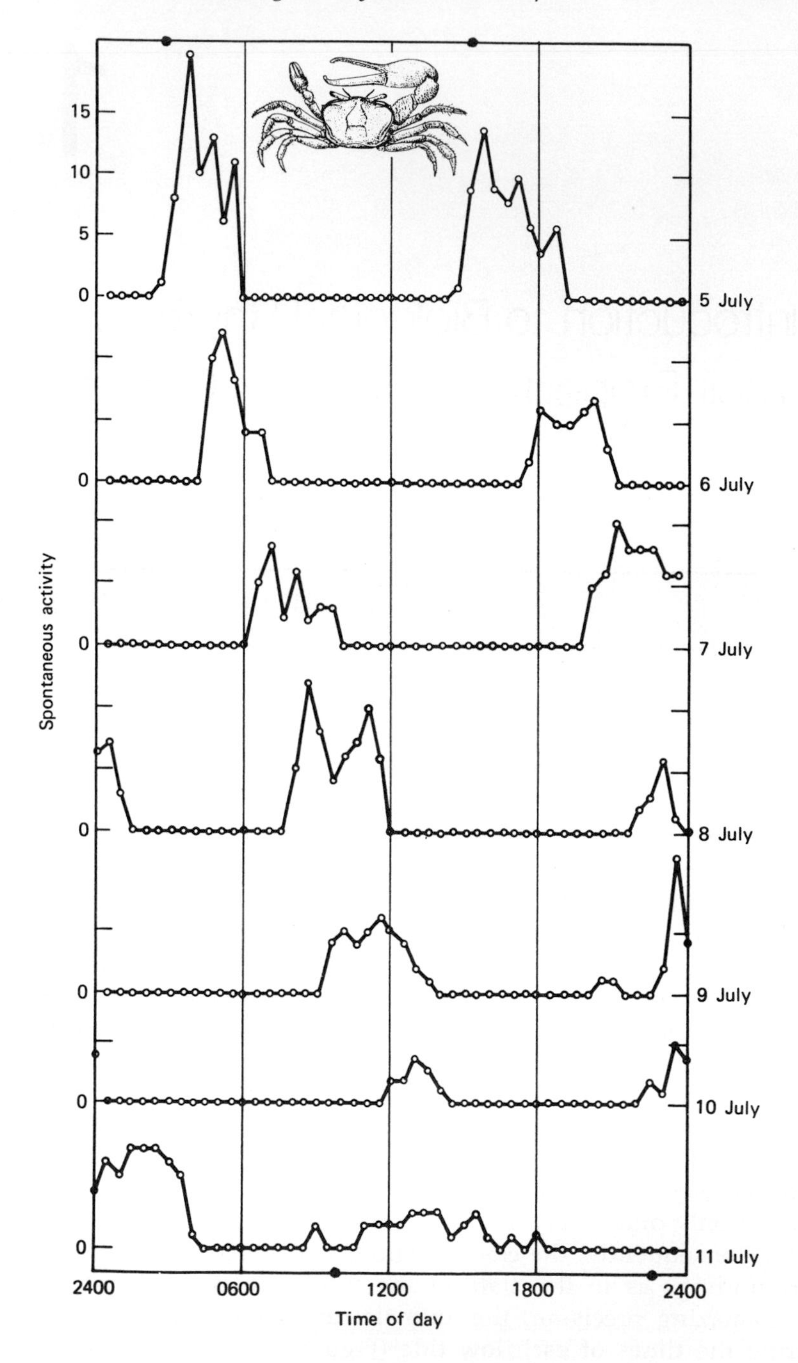
15
10
5
0
Spontaneous activity
5 July
6 July
7 July
8 July
9 July
10 July
11 July
2400
0600
1200
1800
2400
Time of day

ior patterns (which are recorded on a chart outside the incubator), rather than the tide tables of the Geodetic Survey, to plan their field trips to the crab's old home 30 miles across Cape Cod.

How do crabs do it? It is not yet known.

Because the crabs repeat their bursts of activity with such *beat*like precision, the response is called an activity *rhythm* and, as described here, since the response is synchronized with the ebb and flow of the sea, it is a *tidal* activity rhythm. When (as in this case and in those to be described in the rest of the book) the rhythmic behavior continues in the laboratory, the rhythms are referred to as being *free-running* or *persistent*. Because they do persist with such precision in the absence of the most important environmental time cues (such as high and low tides and day–night cycles), it is reasoned that within the bodies of all organisms must be a living horologe—a mechanism that has come to be called a biological clock. The nature of the clockworks is unknown.

The behavior just described is not just an interesting, but rare, oddity of nature. Almost all organisms, both plants and animals, possess some form of clock and are able to tell—if not the state of the tide—at least the time of day. And, in addition, many possess clocks that signal the day of the month and the month of the year. As an example of a daily rhythm, take the bluefish—one of the most popular game fish on the East Coast—which is an organism that displays several daily rhythms. Figure 1-2 depicts the results of a laboratory study performed under controlled conditions, where it was found that the fish are most active and have the greatest tendency to aggregate into schools during the daytime. This "discovery" is just what good bluefishermen have known for years and taken advantage of by planning their fishing excursions during the daytime.

For an example of a monthly rhythm, observe the sea-dwelling worms whose reproductive sprees are commonly associated with the phases of the moon. A dramatic demonstration of this rhythm (and a yearly rhythm also) is acted out each October and November by the Samoan Palolo worm. This animal spends most of its existence in a honeycomb of tunnels within island-bordering coral reefs, but on the day of

---

Figure 1-1 The spontaneous activity rhythm of one fiddler crab in a darkened incubator and isolated from the ebb and flow of the tide. The peaks represent bursts of activity which are approximately centered on the times of low tide on the beach where the crab was collected. As can be seen by the crab's behavior, there are two low tides each day, which occur, on the average, 50 minutes later each day. Because the crab continues to express this pattern in the absence of any environmental clues, it is said to possess a biological clock which governs this temporal aspect of its behavior. From J. Palmer, *Biol. Rev. Cambridge Philos. Soc.* **48,** 377–418 (1973).

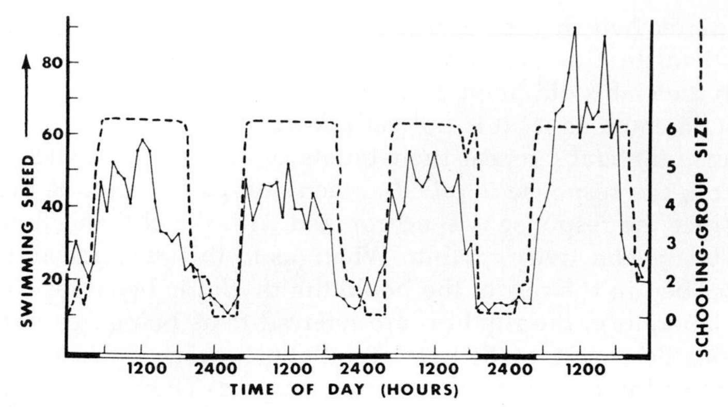

Figure 1-2 The average hourly swimming speeds (solid curve) and degree of schooling (dashed line) of 6 bluefish in the laboratory. The shaded portions of the subtending bar signify the hours of darkness, the unshaded intervals are the hours of light (a convention that will be used throughout the text). Note that the intervals of activity and schooling are greatest during the lighted portions of each of the 4 days shown. This rhythmic behavior will also persist in the absence of light-dark cycles, signifying that it is under the control of a biological clock. Plotted from the data of B. Olla and A. Studholme, *in* "Behavior of Marine Animals" (H. Winn and B. Olla, eds.), Vol. 2, pp. 303–326. Plenum, New York, 1972.

last quarter of the moon, at sunrise, one-foot-long tail segments break off the worm and swim upward to the surface. There, they churn the seawater into a frothy brothel while undergoing a decerebrate breeding dance. Then, as if on a given signal, they explode, liberating a slurry of eggs and sperm into the sea where fertilization takes place. The natives of Samoa, alert to this annelidian rhythm, are prepared in advance for the upwelling of worms, and whole villages set out to sea to scoop up the worms before they self-destruct. Then, while people in the United States consume our Thanksgiving Day turkeys, these natives feast on the very comestible, fresh-baked Palolo.

The golden-mantled ground squirrel, an obligate seasonal hibernator, provides a good example of an annual rhythm. In the natural habitat, just prior to the onset of winter, this animal lowers its body temperature so that it hovers just a degree or two above freezing (a drop of over 30°C below its normal level) and then sleeps. The heart slows to one beat per minute.

The hibernating behavior has been studied in the laboratory for years at a time; daily observations of intermittent arousal being made by the "sawdust" technique, which does not disturb the animal in any observable way. This method capitalizes on the fact that, when hibernating, these little animals remain in one position for many days at a

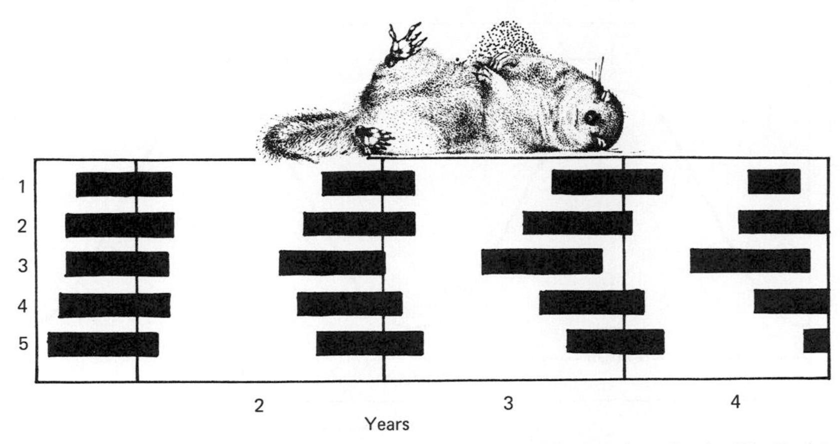

Figure 1-3 The hibernation records of 5 golden-mantled ground squirrels (*Citellus*) isolated from birth in constant darkness and 3°C. All showed hibernation rhythms with circannual periods either longer or shorter than a year. The third animal down from the top of the figure had an almost perfectly repeated cycle of 10 months. The darkened blocks signify the periods of hibernation. Modified from E. Pengelley and S. Asmundson, *Sci. Am.* **224,** 72–79 (1971); *Comp. Biochem. Physiol.* **32,** 155–160 (1970).

time. It is a simple matter to pile a mound of sawdust on their backs (Figure 1-3) and check them daily to see if it has been disturbed. When they do arouse, their first action is to shake the sawdust off. Therefore, the presence of an undisturbed pyramid of sawdust proves that hibernation was uninterrupted.

To demonstrate the control of hibernation by a living horologe, ground squirrels that had been born and raised in constant conditions in the laboratory were placed in darkness at a temperature of 3°C and observed daily for the next three years. As depicted in Figure 1-3, near annual rhythms were described by all the animals in the study.

It is important to notice that the length of the biological rhythms just discussed all match major geophysical cycles [i.e., they are a day (24 hours), a lunar day (the interval between successive moon rises (Figure 3-1) which encompass two full tidal cycles), a month (29.5 days), or a year (365¼ days) in length]. They also share other properties, which will be described in detail in the remainder of this chapter, that make these rhythms uniquely different from all other rhythmic biological processes such as the minute-to-minute heart beat and breathing cycle, the 10 cps α wave in brain activity, the periodic contraction of the intestines, the Sunday morning church attendance, etc. In this text, only the former rhythms, those that approximately mimic the most predominant geophysical cycles, will be discussed.

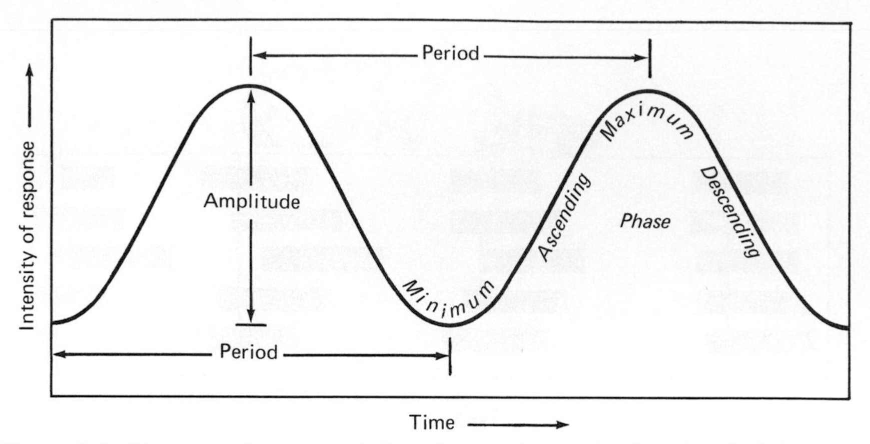

Figure 1-4 Diagrammatic representation of several aspects of cycles. Definitions are given in the text.

RUDIMENTARY TERMINOLOGY

Before delving any deeper into the subject, it would simplify matters greatly to momentarily digress to a discussion of basic terminology and the standard means of picturing oscillatory behavior in physiological manifestations. Organismic rhythms are portrayed graphically as repeating curves in which the changing intensity of a response is plotted against the time of day. In Figure 1-4, two cycles are seen. A *cycle,* used here synonymously with *rhythm* and *oscillation,* is formally defined as a sequence of events that repeat themselves through time in the same order and at the same interval. To facilitate reference, various aspects within a single cycle are individually named. Referring again to Figure 1-4, the *period* is the time interval of one complete cycle, e.g., the time elapsed between consecutive peaks or troughs; the *frequency* is the number of cycles per unit of time; the *amplitude* is a measure of intensity and thus signifies the height of a peak in a cycle; and the *phase* is a term with a dual meaning: (i) it may be an arbitrarily chosen fractional part of a cycle, labeled as ascending, maximum, etc., in the figure, or (ii) it is a term used to describe the positional relationship between two or more cycles, e.g., "the activity portion of the fiddler crab locomotor rhythm is in phase with the times of low tide." Other specific terms will be defined as they first appear and definitions will be repeated in the glossary at the end of the text.

Rhythms are classified generally by their period lengths, e.g., tidal, daily, monthly, or annual. Each general type may then be made more specific by describing the phase, e.g., 24-hour rhythms that peak during the daytime are called *diurnal* rhythms; those in which the maximum occurs at night are *nocturnal* rhythms. Here the focus is mainly on 24-hour rhythms, about which there is the greatest amount of information.

PROPERTIES OF CLOCK-CONTROLLED RHYTHMS

The Circadian Nature of Daily Rhythms

In the natural habitat, daily rhythms are strictly 24 hours in length; as will be described shortly, they are "locked," or "entrained," to this frequency mainly by the daily light-dark cycles generated by the rotation of the earth on its axis. When plants and animals are brought into the laboratory where light and temperature cycles are precluded (called constant conditions), the rhythms often persist, but the periods usually become slightly longer or shorter than 24 hours (Figure 1-5).

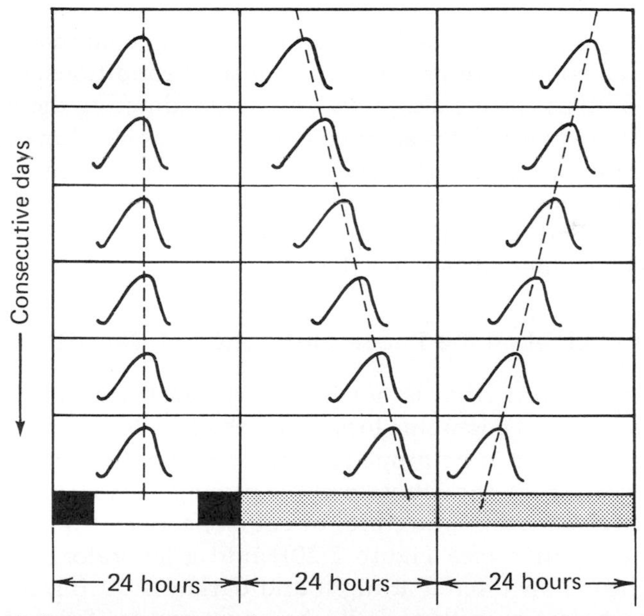

Figure 1-5 A diagram showing that in natural day-night cycles (first column), the period of a solar-day rhythm is strictly 24 hours long. In laboratory constant conditions (second two columns: constant light intensity is signified by gray blocks), the period can become longer or shorter than 24 hours; that is, its *circadian* nature is revealed. After J. Palmer, "Biological Clocks in Marine Organisms." Wiley (Interscience), New York, 1974.

When it was eventually recognized that this change in period length (which ranges between about 20 and 28 hours) almost always took place in organisms brought into the laboratory, the property was dignified with a name—it was called circadian (*circa*, about; *diem*, day + -an), i.e., rhythms *about* a *day* in length. As will be described at the end of the chapter, this property became of prime importance in the construction of explanations concerning the basic nature of the clock governing daily rhythms. However, most scientists did not recognize this importance; in fact, they did not appear to understand that they were dealing with a property of clock-controlled rhythms—one that was only displayed in the unnatural setting of the laboratory. So, they adopted the term as a snappy neologism synonymous with daily rhythm and bandied it about with an air of esoterica. Writings of one parasitologist described the rhythm he was studying as circadian, in spite of the fact that it had a period of *3 days!* How true are the words of Edward Albee: "I don't like labels: they can be facile and lead to nonthink."

The Circa Character of Other Rhythms

When organisms that display tidal, monthly, and annual rhythms in their natural habitat are brought into constant conditions, the periods of these rhythms change also, now only approximating the old natural periods. These new lengths are then generalized as circalunadian, circamonthly, and circannual. By way of example, reexamine Figure 1-1 and you will find that the period of this tidal rhythm has lengthened in constant conditions. On the other hand, most of the animals in Figure 1-3 have periods shorter than 1 year.

The Role of Light on the Phase and Period of Daily Rhythms

The phase of a rhythm is not necessarily restricted to a particular time of day at one geographic locality on the face of the earth. Instead, it can be reset so as to be properly attuned to any time zone. This can be demonstrated in the laboratory by placing a mouse exhibiting a normal activity rhythm [mice are, of course, active at night and sleep during the daytime (see Figure 2-20)] into a laboratory situation in which light is now offered at night and darkness during the day (the same lighting regimen that would be encountered if the mouse had been transported half way round the world). Over the next few days, the animal rapidly alters the phase of its rhythm so that it now sleeps during the new hours of illumination. In fact, a mouse will adjust its sleep time to any part of an artificially altered 24-hour day in which "light-on" is offered. And, if after this adjustment is complete the animal is placed into constant conditions, the new phase relationship obtains,* showing that the clock-controlled rhythm has been completely reset by the treatment with light-dark cycles. Such phase lability is a general property of daily biological rhythms.

Light cycles can also control the period length of an organismic rhythm. For example, the period can be increased by offering abnormally long "days" (say, 13.5 hours of darkness alternating with 13.5 hours of illumination), or decreased by offering abnormally short "days" (down to about 18 hours total) as seen in Figure 1-6 (A,B). However, this treatment does not have a persisting effect on the rhythm; when organisms previously subjected to such unnatural day lengths are placed in constant conditions, the period of their rhythms immediately returns to about 24 hours [Figure 1-6 (A,B)]. Moreover, if more extreme "days" are offered (e.g., 8-hour days, as seen in Figure 1-6C), the organisms "ignore" them and continue to display their natural period of about 24 hours, which emphasizes the very deep-seated nature of the interval of the rotation of the earth in relation to the sun.

THE PHASE-RESPONSE CURVE AND AUTOPHASING

The means by which organismic rhythms *entrain* (i.e., how the phase and/or period are indentured to an exogenous forcing cycle) to light-dark cycles is rather interesting: it is the result of another rhythm, this one being sensitivity to light. The rhythm has been elucidated by maintaining an organism displaying an overt rhythm (e.g., in activity) in constant darkness and then turning on the light in its quarters for a short duration—say for one hour—at some time during the day. The day after this treatment, the organism's activity rhythm is examined and sometimes, but not always, found to have been phase shifted by the one-hour light pulse. Whether or not a phase change is produced depends on what point in the organism's cycle the light pulse was offered. Additionally, when a phase change is produced, its *magnitude* and *direction* also depends on the time that the pulse is given. When the phase changes produced by interrupting (in a different experiment each time) the constant darkness for all the one-hour intervals encompassed by a single cycle of the activity rhythm are

* It should be mentioned that in constant conditions, the new phase setting remains as such only temporarily because, of course, a circadian period is assumed, and the change in period length brings about a change in phase. Therefore, the new, previously acquired phase is gradually lost as can be seen in Figure 1-5.

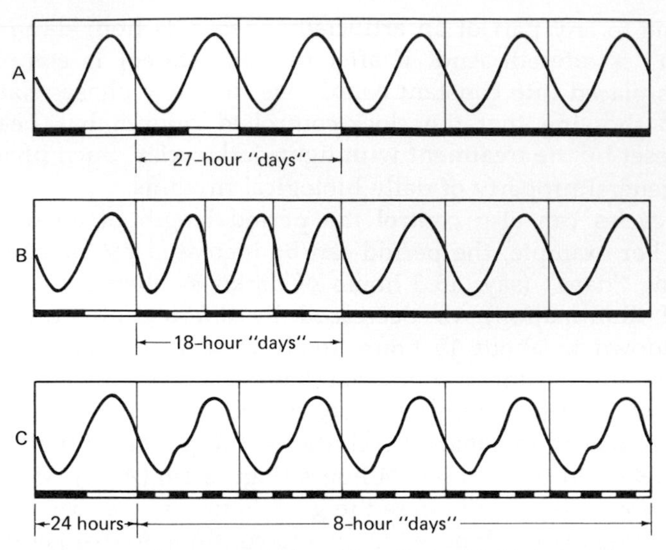

Figure 1-6 A diagrammatic representation of the effects of light-dark cycles on the period of a typical biological rhythm. (A) After many days in a 24-hour light-dark regime (of which only the last day is shown) a rhythm is subjected to two 27-hour "days" (13.5 hours of illumination alternating with 13.5 hours of darkness). The rhythm adjusts to this new regime, but only superficially, for when this treatment is followed by constant darkness, the rhythm instantly reverts to its natural period of about 24 hours. (B) The same rhythm subjected to three 18-hour "days." Entrainment occurs, but again the period reverts back to 24 hours when the rhythm is tested in constant darkness. Note in both A and B that the phase of the circadian rhythm in constant conditions has been set by the last exposure to light (i.e., the first peak in constant darkness comes 24 hours after the last peak in the artificial light-dark cycle). (C) The same rhythm now subjected to acutely extreme "days" of 8 hours. The form of the curve is distorted by the treatment, but the period remains constant at about 24 hours.

plotted, a *phase-response curve* is the result (represented diagramatically in Figure 1-7 or as actual data in Figure 8-2), which depicts a rhythm in changing sensitivity to light. The curve for all organisms is the same at least in "idea of form." It is this *light-sensitive, phase-setting* rhythm that underlies the rapid entrainment of other rhythms to a new light-dark cycle. Its action is described in Figure 1-8, where it is seen that the sensitivity rhythm is locked to the organism's overt rhythms and drags them along to a new desired phase setting when they are exposed to a different light-dark cycle. If an organism is translocated in an easterly or westerly direction across the time zones of the earth, or—in a less expensive experiment—the time that the light is turned on and off in the laboratory is changed, at least a segment of the responsive portion of the light-sensitivity rhythm is illuminated.

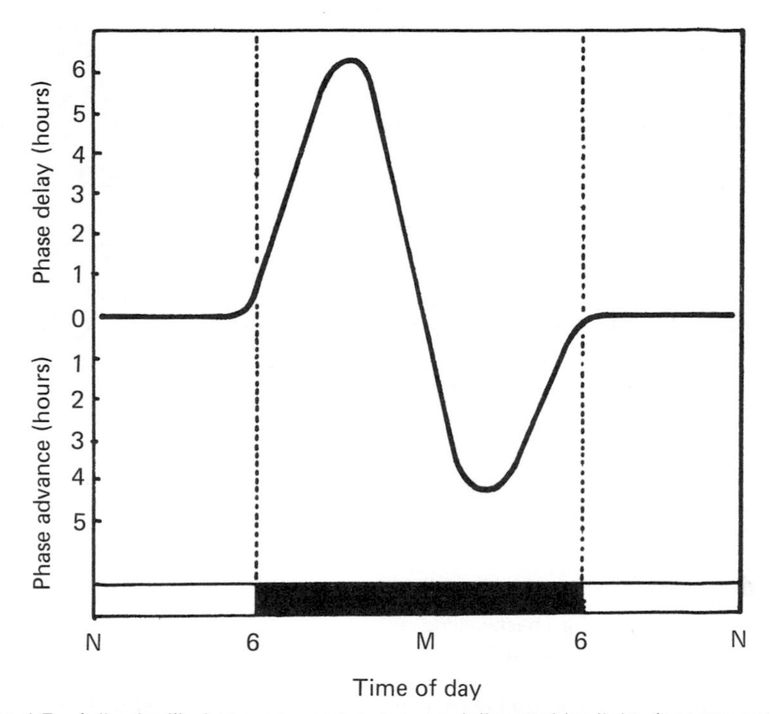

Figure 1-7 A "typical" phase-response curve, as delineated by light- (or temperature-) pulse experiments. The curve represents the amount and direction of phase change produced by short light (or temperature) pulses of a constant intensity and duration, offered at different hours throughout the day, on the overt rhythmic processes of an organism maintained in constant darkness. The black portion of the subtending block signifies the hours during which darkness would normally occur and the open blocks, those representing daylight. As indicated by the curve, pulses offered between 6 A.M. and 6 P.M. produce no phase changes; pulses offered between 6 P.M. and midnight produce delays in phase (the exact change being dependent on the time the pulse was offered); while those given between midnight and 6 A.M. produce phase advances of different magnitude.

If, for example, as indicated in the right-hand portion of Figure 1-8, the light is made to come on at midnight, instead of 6 A.M., that portion of the response rhythm causing phase advances is illuminated and this causes, over the next few days, sufficient advance of both rhythms so that the sensitive portion of the response rhythm is no longer exposed to light. Thus, entrainment to the new light-dark schedule is complete. When the opposite is done, e.g., the light not being turned on until noon and left on until midnight (or the organism is transported westerly, one quarter of the way around the earth), the delay portion of the sensitivity rhythm is illuminated (Fig-

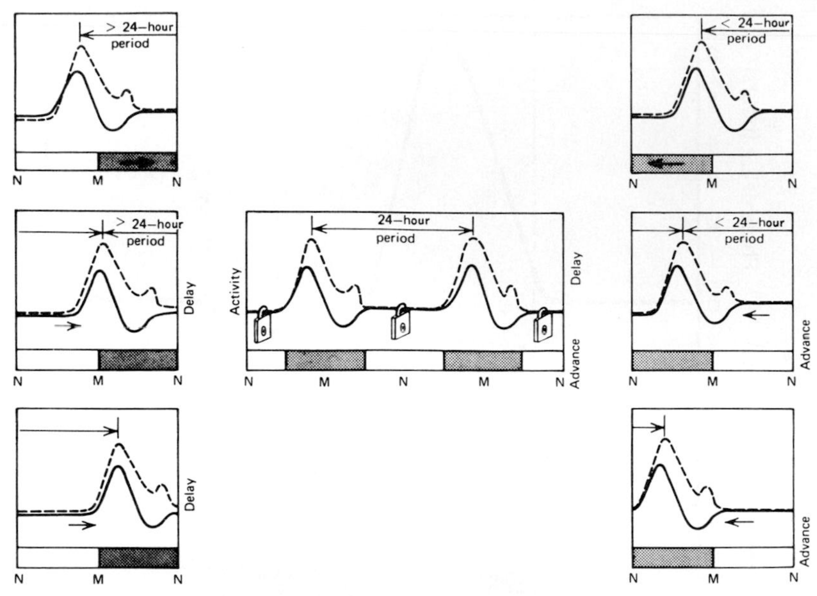

Figure 1-8 The role of a phase-setting rhythm in adjusting a clock-driven rhythm to a light-dark cycle. Center: the dashed curve represents a nocturnal overt rhythm "locked" to an underlying light-sensitive phase-setting rhythm (solid line). Because of the action of the latter, the peaks of the overt rhythm are confined to the period of darkness signified by the shaded portions along the abscissa. To illustrate the way a resetting rhythm functions, the dark portion of a laboratory light-dark cycle is made to begin 6 hours later (upper left-hand corner) than in center diagram, which then exposes the delay portion of the phase-setting rhythm to light. This produces, by the next day (middle left), a delay in the peak of the overt rhythm. These delays are continually caused until the old desired nocturnal phase relationship again obtains (bottom left). The vertical series on the right-hand side of the diagram illustrates a similar phase adjustment, but in this case the dark period has been advanced by 6 hours, thus exposing the advance segment of the resetting rhythm. Consequently, after a series of phase advances, the overt rhythm is again centered in the hours of darkness (bottom right) (N = noon; M = midnight). From J. Palmer, "Biological Clocks in Marine Organisms." Wiley (Interscience), New York, 1974.

ure 1-8, left-hand column) over an interval of a few days, successive phase delays carry the overt rhythm into the proper synchrony with the new light-dark cycle. Thus entrainment is again completed.

The light-sensitivity phase-response rhythm should also play a role in the genesis of the circadian periods when organisms are maintained in the laboratory under constant light (i.e., when the light is left on day after day and held at a constant intensity). In this setting, the entire response portion—both the advance and delay segments—of the light-sensitivity rhythm is exposed to light. This very unnatural lighting situation provides stimuli for a hodgepodge of phase ad-

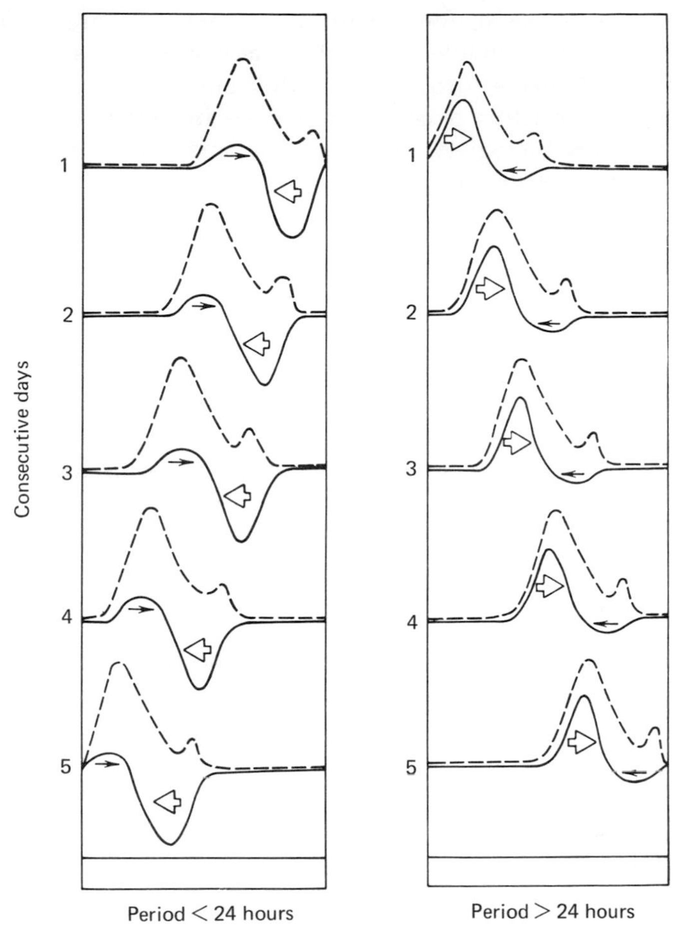

Figure 1-9 The action of a light-sensitive phase-setting rhythm in the laboratory. In constant light, both the advance and delay portions of the resetting rhythm are illuminated and the net amount and direction of "daily" resetting is determined by the relative effects of these two portions. Therefore, the phase change produced is a function of the form of the resetting rhythm. The form on the left produces phase advances, while that on the right produces delays. In the absence of light-dark cycles, the resetting rhythm cannot ever "find" the proper phase relationship with nighttime, so the search lasts eternally, and, as a consequence, produces a *circadian* period. From J. Palmer, "Biological Clocks in Marine Organisms." Wiley (Interscience), New York, 1974.

vances and delays, with the net change each day being a function of the *form* of the phase-response curve (Figure 1-9). Because the light in the laboratory is never extinguished, a similar phase alteration is produced each day; since phase advances produce a mandatory period shortening and phase delays demand the opposite, a circadian period is generated. The inevitable, daily phase adjustment has been termed *autophasing* by Frank A. Brown, Jr., author of Chapter 7.

Another interpretation of circadian periods in constant conditions is favored by many biologists. They feel that this period represents the *natural frequency* of the clock, i.e., the rate at which it normally runs when not entrained to exogenous light and temperature cycles. It is the phase-response rhythm that then molds this fundamental period into a 24-hour one, by causing the driven rhythms to swell or shrink in length to fit into the rotational interval of the earth.

The length of the steady-state circadian period assumed in constant light is additionally a function of the intensity of the ambient constant light. This relationship is portrayed in Figure 1-10, where it is seen that in some organisms (often those that are day active, such as the perching birds) the period length is decreased by higher intensities of constant light; while in others (usually night-active animals like mice, rats, and cockroaches), the period is lengthened. Very high intensities of constant light usually inhibit rhythmic expression.

Light does not play similar roles with tidal, monthly, and annual rhythms. For example, entrainment of tidal rhythms is mainly the result of temperature and pressure cycles, while the entraining stimuli for annual rhythms are not known.

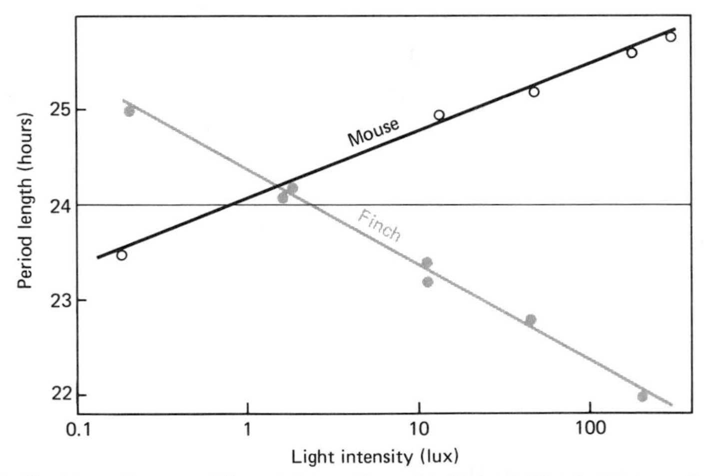

Figure 1-10 The effect of different intensities of constant illumination on the period length of circadian rhythms. The period of the mouse spontaneous locomotor rhythm increases with increasing light intensities, while the period of the activity rhythm of the chaffinch varies inversely with light intensity. Redrawn and modified from J. Aschoff, *Cold Spring Harbor Symp. Quant. Biol.* **25,** 11–28 (1960).

The Role of Temperature on Biological Rhythms

ENTRAINMENT BY TEMPERATURE CYCLES

Temperature cycles can also be used to entrain and set the phase of daily and tidal rhythms. Figure 1-11 is a diagrammatic representation of the ability of low amplitude temperature cycles to entrain a daily rhythm. As seen, the phase of the rhythm can be caused to synchronize with any hour of the 24-hour day. Moreover, when the organism is then transferred to constant light and temperature, the phase remains essentially unaltered (but see footnote on page 9), showing that the treatment has actually reset the clock-timed overt rhythm in the same manner light cycles had done.

As might be expected from the previous discussion of entrainment by light, a temperature-sensitivity phase-response rhythm (Figure 1-7) underlies and produces entrainment to temperature cycles in daily rhythms. Its action can be understood just by substituting in the legend of Figure 1-8 *warm-cold* for "light-dark," *temperature sensitive*

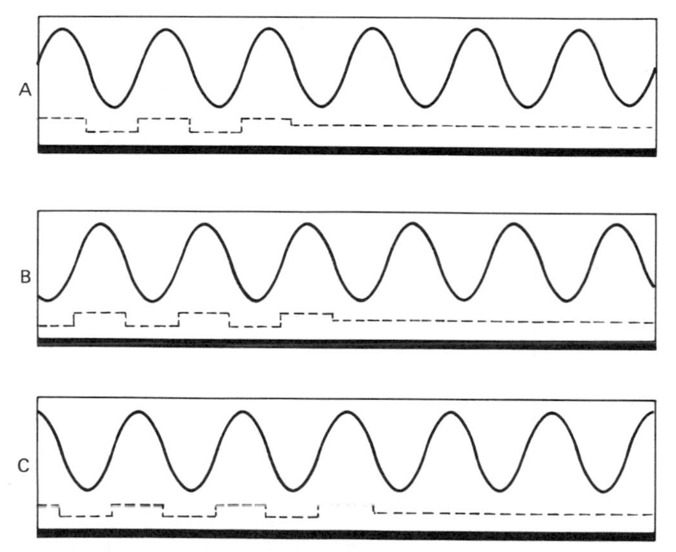

Figure 1-11 Diagrammatic representation of the effect of an 8°C temperature cycle on a biological rhythm. A, B, and C represents three different, but identical, rhythmic systems maintained in constant darkness and subjected to temperature cycles each 8 hours out of phase with one another. The square wave subtending each of the rhythm curves indicates the ambient temperature cycle; the depressed portions of the square wave signify a temperature of 20°C; the plateaus, 28°C. At the time indicated by the straightening of the temperature curves the rhythms were subjected to a constant temperature of 24°C. It is seen that the phase remained unchanged, showing that the clock had been reset by the treatment.

for "light sensitive," *cool temperature* for "darkness," and *warm temperature* for "light." Similarily, the action of this response rhythm in constant darkness can be responsible for the genesis of the circadian period in constant conditions. This form of autophasing can be understood from Figure 1-9 by substituting *temperature sensitive* for "light sensitive," *constant temperature* for "constant light," and temperature cycles for "light-dark cycles."

THE EFFECT OF TEMPERATURE ON PERIOD LENGTH

The rate at which chemical reactions run is very much a function of the temperature of the milieu in which they are confined; the higher the temperature of the reactants, the faster chemical conversions take place. As a rule of thumb, there is at least a doubling of the rate (as shown in Figure 1-12) for every 10°C rise in temperature, but even greater changes are also often recorded. This rate alteration is not only true of reactions carried out in the test tube, but also those that are constantly taking place inside plants and animals.

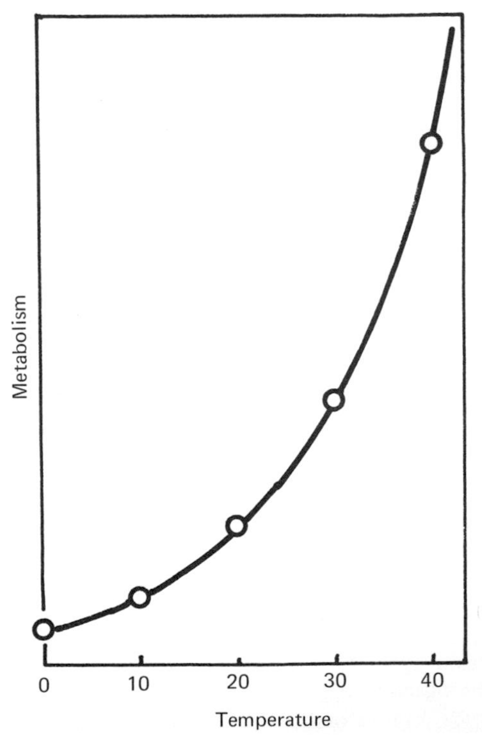

Figure 1-12 The typical exponential relationship between temperature and metabolism. In the case illustrated, there is a doubling in rate for each 10 degree rise in temperature.

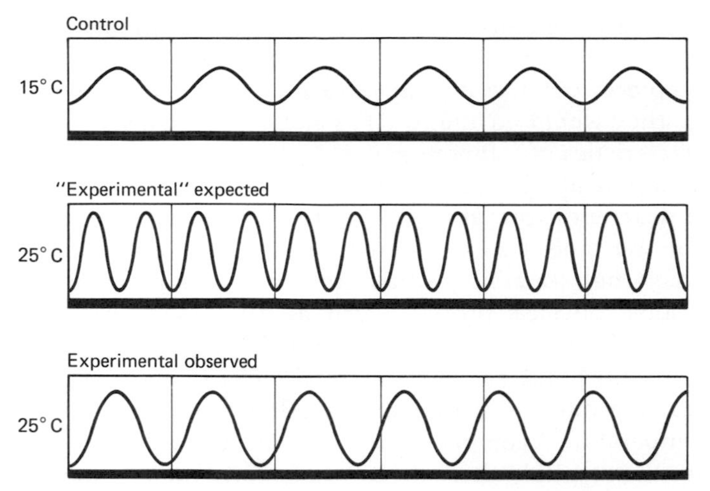

Figure 1-13 A diagrammatic representation of the expected and observed effects of a 10°C rise in temperature on a typical biological rhythm in constant conditions. With this temperature increase, the rate at which a simple biochemical clock runs would be expected to at least double, thereby generating a period in the rhythm that it governs only half as long as that seen at the previous temperature. However, as indicated in the lowest curve, the period is only slightly shortened, demonstrating its so-called virtual temperature independence. Note that while the effect on the period is slight, the amplitude of the rhythm is influenced in the expected way—it is doubled.

Most students of biological rhythms believe that the so-called clock is a biochemical oscillator of sorts. This being the case, unless it is a rather complex system, it should run faster at higher temperatures. Any increase in clock speed produced at the elevated temperature would become evident to the observer as a shortening of the period of his experimental organism's rhythm. The contraction should take place in a predictable way: as at least a halving of the period length for every 10°C rise in temperature of the biochemical clock. Whenever this type of experiment has been performed, the results are always the same (and counter to what some might first expect)—the period of the rhythm is not changed at all or is only altered slightly (usually about 10% of expected) (Figure 1-13). This minute deviation from the expected response was first deemed so unusual that the property has been described as "virtual temperature independence" of the period.*

While this response might be unexpected from a biochemical stand-

* It may be, in fact, that the period length is completely independent of temperature, but that the temperature-sensitivity phase-response rhythm produces the apparent change in period during its endless and futile search for the "proper" phase relationship with constant conditions.

point, from a pragmatic perspective, no clocks—grandfather, cesium, or living—can be sensitive to temperature changes and still function as a timepiece, because if temperature did change the rate at which they ran, they would no longer act as clocks, but instead would take on the characteristics of a thermometer, indicating the ambient temperature by the rate at which they ran.

Three different explanations, all plausible, have been posited to account for the bioclock's continued accuracy under different thermal situations. One idea assumes that the clockworks must be constructed around some physical process, such as diffusion, since the rates of physical processes are known to be much less sensitive to temperature. A second point of view is that the clock does have a chemical basis, but the rates of component reactions are either totally or partially temperature "compensated." As an illustration of this, let's examine the typical alternatives that may occur when enzyme-controlled reactions are exposed to an increased temperature. As seen in Figure 1-14, concomitant with a temperature increase, the reaction rate rapidly rises. Then one of two possible events takes place: (i) the rate remains augmented at the new level, or (ii), with time, the rate decreases toward the fiducial point. It is the latter tendency that is termed compensation and may assume three levels: *partial* (the one encountered most frequently), in which a level intermediate between the newly augmented one and former baseline is assumed; *complete* in

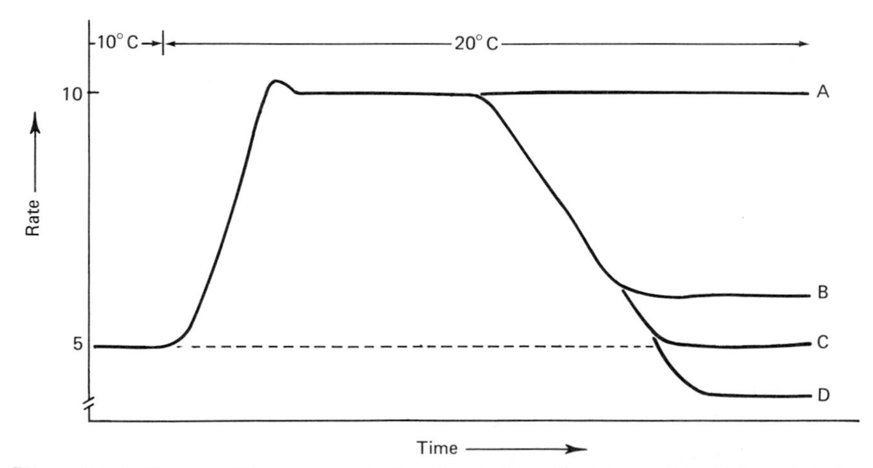

Figure 1-14 The possible responses of a chemical reaction to an abrupt increase in temperature. With the onset of a temperature elevation, a rapid increase in the reaction rate is seen, terminated by a slight overshoot and then a leveling off at a new stabilized state. After a period of time at this new temperature, any of four responses may occur: (A) no change; (B) partial compensation; (C) complete compensation; or (D) overcompensation.

which the pretreatment level is assumed again; and *overcompensation*. Therefore, when a temperature increase produces a slight decrease in the period length of a rhythm, it is thought to represent partial compensation; no change signifies complete compensation; and finding a lengthening of the period at a higher temperature (Figure 2-9) is considered a case of overcompensation.

The third suggestion to account for the absence of a large effect of temperature derives from a major hypothesis on the very nature of the bioclock itself. This line of logic, which states that the major source of timing control for biorhythms does not emanate from an autonomous clock within the organism, but from a rhythmic geophysical entity that pervades all aspects of an organism's physiology, will be introduced at the end of this chapter and is the substance of Chapter 7. Needless to say, if the clock does not reside inside the organism, then changing the body temperature of the organism could not be expected to have any effect on the clock. Again, the small effect usually observed could be a result of the action of the temperature-sensitivity phase-response rhythm.

The Resistance of Rhythms to Chemical Manipulation

Another property of biological clocks is their insensitivity to a great variety of chemical inhibitors, narcotizing agents, growth stimulants, sublethal doses of metabolic poisons, and other types of pulsed or sustained chemical insults. The generalization arising from a great many studies designed to test these categories of substances is diagrammed for one substance in Figure 1-15. Here it is seen that treatment with a sublethal dose of cyanide, which greatly reduces the metabolic rate of an organism and would therefore be expected to act similarly on a metabolic clock, reduces the amplitude of the rhythm, but has no effect on the period. This insensitivity has been found to hold true even when an organism's metabolism was reduced to 5% of normal. Since it is the period of a rhythm that is thought to reflect the rate at which the clock is running, lack of alteration signifies that the clock has probably not been altered by the chemical substance applied. These findings have been described as "expected" by many students in the field of biorhythms, who emphasize that accuracy is the most important attribute of any clock; if the rate at which a living clock runs is altered by each new chemical substance that it encounters in the environment, it would be highly inaccurate and therefore useless. As axiomatic as this statement is, rate insensitivity to chemical (and temperature also) perturbation is not a property of other pacemaker subsystems, such as those controlling heart and breathing rates.

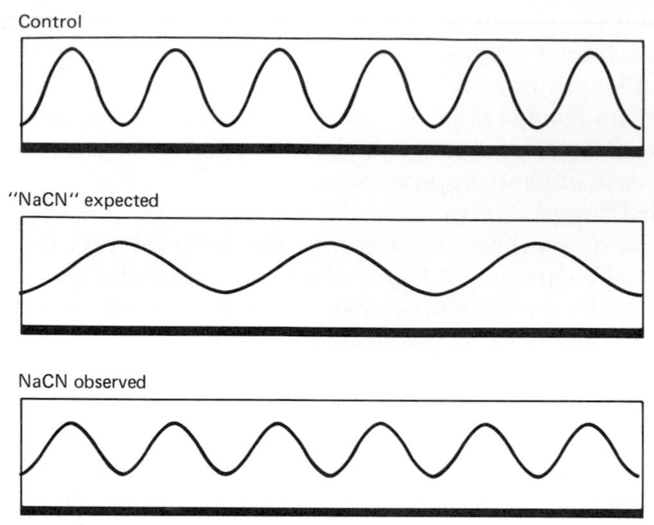

Figure 1-15 Diagrammatic representation of the effect of a sublethal dose of a metabolic inhibitor (sodium cyanide) on the period length of a biological rhythm. Such treatment might be expected to slow down the clockworks of a metabolic horologe, thereby increasing the length of the period (as portrayed by the "expected" curve). When the experiment was performed, it is seen that while the amplitude was reduced, the period was virtually unaffected.

CHROMOMUTAGENIC SUBSTANCES

In spite of the previous conclusion, recently a few substances have finally been found that will consistently produce phase and/or period changes in biological rhythms. The best known of these chronomutagenic (*chrono,* time: *mutatio,* change; *genic,* producing) substances, as they have come to be called, are deuterium oxide (heavy water); the antibiotic, valinomycin; alcohol; and an inorganic salt, lithium. Their actions will be described in Chapters 2 and 8.

Innateness of the Clock

Another major property of biological rhythms is also one that might not be expected: the period of a rhythm is not imprinted on the organism by exposure to day-night cycles of light and temperature, or the seasons. Instead, the capacity for rhythmicity is innate. This fact has been demonstrated by raising animals from birth, and seeds from the time of germination, in noncyclic laboratory conditions. The developing organisms either become rhythmic *de novo* (as already described for the ground squirrel), or they can be made to become rhythmic by subjecting them to a single, nonperiodic stimulus.

The fruit fly (*Drosophila*), that airborne pest found hovering around the summer fruit bowl, provides us with a clear-cut example. In its development from egg to adult, the fly passes through larval and pupal stages, and during the latter, the animal metamorphoses into an adult while contained in a sarcophagus-like case called a puparium. When development is complete, the new adult emerges from the puparium, inflates its neatly folded wings with blood so that they become functional flaps, and flys off. In nature, this emergence (called *eclosion*) takes place only at dawn, and this timing will persist in constant conditions. If, however, batches of eggs are laid and made to develop in constant conditions, the resulting adults eventually emerge at all times of the day, i.e., the population is arrhythmic (Figure 2-24A). It was found that even after 15 generations of arrhythmicity in constant conditions, when the developing larvae or pupae were given one nonperiodic stimulus—a light was turned on and left on (which provided no information about period length)—a rhythm was initiated in the population (Figure 1-16). Therefore, it is quite obvious that the ability

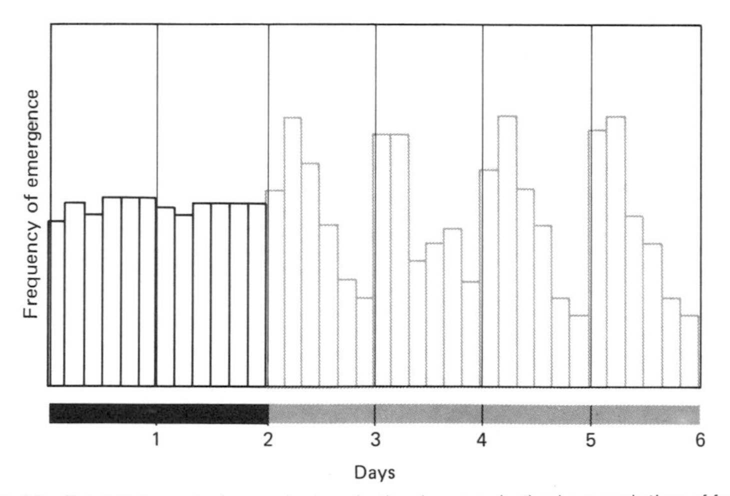

Figure 1-16 Establishment of an eclosion rhythm in an arrhythmic population of fruit flies (*Drosophila*). Flies raised from eggs laid in constant darkness emerge randomly over the 24-hour day from their pupal cases, i.e., they are arrhythmic. However, if the larvae or pupa are exposed to one, nonperiodic stimulus—such as the light being turned on and left on (as indicated by the gray bar beginning at day 2 above)—a 24-hour rhythm is established. Because no information about the interval of a day was passed to the organism by the single stimulus, it may be concluded that this interval is unlearned. From E. Bünning, *Ber. Dtsch. Bot. Ges.* **53,** 594–623 (1935).

to measure off periods of about 24 hours is an innate property of protoplasm.

Not only is the capacity to be rhythmic inherited, the phase and period length have also been shown to be under genetic control. For clear-cut examples, additional experiments with the fruit-fly eclosion rhythm will be described. To study the inheritance of the time (i.e., phase) of emergence, the 12-hour interval centered on dawn was divided into three, 4-hour blocks. For 16 consecutive generations, the flies that emerged during the first and last 4-hour blocks were isolated and bred to members of their own block. For each generation, the number emerging in the early or late blocks were compared to a control population of flies and the progressive selection for "earliness" and "lateness" plotted in Figure 1-17. It is quite obvious that the phase has a genetic basis.

The period of the rhythm displayed in constant conditions has also been shown to be genetically determined. Fruit flies were treated with ethyl methane sulfonate, a substance known to cause mutations. Of the 2000 flies thus treated, two were found to have gene-altered period

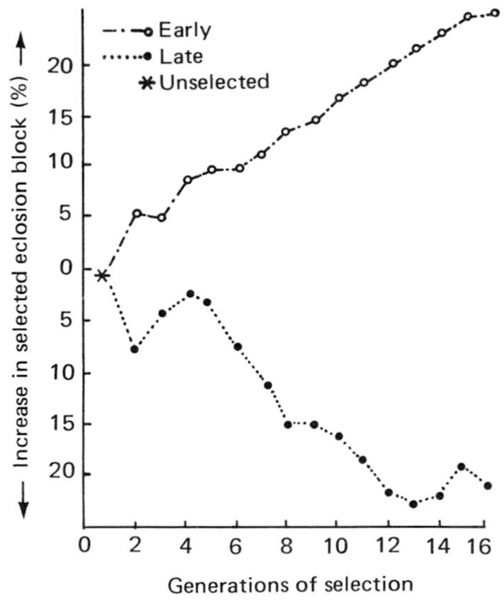

Figure 1-17 Selection for early- and late-phase eclosion strains in the fruit fly. For 16 generations, flies emerging between 6 and 2 hours before dawn were interbred, as were those that emerged between 2 and 6 hours after sunrise. The percentage of early and late emergers in successive generations was then compared to a control population not subjected to the selection routine, and the differences used to produce the above curves. Modified from D. L. Clayton and J. V. Paietta, *Science* **178,** 994–995 (1972).

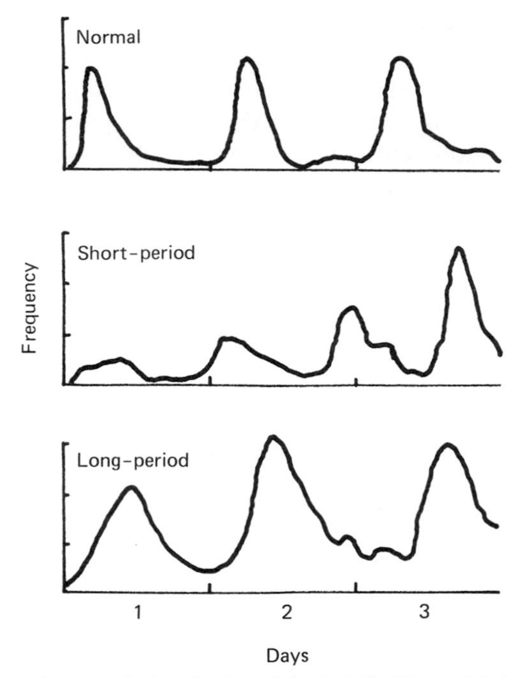

Figure 1-18 The persistent eclosion rhythm of the fruit fly (*Drosophila*). The top curve depicts the rhythm in the wild-type fly, while the lower two curves show the genetically shortened (about 19 hours) and lengthened (about 28 hours) periods of two induced mutants. Modified from R. J. Konopka, and S. Benzer, *Proc. Natl. Acad. Sci. U.S.A.* **68,** 2112–2116 (1971).

lengths in constant darkness: one had a period of 19 hours and the other 28 hours (the wild-type period is 24 hours in constant darkness) (Figure 1-18). All the mutant traits appeared to involve the same functional gene on the X chromosome. Depending on one's interpretation of the cause of the appearance of circadian periods in constant conditions, the mutation can be thought of as changing the basic frequency of the clock period, or a change in the form of the temperature-sensitivity phase-response rhythm. Either alteration would produce the change in the circadian period.

The Nature of the Clock

Rhythms that match the major geophysical periods of the earth and that persist in constant conditions are ubiquitous in their distribution throughout the living kingdom. It is thought that this clock capability arose long ago in some primitive organism, where it was found to have a significant survival value in that it could "notify" its owner in advance of coming periodic environmental events, such as sunrise, sunset, and the flood tide. It was therefore selected for in the evolutionary process and eventually became widely established in the plant and animal phyla.

Because these rhythms are so common, it is somewhat surprising that more is not known about the governing clock that underlies and makes possible their existence. As a matter of fact, not even a single component of the clock has been identified yet. This lack of factual knowledge, however, is in no way indicative of the amount of speculation present on the nature of the clock. Many models have been proposed, but conveniently, most of them can be apportioned into two general categories (Figure 1-19): the *escapement* and *nonescapement* types. (Often, in the literature, these contrasting types are referred to as endogenous and exogenous timers, but the other terminology is preferable since both types are obviously influenced by rhythmic *exogenous* factors, such as light and temperature cycles, and an exogenous clock must have an *endogenous* receiving unit.)

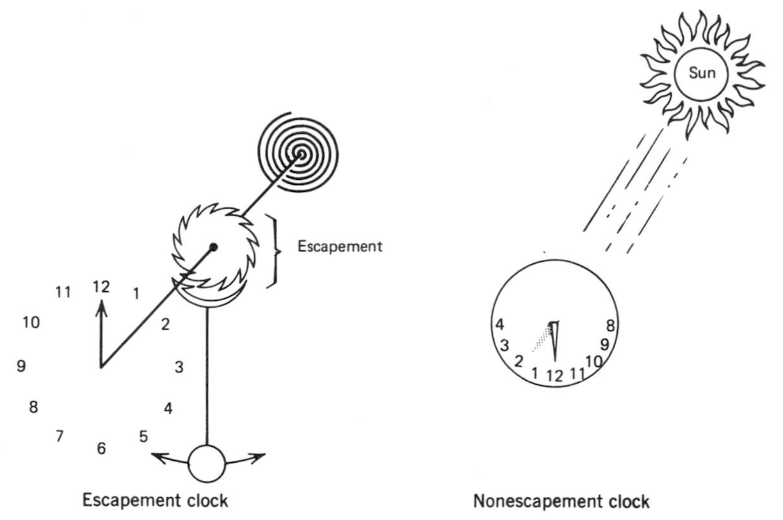

Figure 1-19 Diagrammatic comparison of escapement and nonescapement clocks. In the former, the energy provided by the spring is released in identical packages at uniformly spaced intervals via the escapement mechanism and swinging pendulum. Therefore, this type of clock generates its own interval of time. That is not the case with the sundial, which simply indicates, by the shadow cast by the gnomon, the interval information transmitted to it by an exogenous source. From J. Palmer, "Biological Clocks in Marine Organisms." Wiley (Interscience), New York, 1974.

The Escapement-Type Clock

In a man-made windup clock, the ratchet wheel and pawl escapement (Figure 1-19) parcel out the energy stored in the spring in evenly spaced intervals, therefore, the clock is autonomous. The living version of this type is also thought to be autonomous with a period that usually differs somewhat from its geophysical counterpart. Properties assigned to the "escapement" of this clock are a built-in protective panoply against chemical perturbation and a compensating governor that corrects for changes in temperature. The clock is thought to be a biochemical or biophysical entity, capable of surviving replication during cell division without losing time. Because of the widespread distribution of rhythms throughout the living kingdom, it is thought that this type of clock would have arisen aeons ago. This being the case, over the years it would have had to alter its period to counteract for the slowing down of the rotational rate of the earth (during the last 600 million years the days have become 3 hours longer). Thus far, this clock has not been located within the cell. Chapter 8 will present a detailed account of this general hypothesis.

The Nonescapement-Type Clock

This biohorologe is thought to be more like a signaling type of clock, such as sundials and electric clocks that do not generate their own intervals of time (i.e., they are not autonomous), but instead simply receive time-interval information (from the apparent movement of the sun, and the 60 Hz line current, respectively) and transmute it into time information for the observer. The living version of this clock type is postulated to exist at the cellular level and to be sensitive to changes in periodic geophysical forces that are capable of penetrating the confines of standard constant conditions and organismic structure. The force would be expected to convey period information of 24- and 24.8-hour and 29.5- and 365.25-day intervals to the clock, which, in turn, would use it to mold physiological processes into rhythms. Permeating, periodic forces such as cosmic radiation, geomagnetism, electrostatic fields, and electromagnetic radiation have all been studied for possible causal connections.

Temperature "insensitivity" and the general immutability by chemical manipulations would be expected properties of this type of clock, since the ultimate timing source lies outside the organism. Autophasing in constant conditions would transmute the fundamental geophysical periods into circa ones. The fact that only the rhythms that persist in constant conditions and whose periods are insensitive to temperature and chemical perturbations are the ones whose basic periods match the major geophysical cycles of the earth suggest that subtle geophysical forces indeed may be involved.

Thus far, neither the receiving clock, nor the periodic geophysical force providing the timing information, has been found. Research in this area is exceedingly difficult because investigators must have combined expertise in biology and geophysics. There is also the ominous prospect that the sought-after force may not even be known to physics yet. The details of this model will be discussed in Chapter 7.

THE COUPLING UNIT

In the early days of biological rhythms study, the "clock" was thought to be simply a rhythmic segment in the chain of chemical events underlying an oscillatory process. So, photosynthetic, bioluminescent, and metabolic rhythms (Chapter 2) were "dissected" in search of these rate-altering subunit reactions. The steps were never found, but what surfaced instead was that the clock, whatever its tangibility, was an entity in itself, completely separate from processes such as activity and photosynthesis, but it was coupled to them in

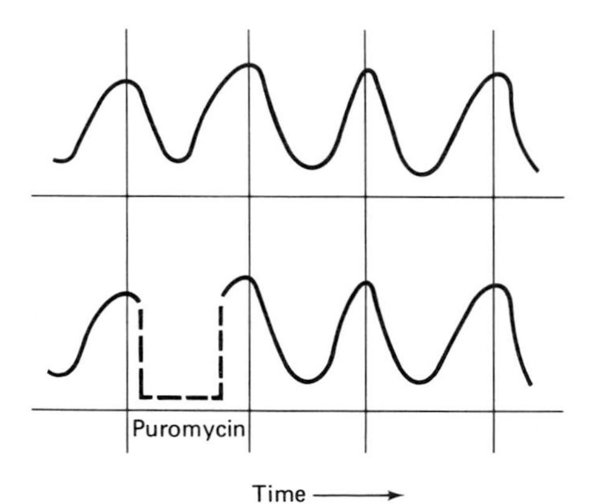

Figure 1-20 A diagrammatic representation of the effect of a single pulse of the antibiotic, puromycin, on a rhythm in bioluminescence (described in Chapter 2). As can be seen, as long as the drug is present, the rhythmic process is stopped, but on removal, the luminescence and the rhythm recommence, and the latter is in phase with the control (upper curve). This indicates that the clock has been running all the while and, therefore, is not a rate determining step in the luminescent reaction. Instead, the clock is a separate entity that is coupled to luminescence and, thus causes it to assume the waveform pattern.

some as yet unknown way which caused them to become rhythmic (Figure 3-14). An experiment showing the existence of this coupling is described diagrammatically in Figure 1-20, where a rhythmic process (which happens to be a rhythm in bioluminescence and is described in detail in Chapter 2) is subjected to an antibiotic, puromycin. The drug, while present, completely inhibits bioluminescence; but, when its action is stopped, the organism again begins to luminesce and the rhythm displayed is in exact phase with the controls not treated with puromycin. Clearly, then, the clock had continued to run unperturbed all through the treatment, which indicates that it is separated from, and only coupled to, bodily physiological processes.

MASTER OR MULTIPLE CLOCKS

Many physiological processes in algae and protozoans are rhythmic, a fact which demonstrates that the unicellular level of organization is sufficient for the expression of rhythmic behavior. Additionally, unicells are known to display more than one rhythm simultaneously, which begs the question of whether a single master clock within their bodies drives all their rhythms (each via a separate coupling unit), or whether each rhythm has its own specific clock. The answer is not yet known.

A similar situation exists in multicellular individuals that display many different rhythms. Their construction raises the additional question of whether the same master, or a republic of individual clocks, is repeated in every cell of their bodies. The animal portions of this group have nervous systems, and both animals and plants have a variety of forms of humoral coordination that could ostensibly alleviate the need for duplicate clocks in each cell. These coordinating systems, however, apparently do not preclude this need, for as will be discussed in detail in the next three chapters, segments of rhythmic multicellular organisms can be removed, maintained alive in separate culture, and the rhythms will continue to be displayed (Figures 2-1, 2-5, 2-8, 8-12). Even more striking, rhythms will persist in a single cell from which the nucleus has been removed (Figure 2-5). Now, complicating matters even more, evidence is beginning to accumulate which suggests that more than one clock may be involved in the control of a single rhythmic process.

In conclusion, there is still not enough evidence available to select one clock type over the other, to identify the coupling system between clock and rhythm, or to know how many clocks are present in a single cell. Both clock hypotheses have their strengths and foibles and, of course, the possibility exists that when at last the clockwork is deciphered, it will be found to encompass a bit of each hypothesis. The last two chapters of this book are devoted specifically to the discussion of these clock mechanisms. They are written by two of the leading figures in the field of biological rhythms, Frank A. Brown, Jr., and Leland N. Edmunds.

SUMMARY AND CONCLUSIONS

1. Organismic physiology and behavior is often rhythmic and these rhythms will persist in the laboratory in the absence of day-night light and temperature cycles. Because they do persevere, it is concluded that they are under the control of a so-called biological clock.

2. Those rhythms that will persist in the absence of the most obvious environmental cycles all have periods that match the prominent geophysical cycles on earth: a solar day (24 hours), a lunar day (24.8 hours), a month (29.5 days), and a year (365.25 days). Often, in the artificial constancy of the laboratory, the periods of these rhythms deviate slightly from the ones displayed in nature and are referred to as circadian, circalunadian, circamonthly, or circannual.

3. The phase of daily rhythms can be set by light-dark and temperature cycles to any time of the day. When the organisms are then placed in constant conditions, this new phase persists.

4. The period length of daily rhythms can be lengthened or shortened by subjecting organisms to light-dark alterations whose combined lengths are longer or shorter than 24 hours. The change thus produced is only illusory, however, since as soon as these organisms are placed in constant conditions, the period reverts back to about 24 hours.

5. Entrainment to light-dark and temperature cycles is brought about by the actions of a light-sensitivity and a temperature-sensitivity phase-response rhythm, respectively. These two rhythms are also postulated to generate, by the process of autophasing, circadian periods in constant conditions.

6. The length of the circadian period in constant light is a function of the ambient light intensity. Very bright constant light inhibits the expression of rhythms.

7. The period of a daily rhythm is altered by only 10% or less by a 10°C change in the ambient constant temperature. This very small change has been referred to as "virtual temperature independence."

Three explanations have been given for this unexpected insensitivity: (i) the clock is based on some physical process since these are known to be less affected by temperature, (ii) a temperature compensating mechanism is built into the clock, or (iii) the clock is of the nonescapement type.

8. The period length (but not the amplitude) of daily rhythms has been shown to be insensitive to alteration by a whole host of chemical substances. Only recently have four chronomutagenic substances been found that will consistently alter the period; these are heavy water, valinomycin, alcohol, and lithium.

9. Rhythms are innate.

10. Almost nothing is known about the fundamental nature of the clock underlying these rhythms. Over the years, two major hypotheses have evolved. The *escapement-type* conceptualization, which envisions an autonomous clockworks, insensitive to most exogenous chemical perturbations, that is temperature compensated and that can weather cell replication without losing time. The *nonescapement* version is a responding-signaling-type clock that does not measure off intervals of time autonomously, but instead responds to one or more periodic geophysical forces that normally penetrate the confines of laboratory constant conditions and organismic structure into the cellular milieu where the clock is located. Since, in this model, the basic timing source is outside the organism, temperature and chemical insults have little effect on the period. The fact that the only rhythms that persist in constant conditions and share the properties described above match the major geophysical cycles of the earth has also been used to support this hypothesis.

There is thus far insufficient evidence to choose either model as a correct one, although most students of the subject choose one over the other to serve as a working hypothesis.

11. The clock is an entity distinct from the processes it causes to be rhythmic. The two are conjoined by an as yet unknown coupling mechanism.

12. The unicellular level of organization is sufficient for the expression of rhythms.

13. In individual organisms—both unicellular and multicellular—that display more than one rhythmic process, it is not known whether all the rhythms are driven by one master clock, or if each rhythm has its own clock. Certainly, all the cells in a multicellular organism have clocks, since isolated fragments from both plants and animals continue to display rhythms while grown in culture. There is even some speculation that several clocks may govern one rhythmic process.

Selected Readings

Aschoff, J. (1963). Comparative physiology: Diurnal rhythms. *Annu. Rev. Physiol.* **25,** 581–600.

Aschoff, J., ed. (1965). "Circadian Clocks." North-Holland Publ., Amsterdam.

Biological Clocks. (1960). *Cold Spring Harbor Symp. Quant. Biol.* **25,** 1–524.

Brown, F. A., Jr. (1972). The "clocks" timing biological rhythms. *Am. Sci.* **60,** 756–766.

Brown, F. A., Jr., Hastings, J. W., and Palmer, J. D. (1970). "The Biological Clock: Two Views." Academic Press, New York.

Bünning, E. (1973). "The Physiological Clock," 3rd ed. Springer-Verlag, Berlin and New York.

Circadian Rhythmicity, (1972). Proceedings of International Symposium at Wageningen. Center for Agricultural Publishing and Documentation, Wageningen, Netherlands.

Harker, J. E. (1964). "The Physiology of Diurnal Rhythms." Cambridge Univ. Press, London and New York.

Menaker, M. (1969). Biological clocks. *BioScience* **19,** 681–689.

Menaker, M., (1971). "Biochronometry." Nat. Acad. Sci. Washington, D.C.

Mills, J. N., ed. (1973). "Biological Aspects of Circadian Rhythms." Plenum, New York.

Palmer, J. D. (1970). Biological clocks. *In* "Encyclopedia of the Biological Sciences" (P. Gray, ed.), 2nd ed., pp. 107–108, Van Nostrand-Reinhold, Princeton, New Jersey.

Palmer, J. D. (1974). "Biological Clocks in Marine Organisms: The Control of Physiological and Behavioral Tidal Rhythms." Wiley (Interscience), New York.

Pengelley, E. T., ed. (1975). "Circannual Clocks: Annual Biological Rhythms." Academic Press, New York.

Pittendrigh, C. S. (1961). On temporal organization in living systems. *Harvey Lect.* **56,** 93–125.

Pittendrigh, C. S. (1974). Circadian oscillation in cells and circadian organization of multicellular systems. *In* "The Neurosciences" (F. O. Schmitt, and F. G. Worden, eds.), pp. 437–458. MIT Press, Cambridge, Massachusetts.

Rhythmic Functions in the Living System. (1962). *Ann. N.Y. Acad. Sci.* **98,** 753–1326.

Scheving, L., Halberg, F., and Pauly, J. (1974). "Chronobiology." Igaku Shoin Ltd., Tokyo.

Solberger, A. (1965). "Biological Rhythm Research." Elsevier, Amsterdam.

Symposium on Biological Chronometry. (1957). *Am. Nat.*, **91,** 129–195.

Ward, R. (1971). "The Living Clocks." Knopf, New York.

Webb, M. H., and Brown, F. A., Jr. (1959). Timing long-cycle physiological rhythms. *Physiol. Rev.* **39,** 127–161.

2

A Survey of Rhythms in Plants and Animals

A great variety of processes in both plants and animals has been found to be rhythmic. This chapter will describe the oscillatory nature of both the most fundamental organismic processes and those exotic ones which have been extensively studied.

PLANTS

Sleep-Movement Rhythms

A record of biological rhythms goes back at least as far as 350 B.C. It may be of interest to note that Aristotle (as the date might have suggested) was not the first person to observe and record a biological rhythm; an officer in the corps of Alexander the Great made the initial observation. The personal diary of this centurion makes mention of the observance of "sleep" rhythms in certain leguminous plants. At dusk, the leaves on these plants lowered to the sides of the stem, where they remained during the night; each morning, they again raised; as if in a pagan gesture to the sun. That the plants could move at all was thought to be quite remarkable; but, when the first experimental work was done on these sleep movements, 2079 years later, by

Figure 2-1 Sleep-movement rhythm in the bean seedling (*Phaseolus*). *Left:* Triple exposure taken at 3 different times during the day and showing the movements of intact leaves. *Upper right:* Detached leaf with petiole inserted into a tube of nutrient. Rhythm persisted for 28 days. *Lower left:* Preparation in which the leaf blade was cut away from the midrib. Rhythm in these petiol-pulvinus-midrib "leaves" persisted for 8–9 days. Courtesy of K. Yokoyama *et al.*, *Life Sci.* **7,** 705–711 (1968).

an *astronomer,* an even more amazing facet was discovered—the sleep-movement rhythm would persist even when the plants were placed in continuous darkness and in a fairly constant temperature (Figure 2-1A).

Even today, botanists enthusiastically study sleep-movement rhythms. Partially responsible for this enduring popularity is the fact that the oscillations can quite easily and automatically be observed in the laboratory. One of the simplest means is to employ time-lapse photography, the results of which can be seen in Figure 2-1A. By using methods like this, the movement of leaves can be studied continuously for extended periods of time, while the plants are maintained in rigorously controlled constant conditions—which eliminates interruptions associated with periodic human observation.

Leaf movement is brought about by two packages of specialized cells—collectively called the pulvinus—one on the upper, and the other on the lower side of the junction between the petiole and the blade. The cells of the pulvinus swell and shrink in size as water moves into and out of them. The flow of water is controlled by the bioclock which governs the pumping of the salt potassium into the pulvinal cells. At dawn or thereabouts, potassium buildup in the internal milieu of the lower pulvinal cells causes water to rush in from surrounding areas by osmosis. The cells then swell to near bursting and, in doing so, act to raise the blade of the leaf to horizontal. A mechanical analogy would be the lifting of the "blade" of a bulldozer by the filling of the hydraulic rams attached to it. Before dusk, potassium is pumped into the upper pulvinal cells and they then become turgid with water. Simultaneously, potassium and water leak out of the cells on the lower side and they become flaccid; the leaves are then pushed down and assume the sleep position. In addition to being the prime mover in the response, the pulvinus is also the photoreceptor that "sees" ambient light-dark cycles. When plants were maintained in absolute darkness, and a microbeam of light (only 0.5 mm in diameter) slowly swept over the leaf blade and petiole, it was discovered that only when the spot of light fell on the lower pulvinus did the leaf rise.

An interesting study on the role of light-dark cycles has been performed on individual leaves of bean plants. A single leaf on a plant was offered a light-dark schedule exactly opposite to the natural one, while the rest of the plant was adequately screened from the cycle. The leaf adopted the phase of this cycle, and even maintained it for awhile, when the whole plant was placed into constant darkness, i.e., it stood erect while its companions "slept." But after a while, the odd leaf would either drift back in phase with the others, or roll up and become inactive.

In fact, a leaf need not even be attached to a plant for the rhythm to persist. Excised leaves were maintained in good health by inserting their petioles through a rubber stopper and into a tube of nutrient (Figure 1-1B). The sleep movements continued, and when followed photographically, the rhythm was found to persevere for as long as 28 days in constant conditions. Even preparations containing only the petiole, pulvinus, and midrib (i.e., the blade was removed) continued, with the customary vigor, to describe the rhythm (Figure 2-1C).

One last display of individual leaf versatility will be mentioned. In the laboratory, at certain intensities of constant illumination, an event takes place which complicates, somewhat, the interpretation of the underlying clockworks. Under these conditions the periods of the sleep-movement rhythms controlled by the primary pulvini (in the leaf axil) and the secondary pulvini (at the junction of the blade and petiole) differ, e.g., at a light intensity of about 6 foot candles, the two periods are 28 and 30.3 hours, respectively. The question is thus posed as to whether there are two clocks, in an individual plant, each of which runs at a different rate, or whether one of the multiple couplings between a single clock and the many rhythms it drives has been altered by the artificiality of the laboratory setting.

The Effectiveness of Chemical Manipulations

As mentioned in Chapter 1, rhythmic organisms have been subjected to literally hundreds of chemical substances in attempts to alter their clockworks. None but a small handful of these compounds has produced any modification. Four substances, alcohol, lithium ions, deuterium oxide, and valinomycin, stand almost alone as known chronomutagenic agents, and their actions have all been demonstrated on the sleep-movement rhythms of the bean and a house plant called *Kalanchoe.* The sleep-movement rhythms of these plants will persist for days even after the plant stems have been severed off at ground level and the cut plants removed to beakers of water. Testing the effect of various substances is done by simply adding them to the water in the beaker, where they freely enter the transpiration stream going up the stem. Agents like alcohol, deuterium oxide, and lithium ions, when administered in this fashion, have all been found to increase the period of the rhythm significantly, and period lengthening is a type of response that would be seen if the clockworks were actually slowed down. The antibiotic valinomycin also produces an apparent change in the clock. When it is administered via cut stems for short intervals during the day, the sleep-movement rhythm is rephased, the amount and direction of the phase change produced being a function of the time of day that valinomycin is offered (Figure 8-21).

Because of the variety of other known effects of these substances on biological systems, a great deal of speculation about their role on rhythms is possible. One of these hypothetical schemes involves their role on membranes. Alcohol, deuterium oxide, and lithium ions all are known to affect membrane permeability and salt distribution between cell and the external milieu. Valinomycin, in low concentrations, is known to form a complex with potassium ions, which enhances their transport through biological membranes. Remembering that the sleep movements of plants are really only secondary displays of the movements of water into and out of the pulvini and that this is preceded by the movement of potassium ions, one may not be surprised that these three agents produce an effect. But, as will be described later, these appear to be universal chronomutagenic substances and produce the same types of alterations in other rhythmic systems, including such nonpulvinal bearers as mice. Therefore, their actions at the cell boundary have been used to develop the "membrane hypothesis"—a tangible version of escapement-type clock—which will be discussed in detail in Chapter 8.

Early in the study of biological rhythms, an as yet unidentified periodic geophysical factor in the environment, one which could penetrate into the laboratory, was suspected of being involved as a timing source. So, plants were taken to the bottom of a 180-meter-deep rock-salt mine in Germany to see the effect of subterranean conditions on the sleep movements. Only the more energetic components of cosmic radiation could penetrate to this level. The rhythm ceased. On return to constant conditions on the surface, the rhythm began again. The plants responded, therefore, in a way that suggested that the underground laboratory must have screened out a geophysical factor important to the rhythm.

PERSISTENCE IN ISOLATED ORGAN PARTS

Rhythmic appendage movements in plants are not limited to leaves alone; petals also show these movements. These flower-opening rhythms differ in phase from leaf sleep movements in that different kinds of plants open and close at different times of day, e.g., the 4 o'clock morning glory does not unfold its petals until the afternoon. In the seventeenth century, many households had what could be called clock gardens, a section of the yard planted with groupings of flowers which opened at different, fairly specific, times of day. It is reported that on a sunny day the timing information provided by the garden was accurate to within a half hour.

That the petal-movement rhythm is also under the control of a bioclock was demonstrated by studying their movements in constant conditions. Especially interesting results have been reported in a semitropical shrub of the Caribbean, the night-blooming jessamine, which opens its flowers each night and emits a powerful fragrance. The odor attracts night-flying insects, which pollinate the flower during its visits. The rhythm in flower opening will persist in constant light or darkness, in both intact plants or those cut and placed in a vase.

Concomitant with the cyclic opening and closing of flowers is a rhythm in fragrance which also persists in constant conditions. Such a rhythm would be expected as merely a consequence of the periodic opening of flowers and the subsequent escape of the volatile odor. However logical this may seem, it is an oversimplification. The substance responsible for the pungent odor emanates from the tips of the petals and if the excised petals are floated on water and placed in constant conditions, the odor rhythm persists until the petals die. The expression of the odor rhythm is therefore independent of the opening and closing rhythms of the flower and, furthermore, does not depend on the intact plant.

Photosynthetic Rhythms

In 1957, biological oceanographers, measuring the primary productivity of oceanic water masses, made an interesting observation: the photosynthetic rate of pelagic phytoplankton varied with the time of day. When successive plankton samples were collected at different times of day, in approximately the same geographic area, it was found that the photosynthetic rate was highest in the morning and lowest in the late afternoon, with a substantial, five- to sevenfold difference between extremes.

These fluctuations could be contributed to a variety of things, such as the use of samples of different size and species composition, fluctuating light intensities, and temperature fluctuations. Some of these variables were eliminated in follow-up studies by using algal populations isolated in large glass carboys and by measuring photosynthetic rates under identical light intensities. The rhythms were found to persist, but, in all cases, the sample populations were maintained in natural day-night conditions, which, of course, can induce rhythms in organisms.

IN GONYAULAX

In response to these field studies, laboratory investigations of the phenomenon were undertaken. The most interesting results have come from studies of pure cultures of algae, in particular, a species of dinoflagellate called *Gonyaulax polyedra* (Figure 2-2), a common photosynthetic plankter on the West Coast of the United States. The photosynthetic rate in this alga was periodically measured throughout the course of the day under identical test conditions. Aliquot samples were systematically withdrawn (both day and night) from a parent population and incubated with radioactive carbon for a standard time interval at the same light intensity. This technique measured the photosynthetic *capacity*, or "ability," of algal cultures to photosynthesize under identical conditions, but at different times of the day. It was found that during the nighttime, when photosynthesis usually does not occur, the photosynthetic capacity was relatively low. During the normal light period, the capacity was greatly increased with a maximum somewhere near midday (Figure 2-3).

Figure 2-2 Scanning electron micrograph of *Gonyaulax polyhedra.* (Courtesy of Beatrice M. Sweeney.)

To elucidate the persistent nature of the rhythm, algae were maintained under unvarying conditions and, when the proper levels of constant illumination were provided, the rhythms were found to persist for several cycles. At high levels of illumination, the rhythm was inhibited; this feature, though not understood, is the usual effect of bright light on the expression of biological rhythms.

Two reasons come immediately to mind as possible causes of arrhythmicity in a *Gonyaulax* population after exposure to constant bright

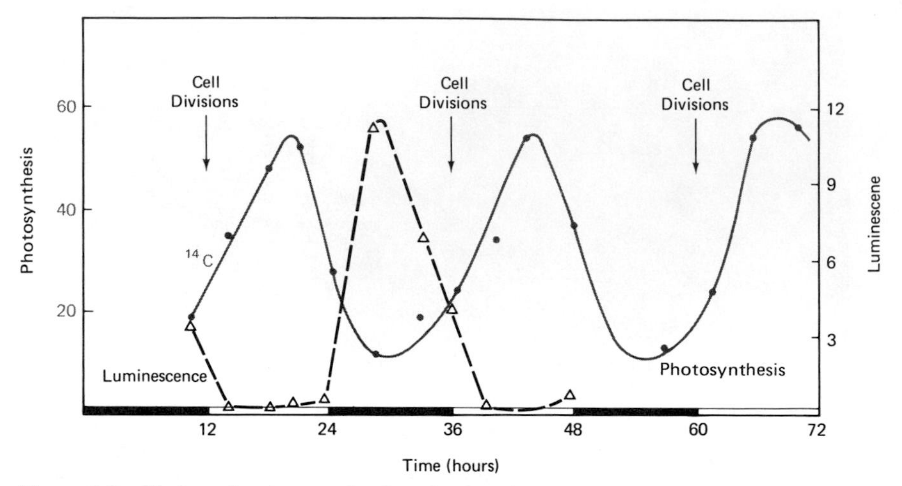

Figure 2-3 Rhythms in photosynthesis, stimulated bioluminescence, and cell division in *Gonyaulax*, maintained in alternating day-night cycles. The 12-hour dark periods are indicated by the shaded bars on the abscissa. The times of peak cell division are indicated by the falling arrows. Modified from J. W. Hastings *et al.*, *J. Gen. Physiol.* **45,** 69–76 (1961).

light: (i) the clock has been stopped; or (ii) all the individuals in the population continue to be rhythmic, but the peaks of each one's rhythm comes out of phase with its neighbor, which cancels out the unified cycle display of the population. To distinguish between these two possibilities, the photosynthetic rate of a single *Gonyaulax* cell was measured in a sensitive Cartesian-diver microrespirometer. Under normal, dim constant light, the rhythm persisted; but, at a bright intensity, it was inhibited (Figure 2-4). Therefore, the rhythm is truely inhibited by constant bright light, which signifies that the clock has either been stopped, or come uncoupled from the photosynthetic process.

IN ACETABULARIA

An interesting experiment, and one which provides some basic knowledge about the timekeeping mechanism of biological clocks, was done using the photosynthetic rhythm of a large, single-celled alga, *Acetabularia*. This plant inhabits the inshore waters in certain coastal habitats and is interesting in that it may grow to heights of up to 10 centimeters, in spite of the fact that it is but a single cell. The vegetative plant contains a rather large nucleus located in the base of the plant, usually in one of the rhizoids (Figure 2-5, top). Because of this

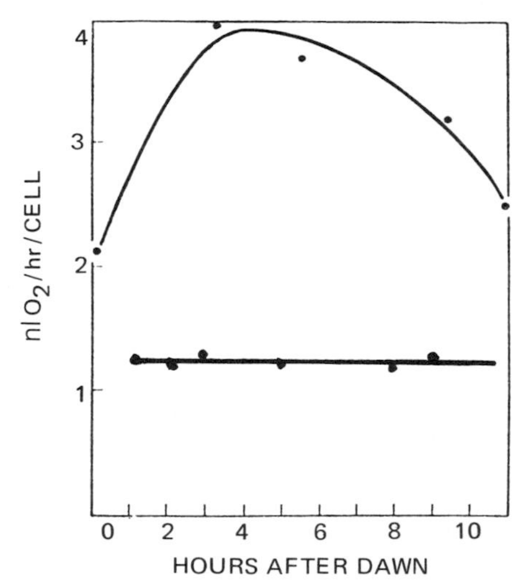

Figure 2-4 The role of constant bright light in the inhibition of the photosynthetic rhythm of a single *Gonyaulax* individual. The dome-shaped curve shows the persistence of the rhythm in dim (50 foot candles) constant light, while the loss of rhythmicity at 800 foot candles in another cell is indicated by the lower curve. Modified from B. M. Sweeney, *Cold Spring Harbor Symp. Quant. Biol.* **25,** 145–148 (1960).

arrangement, the plant cell can be easily enucleated by cutting off the basal tuft of rhizoids. Biologists—cytologists in particular—have systematically exploited this plant in experiments designed to elucidate the role of the nucleus in the life of the cell. When this plant is placed in alternating cycles of light and darkness, it is found to display a rhythm in photosynthesis, one which will persist even in enucleated cells for as long as 40 days in continuous dim illumination (Figure 2-5). This study certainly indicates that the nucleus is not required for the immediate maintenance of timekeeping in *Acetabularia.*

The nucleus exerts its influence over the remainder of the cell largely by the production of messenger RNA, synthetized under the control of nuclear DNA. It is important to recognize that extirpation of the nucleus in *Acetabularia* does not deprive the plant of all its DNA, for this important substance is also present in the chloroplasts and mitochondria, and these structures are also sites of RNA synthesis. These extranuclear nucleic acids may therefore be playing an important role in the timing of the overt photosynthetic rhythm.

While the nucleus appears not to have a determinate role in the origin or maintenance of the photosynthetic rhythm in *Acetabularia,* it could be involved in other aspects of the rhythm, e.g., possibly it may

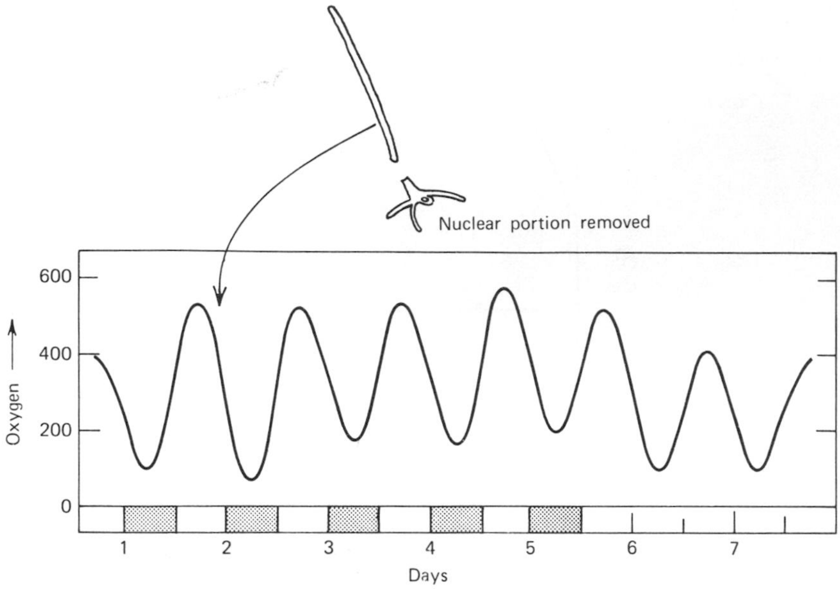

Figure 2-5 Top: The diurnal photosynthetic-capacity rhythm in the single-celled alga *Acetabularia*. The rate of photosynthesis was measured by determining the oxygen evolution from the plant. Each measurement was made under identical light intensities of equal duration all through the day and night. The rhythm persisted even after the cell was enucleated on day 2 and placed in constant conditions on day 6. Drawn from the data of B. M. Sweeney and F. T. Haxo, *Science* **134,** 1361–1363 (1961).

dictate the phase or amplitude of the rhythm. To examine this possibility, a culture of *Acetabularia* was divided into two groups. One group was illuminated from 8 A.M. to 8 P.M. and the other from 8 P.M. to 8 A.M. After a period of 2 weeks, the nuclei were removed from all the plants, scraped clean of cytoplasm, and reciprocally transplanted between the groups. Plants that had been maintained in light between 8 A.M. and 8 P.M. now contained nuclei from plants formally exposed to light from 8 P.M. to 8 A.M. (Figure 2-6A). All plants were now placed in constant illumination and the oxygen evolution measured at 12-hour intervals. The photosynthetic rhythm persisted in spite of this rather drastic treatment; the phase of the rhythm in each plant being determined by the nucleus it contained.

In a further experiment, the plants were kept intact, but the opposing ends of each plant were subjected to opposite photoperiods. After 2 weeks of this treatment, the plants were transferred to constant illumination. The phase of the resulting persistent rhythms (Figure 2-6B) indicated that the nucleus-containing end determines this aspect of the rhythm.

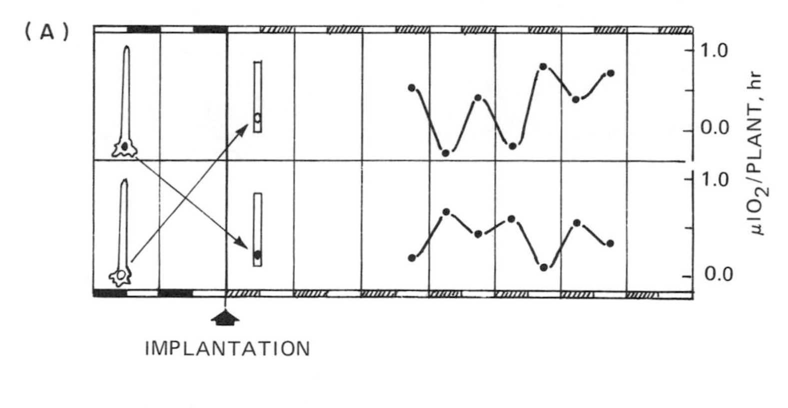

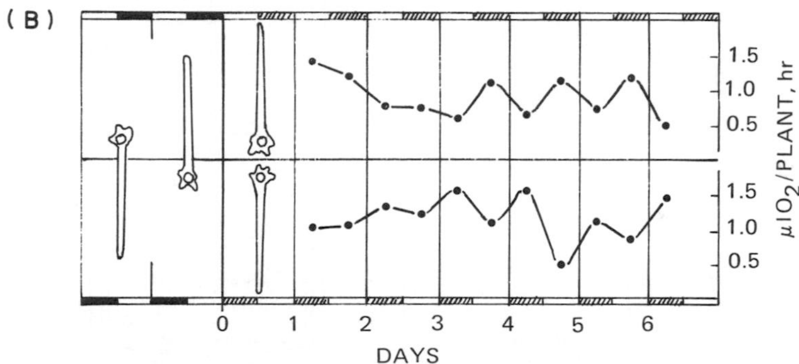

Figure 2-6 The role of the nucleus in the photosynthetic-capacity rhythm of *Acetabularia.* (A) Test plants were maintained in normal and reversed light regimes. The nuclei were then exchanged between plants and the photosynthetic rhythms studied in constant conditions. The phases of the two rhythms were that of the nuclei they contained. (B) Portions of intact plants were subjected to opposite illumination cycles for 2 weeks and the photosynthetic rates then measured in constant conditions. The phase displayed was that of the nuclear, rather than the cytoplasmic end of the plant. Modified from H. J. Schweiger *et al., Science* **146,** 658–659 (1964).

These studies therefore indicate that while the nucleus in *Acetabularia* is apparently not important in the genesis of the rhythm it still participates in some aspects of oscillatory behavior. Since the nucleus is known to exert its influence on the cytoplasm via chemical messengers, this system should prove amenable to biochemical inquiry.

Oxidative Metabolism Rhythms

The metabolic machinery in virtually all organisms uses oxygen to liberate the energy bound in organic molecules. Since oxidative metabolism underlies almost all forms of biological activity, it is an

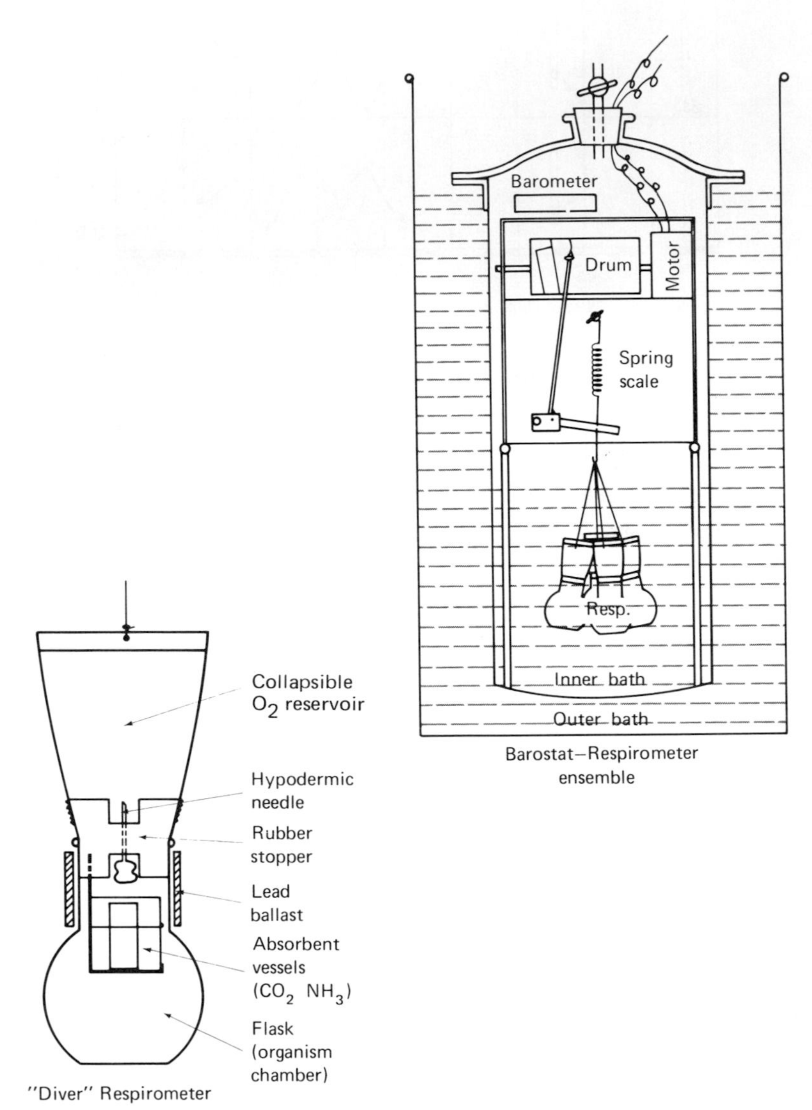

Figure 2-7 Recording respirometer-barostat assembly. As organisms consume oxygen from air in the flask, O_2 is replaced from the collapsible reservoir and CO_2 and NH_3 absorbed by KOH and cupric chloride. The weight of the "diver" increases (buoyancy decreases) as O_2 is consumed with the increase recorded on a rotating drum. The diver is maintained in a hermetically sealed system in constancy of all ordinarily controlled variables as well as of O_2 and CO_2 and pressure. From F. A. Brown, Jr., *Cold Spring Harbor Symp. Quant. Biol.* **25,** 57–71 (1960).

extremely useful and informative process to study. Metabolic rates are usually measured by simply determining the rate at which oxygen is utilized by plants and animals. This measurement is easily carried out over short periods of time by simple manometric techniques; but long-term measurements (several days or weeks) require more elaborate methods. Since, in the study of rhythmic phenomena, continuous long-term measurements are necessary and automation highly desirable—both for convenience and accuracy—a continuous-recording respirometer had to be devised.

One device contrived consists of nothing more than a diverlike reaction vessel submerged in water and suspended from a level system activating a chart recorder. The respirometer "diver" is a small flask in which the organism is placed and is topped with a collapsible plastic bag filled with oxygen (Figure 2-7). As the organism consumes the oxygen in its flask, a fresh supply flows in from the collapsible reservoir. An absorbant within the flask removes the carbon dioxide produced by the organism, so that as the oxygen reservoir collapses, the diver bouyancy decreases, and the diver sinks deeper in the water. The rate of sinking is directly dependent on the rate of oxygen utilization: the greater the oxygen consumption, the faster the sinking rate.

The whole apparatus is placed in an airtight "barostat" and a constant temperature bath. This method of recording is highly desirable as many environmental parameters, such as temperature, light intensity, atmospheric pressure, humidity, oxygen tension, and carbon dioxide tension, can all be held rigorously constant for protracted periods of time.

The potato tuber (*Solanum*) was chosen as the experimental "organism" to be used with this apparatus because it has a generous reserve of stored food and hence eliminated the need of disturbing the constant conditions to add nutrient. Also, potato "eyes" are meristematic areas and, therefore, metabolically quite active. Small cylinders, each bearing an eye, were removed from the potato and used in the divers (Figure 2-8). A single eye could be used for periods lasting up to nine months, but oxygen and fresh CO_2 absorbant had to be renewed at 3- to 8-day intervals. Measurements were made daily for 10 years, and the 1.5 million "potato-hours," and nearly 90,000 consecutive calendar hours of data, are summarized in Figure 2-8. Metabolic rates were lowest during the night and then rose sharply in the early morning to three daytime peaks. So, the potato, which normally inhabits a subterranean environment, possesses a distinct trimodal daily rhythm.

Daily metabolic rhythms have been described in many other plants

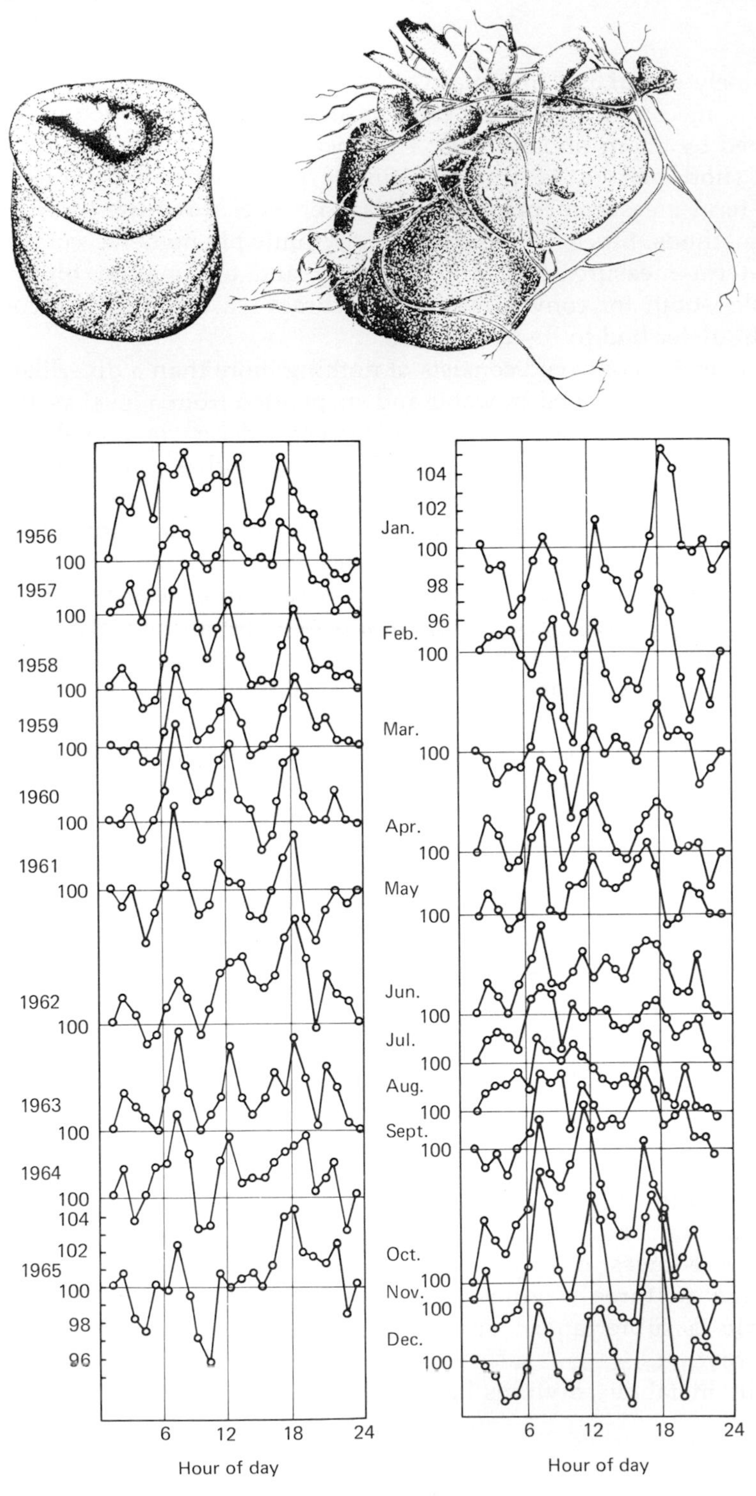

Potato O_2 consumption

also. In one particularly fascinating example, the rhythm was found to persist in dormant onion seeds. In this quiescent state, the nuclear machinery of each cell is virtually turned off. The persistence of the rhythm suggests control by a nonescapement-type clock.

Rhythmic Bioluminescence

At night, luminous "clouds" of light are often seen suspended in the ocean. Early explorers attributed this "burning of the seas" to the presence of phosphorous, which, on slow oxidation, was known to luminesce; and the spectacle was logically called phosphorescence. After 1753, when it became known that phosphorescence was of biotic origin (caused mainly by planktonic organisms), the phenomenon was renamed bioluminescence. It was soon discovered that bioluminescent organisms were stimulated to luminescence by almost any sort of mechanical agitation; wave action, the dipping of an oar, the turn of a propeller, schools of fish, etc., all produced the startling effect. Fishermen are often attracted to schools of sardines by the luminescent disturbances these fish create in the surrounding water.

Over a hundred years ago, it was discovered that when certain of these tiny organisms were brought into the laboratory they could *usually* be made to shine by simply shaking their container. However, such disturbances did not always produce the desired effect, and as early as 1893, from an investigation of a marine dinoflagellate (then known as *Noctiluca,* but now thought to be some other genus), it was found that shaking would elicit the emission of light only during the nighttime. In fact, even when this organism was kept in either constant illumination or constant darkness, it could be made to flash only at the times corresponding to astronomical night. This rhythm in irritability would persist until the death of the organisms—about 1 week.

From a teleological point of view, these results might have been anticipated, for if luminescence is to be useful to an organism, it would be expected that it should have minimal meaning in the daytime—bioluminescence cannot, of course, compete in intensity with

Figure 2-8 Top: Potato plugs used in metabolic studies. The plug on the right had been sealed in a recording respirometer for more than 5 months: it sprouted an abortive rhizome system complete with a "new potato" growing from it. Bottom: Results of a 10-year study of the mean daily respiratory patterns in the potato. The left-hand column shows the patterns for each of 10 consecutive years, plotted as a daily average, for each year. Notice the three recurring peaks each day: 7 A.M., 12 noon, and 6 P.M. In the right-hand column, the same data are used to compute the mean daily curves for all ten Januarys, Februarys, etc. Note the annual modulation in form and amplitude of these curves. From F. A. Brown, Jr., *Scientia (Milan)* **103,** 245–260 (1968).

sunlight. Therefore, one would expect bioluminescent organisms to have evolved some mechanism to prevent luminescence in the daylight; which would conserve the bioluminescent "fuel" for nighttime displays. Indeed, subsequent field observations confirmed this speculation by revealing the existence of daily rhythms in luminescence.

Extensive modern studies have been carried out on the armored marine dinoflagellate, *Gonyaulax polyedra* (Figure 2-2), a common luminescent phytoplankter, often responsible for the "phosphorescent" displays of coastal waters. When these cells are stimulated to luminescence—which is done in the laboratory by bubbling purified air through the culture—each cell responds by producing a blue-green flash of light that lasts about 90 milliseconds. Less than 1 minute of bubbling is sufficient to completely fatigue a small culture. To measure the intensity of the emitted light, a vial containing the organisms is placed in a dark chamber in front of a sensitive photomultiplier tube; light emission is detected, the phototube current amplified, and the response recorded on a chart.

With this technique, it was soon discovered that when cultures were

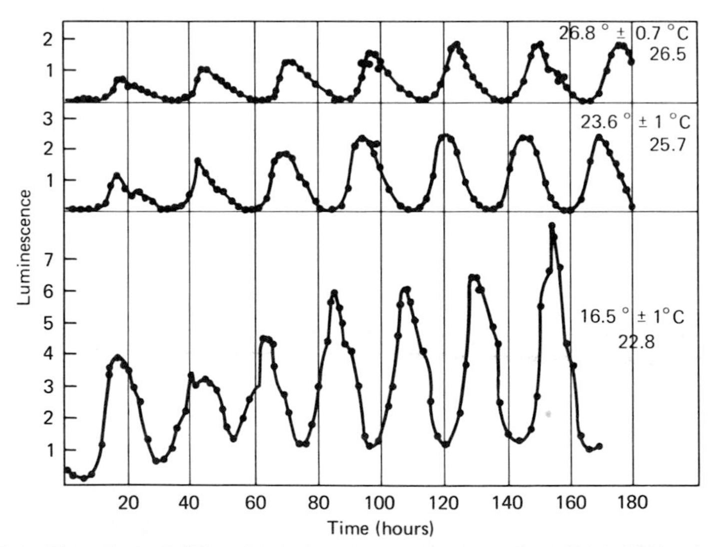

Figure 2-9 The effect of different constant temperatures on the stimulated luminescence rhythm of the dinoflagellate *Gonyaulax*. Prior to the start of this experiment the cells were kept at 22°C in alternating light-dark conditions. At the end of a dark period, they were transferred to constant dim light and one of the test temperatures. Note that the periods lengthen as the temperature increases; this unusual augmentation is interpreted as overcompensation. Modified from J. W. Hastings and B. M. Sweeney, *Proc. Natl. Acad. Sci U.S.A.* **43,** 804–811 (1957).

periodically stimulated to luminescence, 40 to 60 times more light was emitted at nighttime than during the day. Just as had been shown for "*Noctiluca*"—63 years earlier—it was found that this rhythm would persist for as long as 14 days in constant light (Figure 2-9).

This rhythm was subsequently found to be partially a result of a variation in the sensitivity of cells to stimulation: they were more responsive at night. Changes in the concentrations of reactive components of the luminescent system also play a role as was shown by preparing cell-free extracts of *Gonyaulax* luciferase (the enzyme which catalyzes the luminescent reaction) at different times of day; greater activity was found during the nighttime than during the hours of light. Maximal activity of the extracts corresponded to the time when luminescence from flashing was greatest. Extracts prepared from cells maintained in constant conditions also showed the rhythm in activity. When extracts of the substrate luciferin were prepared, it was found that its activity also peaked at night and this rhythm would persist in cells kept in constant dim illumination. However, the phase relationship between the two rhythms is not the usual one found in enzyme-substrate reactions. Usually, an accumulation of substrate acts as the stimulus to induce the enzyme that destroys the substrate; and then, because the enzyme is unstable, it self-destructs. But in the luciferin-luciferase case (as can be seen in Figure 2-10), the peak in enzyme activity *precedes* in time that of the substrate, which indicates that a clock-controlled *de novo* synthesis and destruction must take place.

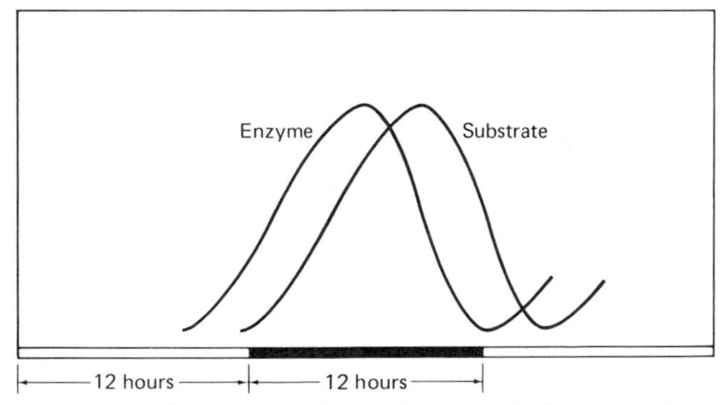

Figure 2-10 Diagrammatic representation of the amount of extractable enzyme (luciferase) activity and the amount of extractable substrate (luciferin) as related to time of day. The cultures were maintained on the light-dark cycle shown. The peak of substrate occurs somewhat later than the peak of enzyme activity. From J. W. Hastings, *in* "The Biological Clock: Two Views" (F. A. Brown, Jr. *et al.*), pp. 61–91. Academic Press, New York, 1970.

AN UNUSUAL TEMPERATURE RESPONSE

The bioluminescence rhythm has also been examined under a melange of constant temperatures. One massive experiment was performed at six different constant temperatures, ranging between 11.5° and 26.8°C. Aliquots of a parent *Gonyaulax* culture, which had been growing at 22°C in alternating light-dark cycles, were pipetted into 1300 test tubes and equal numbers placed, simultaneously, in each of the constant temperatures and in constant dim light. At 11.5°C, the rhythmicity was lost, although the cells remained viable. When these were returned to 22°C, the rhythm returned; the phase was determined by the time at which the cultures were returned to 22°C. The effect on the period between 16.5° and 26.8°C was somewhat unexpected: the period progressively *lengthened* (by about 15% total) with increasing temperature (Figure 2-9). The same results were found when the cell division rhythm (which will be described next) and the photosynthetic rhythm were tested against temperature. These results have been interpreted to mean that temperature overcompensation has taken place.

In addition to the flashing behavior, it was discovered that undisturbed cultures of *Gonyaulax* also emitted a continuous dim luminescence. This spontaneous glow is so dim that it cannot be seen by the dark-adapted eye; in fact, the intensity of the glow from a culture of 10^5 cells is less than the flash from a single organism. The intensity of the glow varies rhythmically over the day and rises to an abrupt and short-lived maximum at the end of each dark period. This rhythm will also persist for a few cycles in constant conditions.

ATTEMPTED CHEMICAL MANIPULATIONS

The glow rhythm is one that has been subjected to attempted manipulation by a whole host of chemical substances. By way of example, those involving antibiotics will be discussed. Actinomycin D, puromycin, and chloramphenicol all inhibit protein synthesis, but each produce this effect differently. When the *Gonyaulax* glow rhythm was subjected to actinomycin D, the effect produced depended on the concentration used. As seen in Figure 8-19, the rhythm and, to a very large extent, bioluminescent glow also were inhibited one or two cycles (depending on concentration) after actinomycin D was added to the culture. Puromycin inhibited all bioluminescence, while chloramphenicol augmented thc amplitude of the rhythms severalfold. Results like these are difficult to reconcile since all three substances are inhibitors of the same category of synthesis, protein production (albeit each acts in its own fashion). The same three substances were found not to act similarly on the photosynthetic rhythm in *Acetabularia*—in fact, in this organism actinomycin D appeared to *inhibit* the first two peaks of the rhythm and permit its expression thereafter, while chloramphenicol *decreased* the amplitude of the rhythm. At any rate, the most important change in constant conditions that one would wish to obtain when applying a chemical insult would be a change in the *length* of the period, for this aspect of a cycle is thought to best represent the rate at which the driving clock is running [of course the period length is also at least partly determined by the form of the phase-response rhythm (Chapter 1)]. None of the above antibiotics had any observable effect on the period.

Three substances have finally been found that will produce phase or period alterations in the *Gonyaulax* luminescent rhythms. These are the same compounds that were found to be effective in lengthening the period of the bean sleep-movement rhythm. Pulsed ethyl alcohol and valinomycin produced phase changes (the degree and direction being a function of the point in the cycle that they were pulsed) in the stimulated luminescence rhythm (Figure 8-22), and D_2O was found to lengthen the period of the glow rhythm in a dose-dependent fashion.

Rhythmic Cell Division

Rhythmic cell division had been reported in 1851, but the persistent nature of this rhythm was not revealed until the turn of the century. The earliest work was done on vascular plants: the rate of cell division being measured by periodically fixing, staining, and sectioning meristematic tissue for microscopic examination. Dividing cells could be found at all hours of the day, but relatively greater numbers of mitotic figures were found at specific times. These rhythms did endure in constant conditions and were apparent in the apical meristems of both roots and stems.

Contemporary with the work on cell division in terrestrial vascular plants was the discovery by oceanographers that daily rhythms in cell division were a natural feature of certain of the marine dinoflagellates. *Gonyaulax,* the alga which has been found so useful in the studies of other rhythmic processes, also displays a cell division rhythm. The daughter cells of this organism adhere to one another for about 30 minutes after cell division (Figure 2-11). Therefore, by periodically counting the number of cell pairs in a population, variations in the rate of cell division could be determined. When cells of this species

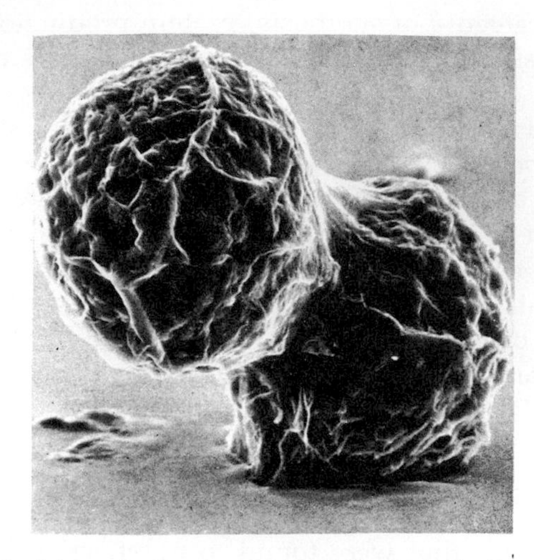

Figure 2-11 Cell division in *Gonyaulax* (×1500). From G. Dürr and H. Netzel, *Cell Tissue Res.* **150,** 21–41 (1974).

were maintained in alternating, 12-hour periods of light and darkness, it was found that about 85% of all of the cell division that takes place in any 24-hour period occurred during a 5-hour period spanning the end of the dark period and the beginning of the light period. The rhythm would persist for as long as 14 days in constant dim light (Figure 2-3).

It is possible, of course, that a persistent cell division rhythm represents nothing more than a synchronization (by pretreatment with alternating periods of light and darkness) of cell division in the culture. This, however, is not the case; for, in all instances, the generation time was found to be longer than the period of the cell division rhythm. Moreover, even at the peaks of the cell division rhythm, never more than about 20% of the cells in a culture were found in a paired condition. This means that the generation time of an *individual* cell must be equal to an even multiple of the period of the cell division rhythm of the culture (i.e., multiples of a period of about 24 hours), which means that once a cell is mature and capable of dividing the biological clock dictates the actual time of the division.

Four different rhythms have been thus far described in *Gonyaulax;* this raises the question of whether a single master clock governs all of them, each via its own coupling mechanism (somewhat like the situation portrayed in Figure 3-14), or whether each rhythm has its own personal clock. There are several observations that seem to endorse the master clock scheme. First, when all four rhythms are tested at different constant temperatures, all show overcompensation to about the same extent; second, the phase relationship between the rhythms does not change during long sojourns in constant conditions (as would be expected if each was driven by its own clock, any of which ran at a slightly different rate), and third, all four rhythms are shifted in the same direction and the same amount by a single light-perturbating stimulus. All of these responses would be expected if just one clock was in charge.

Rhythms in Growth

As early as 1879, the growth in the length of stems in a variety of plants had been found to be rhythmic, and some of these rhythms would last for as long as 14 days in constant conditions. Similar rhythms in root elongation were reported in 1904; and, in 1920, it was shown that maxima in root-elongation rhythms alternated with maxima in root-cell-division rhythms.

In these early studies, elongation rate was measured by a human observer, whose periodic observations necessarily interrupted constant conditions. Nowadays, these less sophisticated methods have been replaced by infrared time-lapse photography, a technique that produces exact measurements with the added benefit of automating mensuration and thus allowing relatively uninterrupted constant conditions to be maintained. When the camera was used to record the hourly growth rates of oat (*Avena*) coleoptiles, elongation under certain conditions was found to be rhythmic. Seeds germinated and maintained in red light showed no rhythm in growth; but, if the seeds were first germinated in red light and then transferred to constant darkness, a *24-hour* growth rhythm became apparent (Figure 2-12). The change from red light to darkness not only initiated the rhythm, but also determined its phase: when the change to darkness was made at various times during the day, the first peak of the growth rhythm always occurred about 16 hours after this change, irrespective of the time of day that the change was made.

Other embryonic structures of the oat seedling also elongate rhythmically, namely, the primary leaf. The first leaf develops within the protective cavity formed by the coleoptile; but, since the coleoptile is somewhat transparent, the growth rate of the temporarily confined primary leaf can still be measured. The growth-increment rhythm in this organ differs slightly from the growth rhythm in the coleoptile, in that the period is not 24 hours, but 24.75 hours long (Figure 2-12).

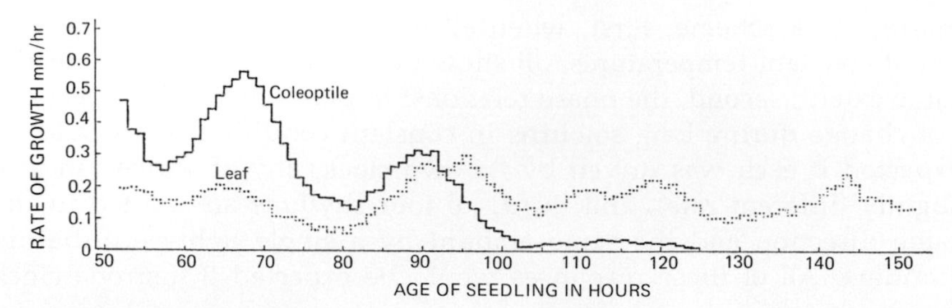

Figure 2-12 Growth rhythms in the oat *Avena*. Solid line, growth of coleopitle; dotted line, growth of primary leaf. The seedling was transferred from continuous red light to darkness at hour 52, and this treatment initiated the rhythms. The period of the coleoptile rhythm is estimated to be 24 hours, while that of the primary leaf, 24.75. Modified from N. G. Ball and G. B. Newcomb, *J. Exp. Bot.* **12,** 114–128 (1961).

Here, then, is a striking example of multiple rhythms, each with a different period, occurring simultaneously in the same plant. This again introduces the problem of postulating whether these rhythms are under the jurisdiction of separate clocks or whether a single clock can drive both rhythms, a problem discussed in Chapter 1.

Rhythms Associated with Reproduction

Many of the classic studies of these types of rhythms have been done with the fungi. The temporal aspects of spore discharge in *Daldinia concentrica*, one of the Ascomycetes, have been thoroughly investigated. The fruiting body of this common fungus has the appearance of a distasteful black ball and may be found living saprophytically on the trunks and limbs of ash trees. The outer surface of the fruiting body is pitted with thousands of flask-shaped cavities, each lined with numerous asci (the sporangia). In dry weather, the turgid asci elongate—one by one—until the tips reach out of the cavity. This done, the ascus bursts violently, shooting its spores into space. In nature, spore discharge is nocturnal and the average sized *Daldinia* may liberate as many as a hundred billion spores overnight.

The periodic nature of spore discharge has been studied in the laboratory in a promethean and unique manner (Figure 2-13A). Fruiting bodies of *Daldinia* were mounted on a model railway flatcar and the toy slowly pulled along a short length of track by a motor-driven winch. During the times of spore discharge, the spores spewed forth into the air and, being sticky, adhered to glass microscope slides suspended over the railway. Each slide was so spaced that it was in line of fire for only a 2-hour interval. At the end of each day, the spores

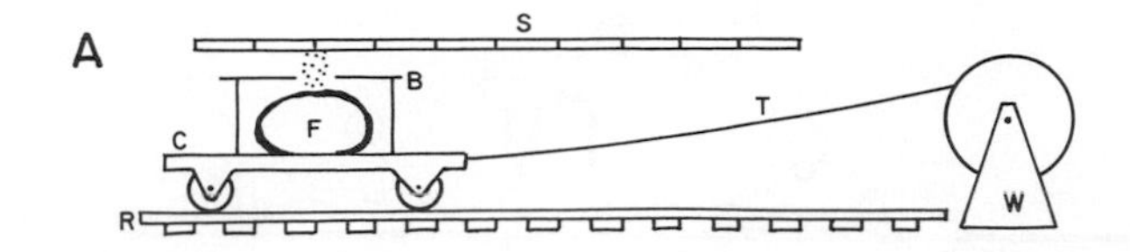

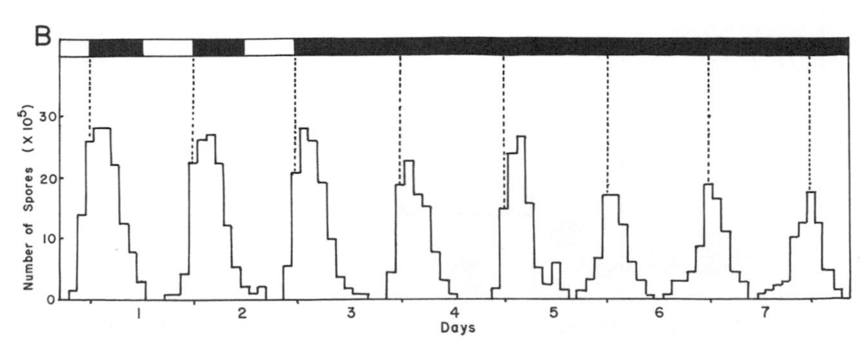

Figure 2-13 (A) Diagram of device for automatic collection of fungal spores. Fungus, F, riding on model railways flatcar, C, is pulled along tracks, R, by a thread, T, attached to a slowly turning winch, W. Spores are seen shooting out of the slitlike opening in the container, B, and, as the spores are sticky, adhere to the glass slides, S, suspended above. Each slide accumulates spores for a 2-hour interval. (B) Rhythmic spore discharge in *Daldinia.* First 2 days in alternating night and day conditions; remaining 6 days in constant darkness. Both drawings modified from C. T. Ingold and V. J. Cox, *Ann. Bot. (London)* **19,** 201–209 (1955).

were washed off the slides and counted and, in this way, spore discharge could be followed continuously for the reproductive life of the fruiting body.

Figure 2-13B, portrays spore discharge in natural, alternating cycles of light and darkness for two days, and then in constant darkness. In the dark, the rhythm would persist for about 12 days, but then, although spore discharge continued, the rhythm was lost. Exposure to three cycles of natural alternating day-night conditions reestablished the rhythm, which would again persist in constant darkness.

Reproduction rhythms have been demonstrated in other fungi also. In 1953, a mutant strain of the bread mold *Neurospora* was found to produce distinct alternating dense and sparse mycelial regions when growing down long cylinders called race tubes (Figure 2-14). Each dense region was found to be composed of a tuft or *patch* of aerial growth, on which orange conidia were produced. Conidia are the spore-forming structures of this fungus. In constant temperature, humidity, and darkness or red light, one conidia "stand" is produced

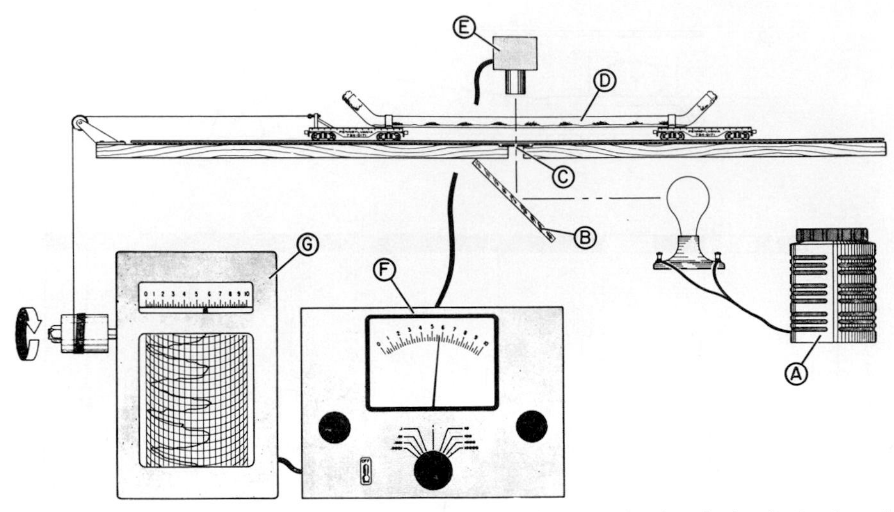

Figure 2-14 The means of studying and recording the reproduction rhythm in the bread mold *Neurospora.* A, variable intensity light source; B, mirror; C, slit to permit the passage of only a narrow band of light; D, race tube; E, photometer sensing unit; F, signal amplifier; G, graphic recorder. The nutrient media in one end of the race tube is inoculated with bread mold which is allowed to grow down the tube for the next few days. At approximately daily intervals, it sends up tufts of conidia. At the end of the study period, the tube is mounted on toy railroad flatcars and pulled by a winch over a light source. The amount of light passing through the fungal culture is measured by a sensitive photocell and the amount transmitted is recorded as a waveform on a chart. From M.L. Sargent *et al., Plant Physiol.* **41,** 1343–1349 (1966).

at approximately daily intervals. The mutant has been appropriately named patch.

When the organism was subjected to constant temperatures ranging between 24° and 31°C, the rate at which the mycelia grew down the race tube almost doubled at the higher temperatures, but the conidia continued to be produced at approximately daily intervals. That is, the period is temperature "independent."

GENETIC BASIS

Comparative studies indicated that this rhythm appeared to be unique to only this strain and, quite naturally, prompted a genetic study in which the patch mutant was crossed with the wild type, nonrhythmic form. The zonation phenomenon was found to segregate in a 1:1 ratio, indicating that the *pattern* was inherited as a single gene. Since this initial investigation, other rhythmic mutants of *Neuros-*

pora have been isolated. They have been named wristwatch, timex, etc., and like patch, in backcrosses with the wild type, gave segregation ratios of 1:1, again indicating that the mutations are at a single gene locus.

A rather interesting facet of the genetic control of the rhythm of one of these strains was recently disclosed. In all previous studies, the ends of the race tubes used had been loosely stoppered with cotton, but now, when rubber plugs were substituted, the rhythm was found to stop. Thinking that the impervious stoppers were exerting their effect by precluding the entrance of oxygen, this gas was pumped through the stoppered tubes. The rhythms returned. However, subsequent experiments showed that oxygen per se was not the necessary element, because blowing almost pure nitrogen through the tubes produced the same effect. It was eventually discovered that the inhibition of rhythmicity was caused by the accumulation of carbon dioxide given off by the fungi and trapped in the confines of the small-diameter tube. In fact, when tubes containing the wild type were aerated, they too became rhythmic! Carbon dioxide concentration as low as 0.125% was found to prevent the expression of the rhythm in the wild type, while the mutant could withstand up to 30% carbon dioxide and still remain rhythmic. Therefore, the 1:1 segregation ratios previously found apparently reflect only genetically determined tolerances of the various strains to carbon dioxide.

Rhythmic "Phototaxis"

PHASE-RESPONSE CURVE IN SINGLE-CELLED ORGANISM

Probably the first conclusive demonstration of a persistent diurnal rhythm in a unicellular organism was obtained in 1948, when it was shown that *Euglena gracilis* responded to light in a rhythmic manner. As this story unfolds, it will be seen again that all the important features of biological clocks occur at the unicellular level of organization; the secret of the organismic timing mechanism is not locked into the complexity of hormonal or nervous systems of multicellular life.

Euglena gracilis is an interesting organism which, because of its plant- and animal-like characteristics, has been somewhat of an enigma to systematists. The organism has chlorophyll and can undergo photosynthesis, but, unlike higher plants, also requires two of the B vitamins and a few amino acids to survive. When placed in darkness, the chloroplasts fragment and are no longer visible, and nutrition becomes heterotrophic. At the base of the flagella is a swelling

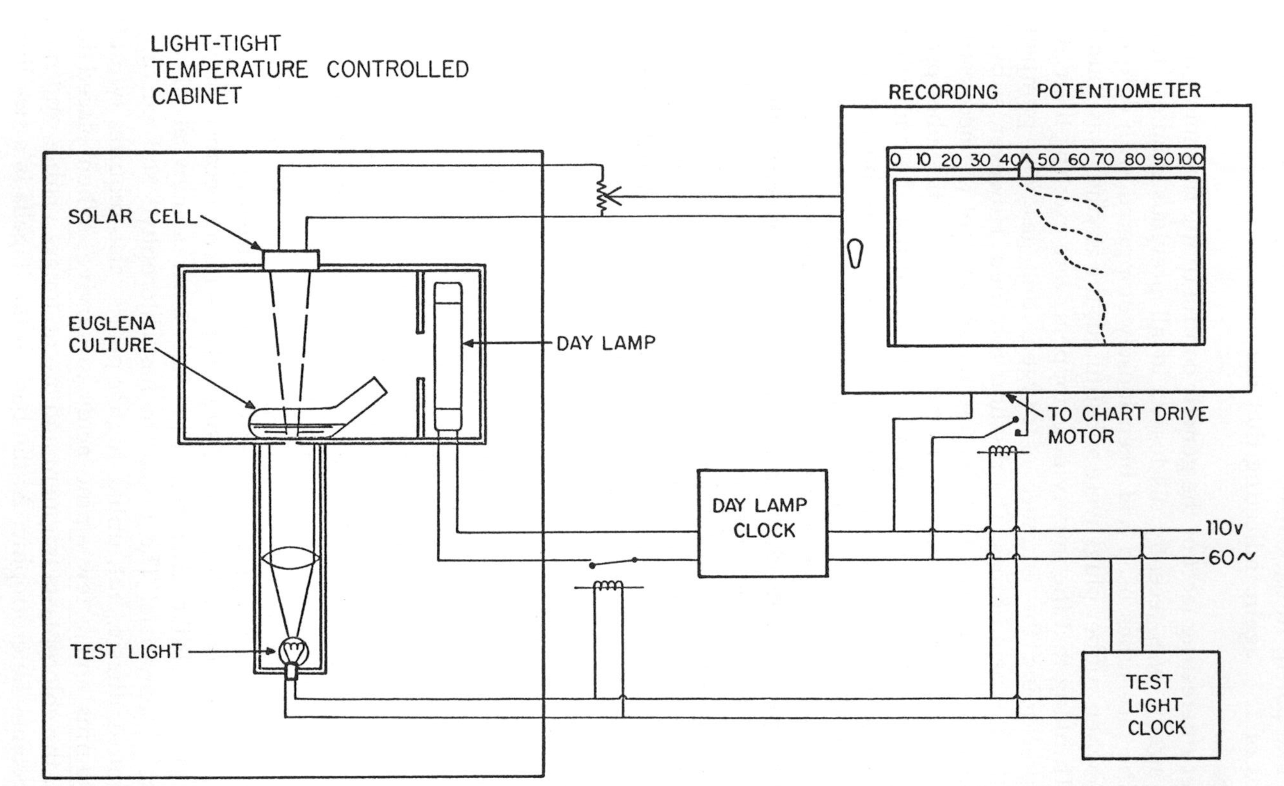

Figure 2-15 Schematic of system used to assay the *Euglena* "phototactic" rhythm. A beam of light is projected through the culture at 2-hour intervals and may or may not attract the cells to it depending on the state of the rhythm. When they do accumulate in the shaft of light, their bodies reduce the intensity falling on the solar photocell and the reduction in current thus produced is recorded on the potentiometer. The day lamp is used to produce a light-dark cycle when required. From V. C. Bruce and C. S. Pittendrigh, *Proc. Natl. Acad. Sci. U.S.A.* **42,** 676–682 (1956).

called the paraflagellar body that functions as a light receptor enabling the organism to swim toward or away from a light source (phototaxis).

To test the phototactic behavior of this organism, a culture was placed in darkness, and, for a period of 15-20 minutes once every 2 hours, a narrow beam of light passed through the culture and allowed to fall on a phototube. The beam served a dual function: first, it attracted the *Euglena* if they were in a photopositive state; and second, it served to measure the numbers of cells attracted to and "trapped" in the beam, since their bodies screened out some of the light falling on the photocell and the current output change caused by this attenuation recorded on a chart (Figure 2-15). It was found that the response to light was rhythmic: the cells being attracted during the middle of the day and showing little or no response at night (Figure 2-16, top). The rhythm would persist for about a week in "constant" darkness and the period was virtually unaltered by constant temperatures ranging between 17° and 33°C. Conditions were, of course, not really constant, as darkness was necessarily interrupted every 2 hours by the test beam. These interruptions were not only necessary to measure the phototactic response, but also to provide the cells with light for photosynthesis. Constant, as used here, means the absence of any 24-hour cycles of light or temperature.

The effect of light pulses in shifting the phase of this rhythm has been thoroughly studied. Because periodic light flashes are necessary to assay this behavior, the phototactic rhythm is not an ideal one for a

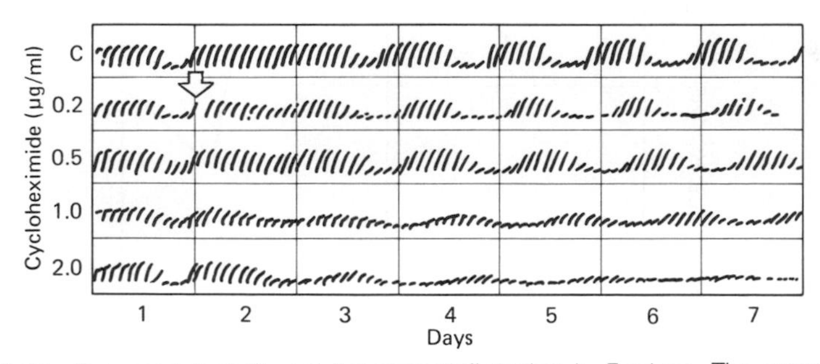

Figure 2-16 The persistent diurnal "phototactic" rhythm in *Euglena*. The curves represent original strip-chart records of four cultures in constant darkness at 25°C. Each rising line corresponds to a decreasing output from the photocell (an increasing concentration of cells in the test-light beam). The top curve, C, is from a control culture; the others, starting at the point indicated by the falling arrow, were treated with cyclohexamide (an inhibitor of protein synthesis). Concentrations (μg/ml) of the antibiotic are given at the left of the figure. Modified from J. F. Feldman, *Proc. Natl. Acad. Sci. U.S.A.* **57,** 1080–1087 (1967).

detailed analysis of phase-shifting responses. However, because *Euglena* is a rhythmic single-celled organism, it was important to test it in this manner to see if it responded to light pulses in a way similar to multicellular organisms.

In a systematic study, single 4-hour light pulses were offered at different segments of the phototactic rhythm to cells maintained in constant dark, and the cultures were observed for several days thereafter for subsequent phase shifts. It was found that quite constant phase changes arose as a function of the time in the cycle during which the light pulses were given. A phase-response curve (Figure 2-17) was constructed that represents an underlying rhythm in the resettability of the phototactic rhythm. The curve indicates that no phase shift occurred if the light pulse was given between hours 0 and 4; increasingly greater delays were produced up to hour 16; and a pulse given at hour 20 caused a 2-hour advance in phase. In other words, the most profound phase changes were caused when a light pulse overlapped, or

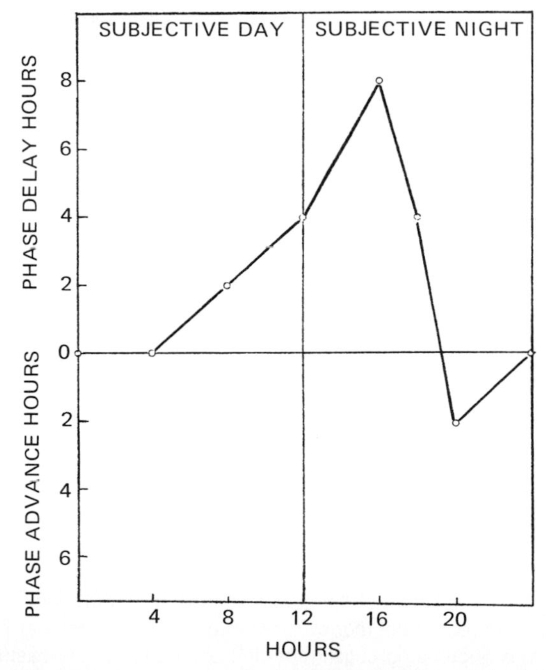

Figure 2-17 The phase-response curve for *Euglena*, for 4-hour light pulses given on the second day of constant darkness. Each point represents the time of the beginning of the light pulse. Zero hour, time when lights would have come on in previous light-dark cycle; hour 12, time lights would have been turned off previously. Modified from J. F. Feldman, *Proc. Natl. Acad. Sci. U.S.A.* **57,** 1080–1087 (1967).

coincided with, the night phase of the rhythm (i.e., when the phototactic response was minimal).

THE ROLES OF CYCLOHEXIMIDE AND DEUTERIUM OXIDE

The phototactic rhythm in *Euglena gracilis* has been examined under treatment by a wide variety of chemical substances such as KCN (a respiratory inhibitor), phenylurethane (a mitotic inhibitor), nucleic acid components [2,6-diaminopurine sulfate (the adenine growth factor analogue), 2-amino-4-methyl pyrimidine (the pyrimidine and nucleic acid analogue), adenine, guanine, thymine, cytosine, and uracil], gibberelic acid and kinetin (growth factors), various pH's, and several culture media, without consistently affecting the phase or period. One substance has been found effective. Various concentrations of cycloheximide, an inhibitor of protein synthesis, were added to late logarithmic-phase *Euglena* cultures (Figure 2-16, arrow) on the second day after the cultures were placed into constant darkness. This produced a longer steady-state period than in control cultures. In cultures treated with 0.2 μg/ml cycloheximide, the period increased to 25.0 hours and lengthened with increasing dosages to 29.6 hours at 2.0 μg/ml. As long as the drug was present, the period remained lengthened; but when the inhibitor was washed out, the period returned again to approximately 24 hours, showing the reversible nature of this treatment.

These results conclusively demonstrate the very positive effect of cycloheximide on the *Euglena*-phototactic rhythm. It must be pointed out, however, that it is still not yet possible to conclude that protein synthesis per se is an integral part of the clock mechanism, since cycloheximide may have other effects also. In fact, protein synthesis, as measured by amino acid incorporation, was not completely stopped by this treatment: inhibition ranging between 19 and 77% for cycloheximide concentrations of 0.2 to 2.0 μg/ml. Additionally, cycloheximide exerts an influence only when cells are well nourished and rapidly dividing; in stable cell culture under nutrient limitation, the period is not influenced.

Both the period and phase of the *Euglena* phototactic rhythm can also be altered by protracted exposures to deuterium oxide (D_2O). Cultures adapted to 20% D_2O showed no alteration in period, but the period of those maintained in 25% D_2O for 3 weeks and then in 45% for a month increased to 27.0 hours, and one culture living in 95% D_2O increased its period to 26.6, while a replicate increased to 27.7 hours. Obviously there was no correspondence between concentration of deuterium and the period-lengthening. When cultures were re-adapted back to H_2O, the period of the persistent phototactic rhythm returned to about 24 hours again.

Unfortunately, the cultures used were too small to determine the actual amount of deuterium incorporated into the cell. Also, the D_2O concentrations stated represent the starting concentrations; there was certainly some exchange of deuterium with atmospheric hydrogen during the long intervals involved in the study (possibly this is why "different concentrations" did not produce different periods). It is quite clear, however, that some isotopic substitution did take place and produced a definite lengthening of the period. The effect of D_2O on rhythmic processes is more distinct when used on mammals and will be discussed later in the chapter.

Other Plant Rhythms

In addition to the rhythmic processes that have already been described in detail, almost all other common plant processes—in a wide variety of plants—are known to be also under biological clock control and thus exhibit persistent rhythms. These include rhythms in nuclear volume, turgor pressure (and in the carrot, this rhythm even persists in cells isolated in tissue culture), starch deposition in leukoplasts, influence of light on the rate of chlorophyll synthesis, heat resistance in leaves, rate of transpiration (this may be an indirect result of a rhythm in the size of stomatal aperture), rate of guttation, rate of negative exudation, sensitivity to auxin, shape of chloroplasts, activity of a variety of enzymes, protoplasmic viscosity, cytoplasmic streaming, migration of plastids, permeability, sensitivity to herbicides, and so on.

ANIMALS

Spontaneous Locomotor Activity

Unquestionably, the most intensively studied rhythm in terrestrial vertebrates has been the one in locomotor activity. There are two reasons for this: it is easy to make long-term observations, because, as will be described in the next paragraph, the animal does all the work and the measurements can be simply automated, thus eliminating the need to interrupt otherwise constant conditions and upsetting an experimental organism by the observer's presence. Although any of a hundred examples could be given, the method used in recording the activity of small mammals will be described here.

METHOD OF STUDY IN MAMMALS

In a room where the temperature and lighting can be controlled, individual animals are provided with small "residence" cages furnished with a 2-week (or more) supply of food and water and built with a floor made of grating so that urine and feces drop through. Affixed to this self-cleaning residence is an exercise wheel into which the animal may step and run in place whenever the mood so strikes. This is often done willingly; the mice in my lab, for example, sometimes rotate their wheels sufficiently during one night to have covered 9 linear miles had they been unrestricted. To transduce these movements into a permanent record, with each revolution of the wheel, a cam projecting excentrically from the wheel margin closes a microswitch which, in turn, produces a pen deflection on a chart recorder (Figure 2-18). With a setup like this, the investigator needs to disturb the constancy of the mice's environment only once every few weeks to renew provisions.

The data obtained in this manner are customarily treated in one of three different ways, each of which is illustrated in Figure 2-18. The simplest way is to cut the chart into 24-hour lengths (Figure 2-18, top) and align consecutive segments one beneath the other, which produces a final figure such as seen in Figure 2-19. This is a particularly useful method of data display to use when one wants to learn the period length of a rhythm in constant conditions; it also uses the direct pen tracings and therefore, is quick and easy. A second way of data analysis is to count up each hour's worth of spikes and plot a daily curve (Figure 2-18, bottom), or combine several daily curves into an average daily one (Figures 2-20B, right-hand side; 3-6). This is really the best method to use, because with it you can determine the period, phase, form, and amplitude of a cycle—however, it is most tedious and time-consuming to produce. The last method, which will be called array analysis is best used for emphasizing trends in data that are otherwise not as "clean" or precise as those shown in Figures 2-19; 7-33. With this method, as is indicated in the center of Figure 2-18, the raw data are reduced to, and presented as, symbols—more specifically, as black squares. The technique is as follows. Hourly activity values are determined for each day's worth of data. A graph is then constructed in which each day is represented as an unshaded horizontal bar divided longitudinally into 24 subdivisions (one for each hour of the

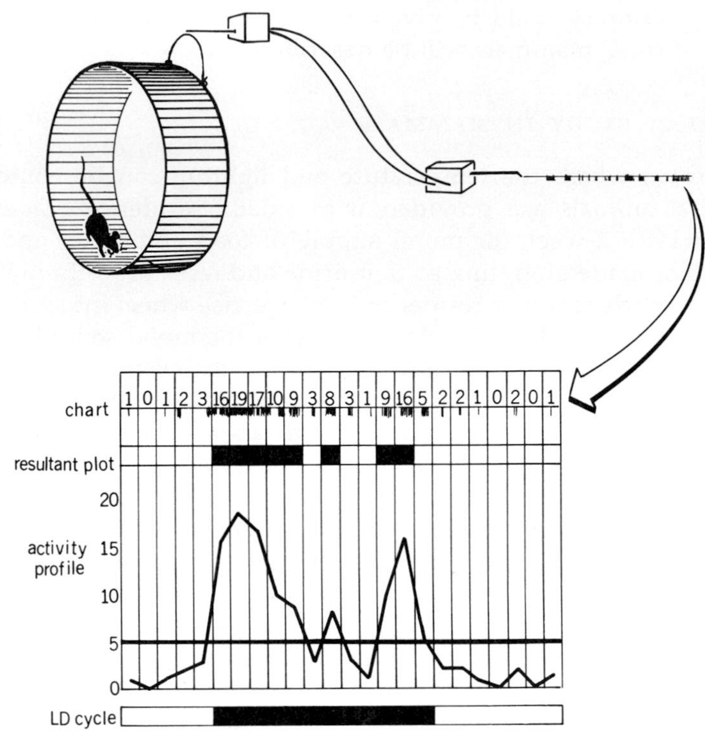

Figure 2-18 Diagrammatic representation of recording and representing locomotory rhythms. Much of the detail of the figure is explained in the text. To produce "array analysis" figures such as 2-20, the individual spikes for each hour are summed (top of figure) and an hourly average for the day calculated (represented by the horizontal line slicing the lower curve). The day is then represented as a row of 24 open blocks, and all blocks in which the hourly activity values are equal to, or greater than, the daily mean are darkened. A day's activity pattern is thus represented as the bar labeled "resultant plot" above. From H. B. Dowse, Ph.D. Thesis, New York University, 1971.

day), and all hourly activity values for that day that equal, or surpass, the daily mean, represented by blackening in the squares on the bar which corresponds to those hours (Figure 2-18, bottom). Thus represented, consecutive days are plotted one beneath the other. The net result is that minor fluctuations (i.e., those below the daily mean) are "filtered out," while the times of maximal activity are boldly emphasized along an otherwise unshaded bar (Figures 2-20; 3-3).

Locomotor rhythms in small mammals such as flying squirrels (Figure 2-19), field mice, and laboratory mice (Figure 2-20) are ideally suited for study. The rhythms of these animals will persist for months in constant conditions, and some studies have been extended to

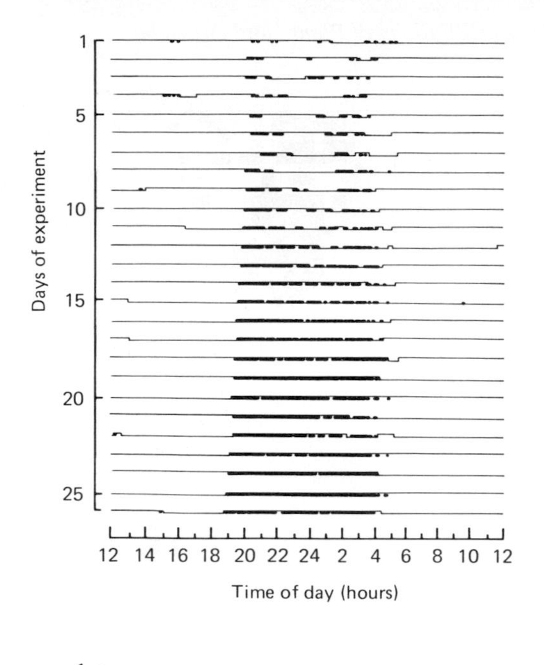

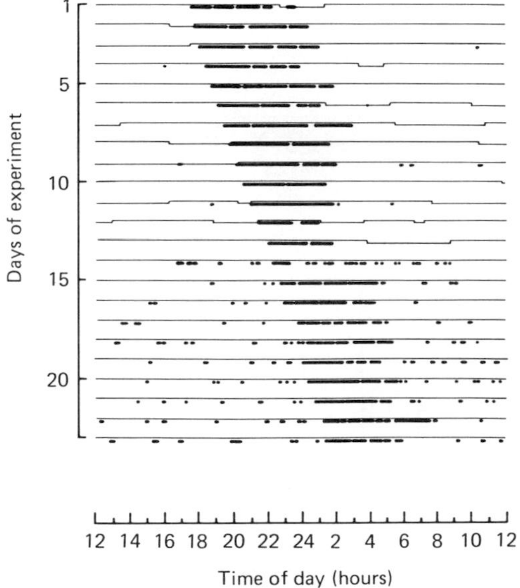

Figure 2-19 The spontaneous locomotor activity of two flying squirrels (*Glaucomys*) recorded in running wheels in constant darkness. The nightly bursts of activity are often so intense that individual spikes blend together to produce a solid band. The circadian period of the upper animal averaged about 23 hours and 58 minutes and that of the other, 24 hours and 21 minutes. Note the extreme precision of the daily onsets of activity, especially in the animal that produced the upper graph. From P. DeCoursey, *Z. Vergl. Physiol.* **44,** 331–354 (1961).

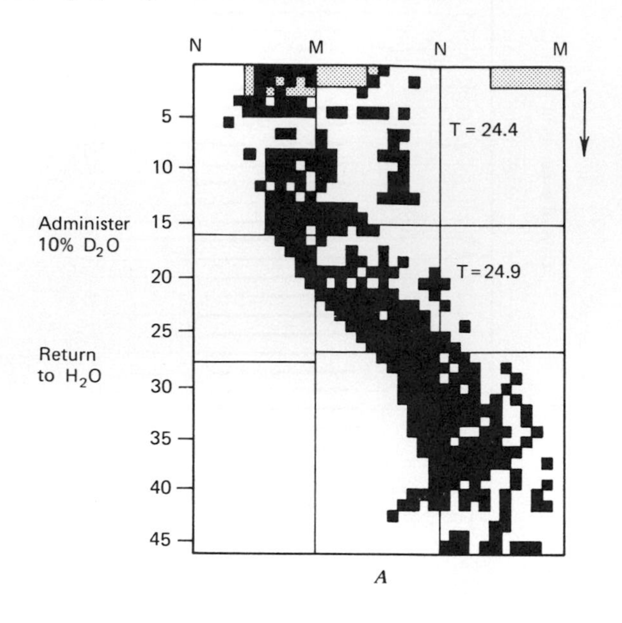
N
M
N
M
T = 24.4
Administer
10% D_2O
T=24.9
Return
to H_2O
5
10
15
20
25
30
35
40
45

A

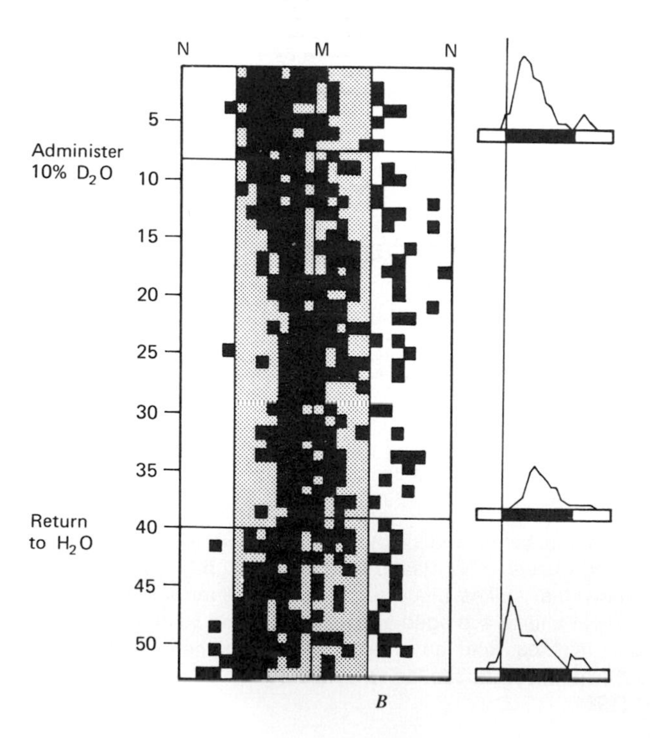
N
M
N
Administer
10% D_2O
Return
to H_2O
5
10
15
20
25
30
35
40
45
50

B

longer than a year. Additionally, individuals within a species are veritable virtuosos as far as the precision of the periods of their rhythms is concerned, which makes them, of course, ideal subjects for experimental manipulation. Thus, the laboratory mouse was chosen for the following study.

THE ROLE OF D_2O ON THE PERIOD, PHASE, AND ENTRAINMENT

As mentioned before, one of the first substances found to have a consistent effect on organismic rhythms was deuterium oxide, also called heavy water. The presence of its known effect on other organisms was tested for in mice. It was administered to them in the simplest possible way—by adding it to their drinking water. It concentrations up to 30%, no deleterious effects were noted; but, above this level, it produced what might be best described as deuterium "intoxication," and the mice did not fair well in their running wheels under its influence.

Heavy water consumption was found to produce a clear-cut response, it lengthened the circadian period of their rhythm (Figure 2-20A) in constant conditions. And, the higher the concentrations that the mice drank, the more the period was lengthened. D_2O also produced an effect in mice maintained in light-dark cycles. In low concentrations, it caused a change in the phase relationship between the illumination cycle and peak activity (Figure 2-20B), or, at higher concentrations of heavy water, the light-dark cycles were no longer able

---

Figure 2-20 (A) Representative example of altering the period of the mouse (*Mus*) locomotor rhythm with D_2O. The dark blocks represent the times of major daily activity; the stippling signifies the hours of darkness. The ordinate shows consecutive days; for graphic clarity, the abscissa has been expanded to a width of 72 hours. At the point signified by the falling arrow, the mouse was switched from a light-dark cycle to constant dim light, and the period of its persistent rhythm determined (to be about 24.4 hours) over the next 13 days. For the next 12 days, it was permitted to drink only 10% D_2O, which lengthened its period to 24.9 hours. The mouse was then given regular water again, and this caused the period to revert to a value close to the starting one. (B) A representative example of altering the phase of the mouse locomotory rhythm with D_2O. The mouse was maintained in cycles of 12 hours of light alternating with a similar length of time in darkness throughout the study. On day 9, the mouse was given 10% D_2O, a concentration that produced a 2–3 hour phase delay (see mean-daily form-estimate curves on right ordinate). A few days after return to regular water the old phase relationship with the light-dark cycle was reestablished. Excerpted from H. B. Dowse and J. D. Palmer, *Biol. Bull.* **143,** 513–524 (1972).

to entrain the rhythm.* Both of these alterations—period lengthening and phase change—are what would be expected if deuterium was acting at the level of an escapement-type bioclock.

Unfortunately, since a consistently effective substance has finally been found, deuterium does not have one, or only a few, specific effects on biological systems. Instead, it is known to produce a whole host of changes ranging from reduction in the rate of many biochemical reactions, through decreased solubility of important respiratory gases, to increased acidity and a reduction in ion mobility. Therefore, it is not known how deuterium effects the clock. There is also the possibility that it is not truly disrupting the clock per se, but is instead altering the coupling mechanism between clock and rhythm. For example, with deuterium, or anything else that causes a slippage in the coupling mechanism between clock and rhythm (as portrayed in Figure 3-14), the period of the latter would be lengthened without the rate at which the clock runs being altered.

IN BIRDS

Birds have also been favorite objects of study for a variety of reasons, but, in particular, because their activity is easy to monitor and they bear up well in captivity. Their movements within a cage are recorded with ease by attaching a microswitch to a hinged perch. Subsequent spontaneous hopping displays a diurnal pattern which will persist for many months at a time in constant dim light. The form of this activity rhythm in some migratory birds changes during the spring and fall of each year. Many species of sparrows and warblers migrate mainly at night and rest and feed during the hours of daylight. If they are captured during peak migratory times and confined to small cages, their "desire" to migrate is sublimated into *Zugunruhe*, or "nocturnal restlessness," which is a tendency to flutter around in the cage each night. Thus, at these times of the year, the activity rhythm is characterized by two peaks per day, one in the daytime and the other at night (Figure 2-21). I introduce *Zugunruhe* behavior here because its existence has made possible many of the pioneering studies on bird orientation which will be described in Chapter 5.

Birds, too, have been conscripted as subjects in heavy-water imbibition experiments, and have been found as appropriate as small

* Entrainment could be reestablished in two ways, either by increasing the intensity of illumination in the light portion of the cycle, or by increasing the total length of the illumination cycle beyond 24 hours.

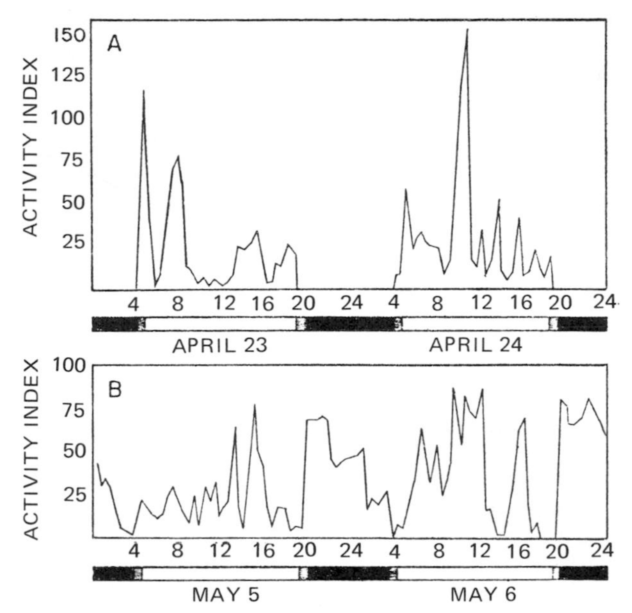

Figure 2-21 (A) The perch-hopping activity of a caged white-crowned sparrow in day-night conditions in the early spring. The bird is active only during the hours of daylight. (B) The activity of a second sparrow captured later in the season while it was migrating. This bird shows a nighttime peak in addition to a daytime one. The "nocturnal restlessness" is called *Zugunruhe* and occurs during the two migratory intervals each year. From J. C. Welty "The Life of Birds." Saunders, Philadelphia, Pennsylvania, 1962.

mammals. They voluntarily consume large quantities of D_2O when it is offered, and concentrations under 30% seem to have no harmful effects. Heavy water has the same effect on their rhythms as it does on all others: it lengthens the period [the degree of lengthening being a function of the concentration consumed (Figure 2-22)] in constant conditions and delays the phase of the rhythm in light-dark cycles. On return to regular water the phase and period assume the pretreatment values.

Extraoptic Entrainment. Bird rhythms may be entrained to light-dark cycles just as can all clock-controlled oscillations. Curiously, not only are the eyes involved in this response, indirect illumination of the brain tissue is also effective. It was found that when house sparrows were surgically blinded and placed in light-dark cycles in which the intensity of the lighted portion was very weak the birds (as might be expected) did not entrain. But, if the feathers were plucked from the already unfortunate bird's head, so that light could penetrate to the

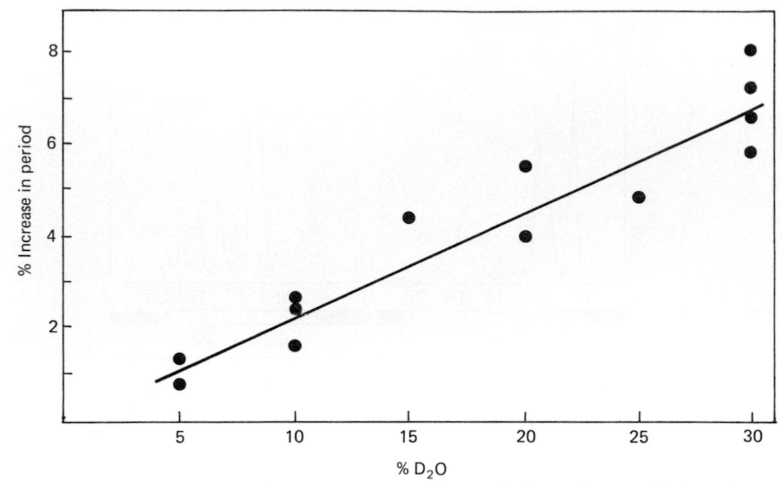

Figure 2-22 The effect of different concentrations of deuterium oxide on the perch-hopping rhythm of the African waxbill (*Estrilda*). At the highest concentrations used, the period was increased by about 7%. From the data of J. D. Palmer and H. B. Dowse, *Biol. Bull.* **137,** 388 (1969).

skull (the feathers reduce the amount of light falling on the skull by 100 to 1000 times), and then on into the brain, the bird immediately entrained to the light-dark cycle (Figure 2-23). Then, however, when India ink was injected under the scalp of these bald birds and allowed to spread over the skull, they again lost their sensitivity to light. Certainly, therefore, the illumination of the brain tissue alone is sufficient to produce entrainment.

Lying embedded on the top of the brain of vertebrates is the pineal organ, a structure known to function as a third "eye" in lizards and some other lower vertebrates. It is present in the sparrow and quite naturally became an immediate suspect as the extraretinal light receptor for entrainment. However, when it was surgically removed from blinded birds, entrainment continued. To date, the elusive photoreceptor has not been found.

The Eclosion Rhythm in *Drosophila*

As briefly described in Chapter 1, in nature adult fruit flies (*Drosophila*) emerge from puparia during a short interval just after dawn. The adaptive significance of emergence at this time—during the coolest, most humid hours of the day when the evaporating power of the atmosphere is at its lowest—is quite obvious: emerging flies still have a soft, permeable cuticle and lose water by evaporation more than twice as rapidly as they do a few hours later when the cuticle has hardened. Also, should an imbalance in water economy obtain, the wings fail to expand properly. (Incidentally, *Drosophila* means "lover of dew.") The eclosion rhythm has been studied in the laboratory in the following manner.

Mature adults of a known age were allowed to lay eggs and the larva made to pupate on a cotton plug, which was subsequently transferred to a funnel-topped canister. In early studies, teams of undergraduate students inverted the canisters hourly and shook out the adults that had recently emerged; later human labor was replaced by an automated shaker that periodically dumped recently eclosed adults into a vial of detergent for subsequent counting. If adults were raised from eggs in constant darkness and allowed to emerge in this condition, eclosion was found to be arrhythmic with some adults coming forth at all times of the day (Figure 2-24A); it will be remembered, however, that transfer to constant light (Figure 1-16) is sufficient to initiate the rhythm. Adults in cultures maintained in alternating light-dark conditions emerged during the first few hours after "dawn" (i.e., after the time of light on); and this rhythm persisted when switched to constant darkness (Figure 2-24B).

In each experiment, the eggs were laid over a 2-day period, yet emergence commenced about 17 days later and lasted about 8 days. Obviously, the flies develop at different rates and would therefore be expected to emerge at all times of the day depending on their individual developmental rates. It is, therefore, somewhat surprising that emergence is confined to a 6-hour slot in the morning. As an explanation of this, it is postulated that the clock controlling the rhythm enforces for each day an 8-hour "forbidden period" on the fly which prevents it from emerging even if it is developmentally ready. Flies ready for emergence at this time are required to wait for the next "allowed period." Support for this proposal is indirectly provided from the shape of the frequency distribution (Figure 2-24) of the daily emergence, which is typically skewed to the right, suggesting that all flies made to hold over during a forbidden period emerged on immediate removal from this restraint.

The fruit fly can be entrained to light-dark cycles, which indicates the presence of an underlying phase-response rhythm in sensitivity to light. Its presence and form were elucidated by experiments such as the one now described. Each of 12 rhythmic cultures, all synchronized to the same phase and then transferred to constant darkness, were subjected to 4-hour light pulses each given at a different time of the

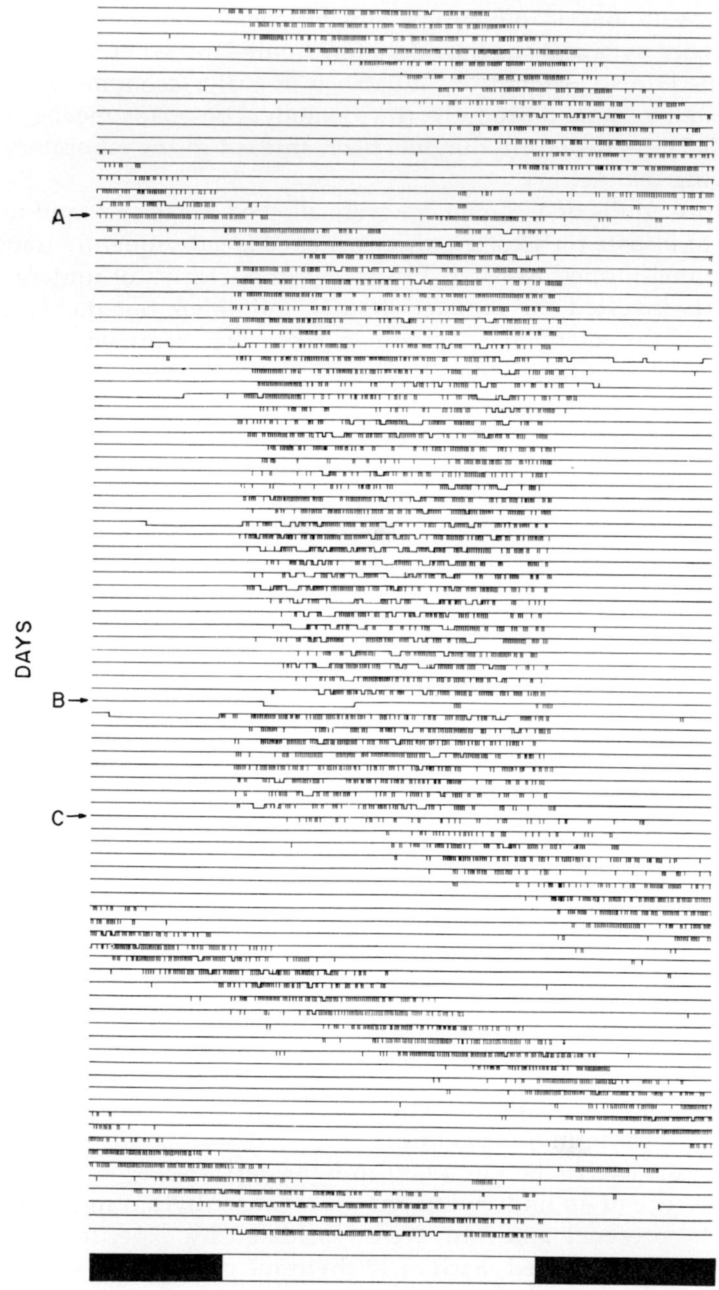

A→

DAYS

B→

C→

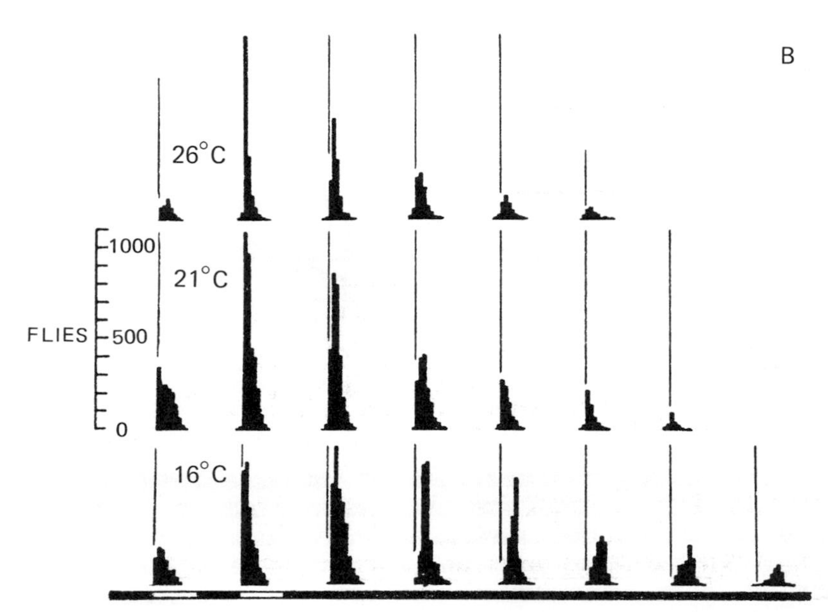

Figure 2-24 The eclosion rhythm of the fruit fly, *Drosophila.* (A) An aperiodic population raised from eggs in constant darkness. (B) The phase relationship of the rhythm to a light-dark cycle and its persistence in constant darkness and different constant temperatures. The cultures were raised at each of the three temperatures and transferred to constant darkness prior to the third peak of eclosion. At 16°C the period is slightly lengthened to about 24.5 hours. From C. S. Pittendrigh, *Proc. Natl. Acad. Sci. U.S.A.* **40,** 1018–1029 (1954).

Figure 2-23 The nonvisual entrainment of the activity rhythm of a blinded house sparrow maintained in light-dark cycles. As can be seen, during the first 17 days, the bird displayed a circadian frequency, signifying that the illumination cycle was not being perceived. On the day indicated by the upper arrow (A), the feathers were plucked from the bird's head, permitting light to penetrate through the skull into the brain—entrainment followed. After 30 days the feathers had regrown sufficiently so that the entraining effect of light was beginning to be lost. Replucking (at B) restored it. At arrow C, India ink was injected under the scalp making the skull opaque to light, and the rhythm became free running again. Therefore, illumination of brain tissue produces entrainment. Modified from M. Menaker, *Proc. 76th Annu. Conv. Am. Psychol. Assoc.* pp. 299–300 (1968).

day (Figure 2-25). On the first day after the light pulse was given, three responses were observed: (i) virtually no change as seen in the top 5 cultures in the figure, (ii) varying degrees of phase delays as seen in cultures 6 through 8, or (iii) phase advances (cultures 9 through 12). Then, on the next day further advances and delays occurred. And, in some of the cultures, even further phase changes took

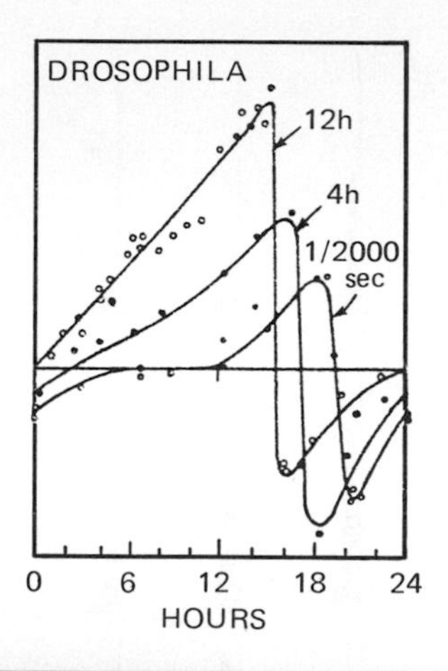

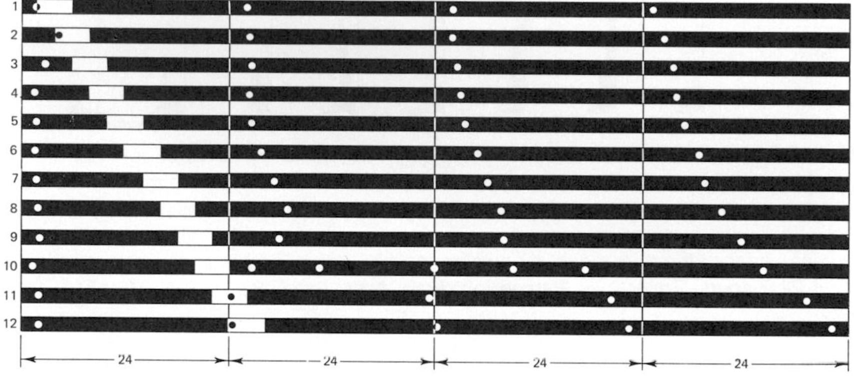

Figure 2-25 The action of light-pulse interruption of the fruit-fly eclosion rhythm in constant darkness. Top: Phase-response curves produced by 12-hour, 4-hour, and 1/2000-second light pulses. Bottom: The effect of single 4-hour light pulses, given at different hours of the day, to 12 separate fly cultures (numbered 1 through 12 on ordinate) previously synchronized to the same phase by light-dark cycles. The circles signify the means of the eclosion peaks. The voids in the solid horizontal bars signify 4-hour breaks in otherwise constant darkness. The phase changes of each culture may be followed by scanning the figure from left to right. Top: from C. S. Pittendrigh, *Cold Spring Harbor Symp. Quant. Biol.* **25,** 159–184 (1960). Bottom: plotted from the data of the same author as published *in* "Perspectives in Marine Biology" (A. A. Buzzati-Traverso, ed.), pp. 239–268. Univ. of California Press, Berkeley, 1958.

place on the next day until all cultures eventually "settled" at a time of day approximately equal to the times of the onset of the 4-hour light signals offered previously. These daily changes in phase that take place until the final "steady-state phase" is reached are called *transients*. By using the final phase adopted, a phase-response curve can be drawn, several of which are shown at the top of Figure 2-25. Different light intensities and pulse intensities produce different response curves; as is seen in the figure, even high intensity pulses—as short as 1/2000 of a second—are sufficient to produce orderly phase changes.

ROLE OF TEMPERATURE ON PERIOD LENGTH, ENTRAINMENT, AND PHASE-RESPONSE CURVES

The relative period stability of the eclosion rhythm has been compared in constant temperatures of 16°, 21°, and 26°C. Separate cultures of flies were kept at each of the test temperatures and a normal light-dark regime all during their developmental stages and then switched to constant darkness when adults began to emerge. As seen in Figure 2-24, the only change in the period was produced at 16°C, where it was lengthened by about 30 minutes.

Entrainment of the rhythm to temperature cycles was found to be possible (Figure 2-26), which suggested, of course, that a temperature-sensitivity phase-response rhythm must be present. The search for it proceeded in the following way. *Drosophila* pupae were glued on a brass holding plate, which, in turn, was bolted to a solid brass "mounting" plate. The temperature of the latter was controlled by circulating water from a temperature bath through milled channels in it.

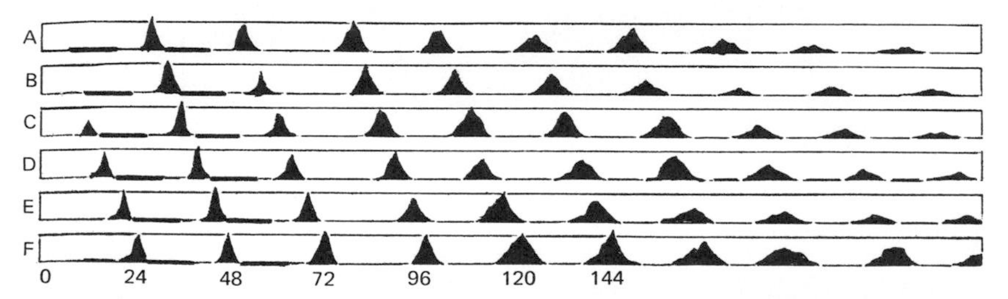

Figure 2-26 Distribution of eclosion peaks during and after entrainment of six fruit-fly cultures to a 24-hour temperature cycle (12 hours at 28°C alternating with 12 hours at 20°C) in constant darkness. The phase of the temperature cycle was systematically varied with respect to local time. The last two 12-hour high temperature pulses are indicated by heavy bars along abscissas. From W. F. Zimmerman *et al.*, *J. Insect Physiol.* **14,** 669–684 (1968).

Temperature variations on the holding plate were less than 0.1°C and temperature changes of 10°C could be produced within 50 seconds. Mounted over the holding plate was a funnel-shaped Lucite container that prevented the escape of newly emerged adults until they could be automatically shaken out into individual detergent vials at 30-minute intervals. Once a day, the flies in each vial were counted.

This device was used to examine the effect of temperature pulses on the eclosion rhythm. Twelve-hour low-temperature pulses (28° to 20°, and then back up to 28°C) were offered in constant darkness at all hours during the cycle. The treatments produced both phase advances and delays; however, the final steady-state change required several days to become established, the interim being filled with transient states. The response curve produced is portrayed in Figure 2-27, where it is seen that the size and direction of the phase change de-

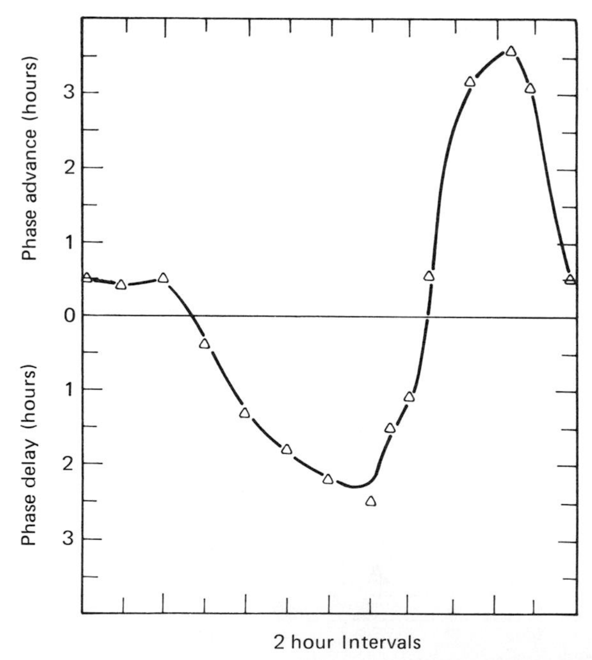

Figure 2-27 The temperature-sensitivity phase-setting rhythm of the eclosion rhythm of the fruit fly. The curve describes the steady-state phase advances and delays produced in the eclosion rhythm 4–5 days after subjecting the pupae to 12-hour low-temperature pulses (28° to 20° and then back up to 28°C) begun at the times indicated on the abscissa. Clearly the magnitude and direction of changes are a function of the time at which the low pulses are given. Modified from W. F. Zimmerman *et al.*, *J. Insect Physiol.* **14,** 669–684 (1968).

pended on the point in the eclosion cycle that the temperature pulse was given.

An experiment that emphasized the importance of an interval of 24 hours in the lives of animals was carried out on the fruit fly. A population was reared in natural light-dark cycles until the first day of adult life, when it was divided into four subpopulations. One group continued living in a 24-hour day cycle, another in a 21-hour "day" cycle (10.5 hours of light alternating with 10.5 hours of darkness), another in a 27-hour "day" cycle (13.5 hours light; 13.5 hours dark), and the last group in constant light. The experiment was repeated four times, and in all cases those flies in the 24-hour day cycle outlived the others by a significant length of time.

The Time Sense of Bees

The time-sense of bees was discovered by a Swiss physician who was plagued each morning at the breakfast table by honey bees who came to gather his marmalade. He observed that they continued to come each morning even when no food was present, nor was the odor of food in the air. He attributed their annoying, repeated visits to a *Zeitgedachtnis* or memory for time. A contemporary naturalist observed that bees frequented a buckwheat field only in the early morning, during the hours that buckwheat flowers secrete nectar. Even if bad weather kept them from leaving the hive for several days, they still returned *on time* to "feed" when clear skies returned. He concluded that these insects probably had a build-in *Zeitsinn* or time sense.

The first experimental observations on the time sense of bees was made in 1929. Colored spots were painted on the backs of 20–30 nectar-collecting bees for individual recognition, and they were then offered sugar water at a specified feeding stand between 10 A.M. and noon each day. Bees visiting this feeding station before 10 and after noon found it empty. After 6–8 days of this training, the frequency of bees visiting the feeding dish was highest between 10 A.M. and noon. The station was now left empty, and an observer recorded the arrival time of each individual bee; it was found that the greatest number of bees arrived to examine the empty bowl during the learned interval. The most industrious ones explored an even wider area in search of their accustomed sugar; they licked all glittering objects such as the watch and pencil of the note-taking observer. When the experimental colony was removed to a hive in a training area called a bee room in the laboratory and maintained in constant conditions, the bees continued to appear punctually at the feeding station at the predetermined time (Figure 2-28) for 6–8 days. The bees could be trained to collect at any time of day.

It was subsequently found that it was impossible to train the bees to a schedule that demanded one visit every 19, 27, or 48 hours. This was true even if the bees were hatched in an incubator in constant conditions; the training interval had to be approximately 24 hours. This aspect of their behavior appears to be inseparably linked to their source of nutrition: flowers have a daily rhythm in nectar secretion, offering it only during certain periods of the day. It has been shown that once a bee has learned what time of day nectar emanates from a particular species of flower, they visit only then and rest in the hive during the intermission.

The period of the feeding rhythm, like all other clock-controlled ones, was found to be "independent" of temperature (between test temperatures of 18° and 35°C).

In an attempt to learn if bees receive their timing information in the laboratory from some penetrating rhythmic geophysical force, they were taken into the depths of the same salt mine in the Bavarian Alps used for the plant sleep-movement rhythm experiment described earlier. In a bee room, 600 feet below the surface of the earth, the time sense was found to persist, with the foragers visiting the feeding dish at the correct time. A past student of mine, H. Burgess Dowse, has visited this mine and reports that the air is "heavy with salt"; a condition that might be expected to disrupt the pulvinal apparatus of a bean leaf (thus destroying the overt rhythm), but not to affect a bee's time sense.

Recently, additional experimentation has been carried out to test for the possibility of a penetrating celestial influence on the bee clock. Identical bee rooms were constructed, one in Paris and the other in New York. Because of the difference in longitudes, local time differed by 5 hours between the two cities. Forty bees were trained to feed between 8:15 P.M. and 10:15 P.M. in constant conditions in the Paris room and, after the last training period, packed in a box and flown overnight to New York where they were set up in the identical bee room. The bees began feeding at their usual Paris summer time (which began at 3:15 P.M. New York time). Had they begun feeding at 8:15 P.M. New York time, it would have signified a 5-hour phase change to local time and, since the bees were in constant conditions, strongly suggested that they were responding to a time-giving geophysical variable penetrating into the test conditions. A reciprocal experiment was performed with training in New York preceding translocation back to Paris. The results were similar.

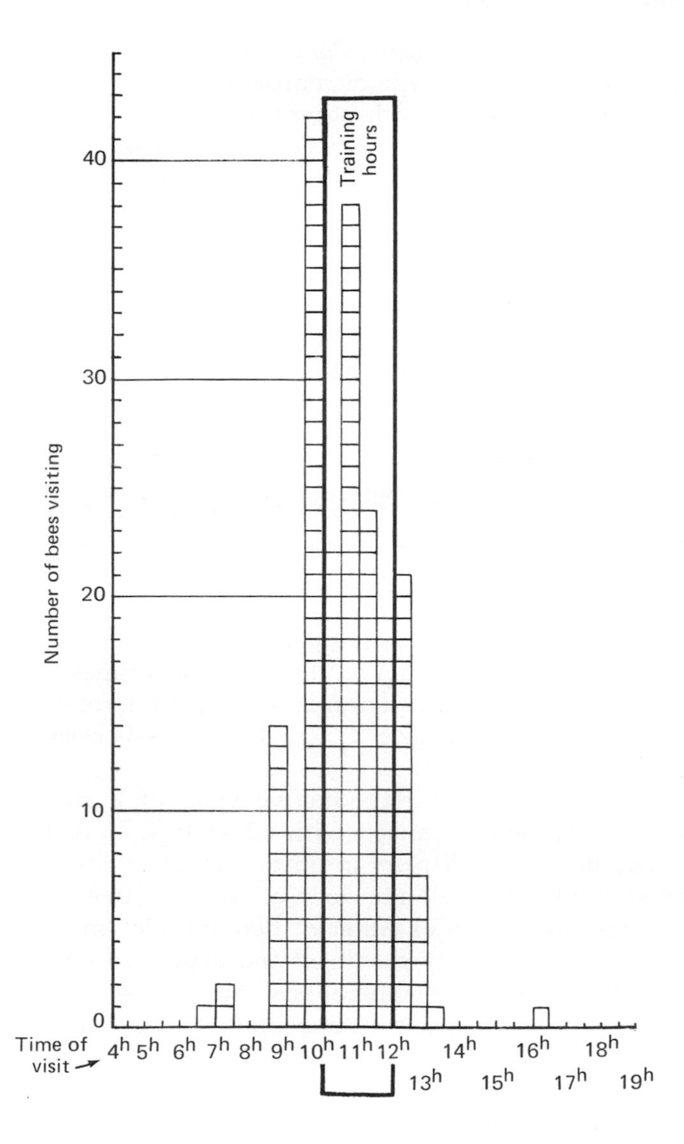

Figure 2-28 The "time sense" in bees. Bees were trained to come to a feeding dish between 10 A.M. and noon for several days and then food was no longer provided. The bees continued to visit the empty dish each day at the predetermined time—even in constant conditions. The graph shows the response on the first day after termination of the training period. The bold rectangle indicates the previous training time. From M. Renner, *Nat. Hist.* **68,** 434–440 (1959).

Rhythmic Sex Reversal

Like all the ciliate protozoans, *Paramecium* undergoes a form of sex called conjugation. In this type of reproduction, two cells conjoin and simultaneously inseminate each other with nuclear material. Usually, there are more than just two sexes, each gender of which is called a mating type and designated by a roman numeral.

Syngen 2 of this species is rather unique in that it undergoes a daily rhythm in sex reversal: in a light-dark cycle consisting of 6 hours of light alternating with 18 hours of dark, the animal is mating type IV during the last 3 hours of light and at least the first 12 hours of darkness, but mating type III during the first 2 hours of light (see Diagram 2-1); at other times, transition is taking place and the mating response is variable.

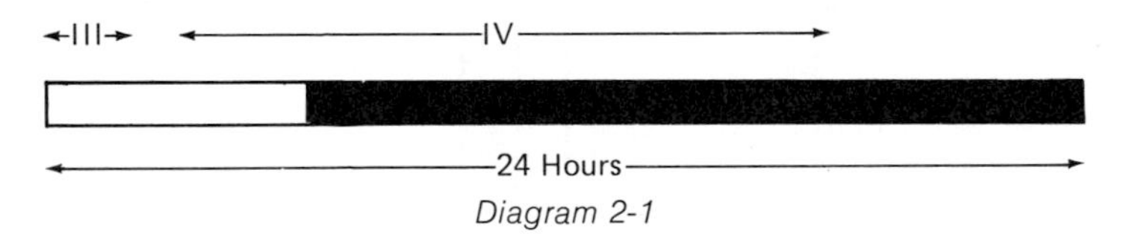

Diagram 2-1

The rhythm will persist in constant darkness with a circadian period slightly shorter than 24 hours, i.e., the switchover times come progressively earlier each day. This rhythmic trait has been found to be under the control of a dominant allele "*C*" (cycler) and only cells homozygous for the recessive are acyclic.

To assay the changing reproductive state of a rhythmic clone, members are periodically mixed with either type III or IV arhythmic clones. The union of cell pairs indicates which of the two types are complimentary to the rhythmic clone at any one time. The rhythm is based on the production of either of two specific "mating-type substances" which are cyclicly produced and expressed out onto the cell surfaces. When present, complimentary cell types agglutinate in preparation for conjugation. The substance then disappears and is replaced by the other mating type substance.

Conjugation does not normally take place until the food supply in a culture has been exhausted. However, up until this time, the cell numbers increase rapidly by asexual reproduction. To demonstrate the role of cell division on the persistence of the mating-type reversal rhythm, 8 individuals from a clone maintained in normal light-dark cycles were transferred to a small flask and placed in constant darkness. By the sixth day the population had increased to 121,800 cells, the nutrient had been consumed, and the cells become sexually reactive. Rhythmic sex reversal was found to be present, with the switchover time coming only slightly before the time of established switchover in control cultures in light-dark cycles. This finding is very interesting for two reasons. First, the phase is seen to be set 7 days before the rhythm was expressed, showing that the clock had continued to run even though no overt rhythm was displayed. Second, note that only the original 8 individuals had been exposed to a light-dark phase-setting cycle—the 121,792 offspring had known only constant darkness—yet all were rhythmic. This means that the periodicity had been accurately passed on through 13 cell divisions (more than 2 per day)—a remarkable feat considering the morphological and functional disruption accompanying cell division!

Rhythmic Cell Division

IN PARAMECIUM

As just described, the mating-type reversal rhythm is under the control of a dominant gene, *C* (cycler). Homozygous dominant (*CC*) or heterozygous (*Cc*) individuals are rhythmic, while *cc* individuals remain as one mating type (either III or IV) eternally. But, when both cyclic and acyclic members of this strain of *Paramecium* were tested for other rhythmic processes, *all* were found to possess one in cell division. To describe the rhythm, 118 cells were removed from parent populations and each one placed alone in its own minute container in constant conditions. Each time one divided, the time of day was recorded and one of the new daughter cells discarded. Cell division occurred over all hours of the day, but the greatest preponderance of fissions took place during the same few hours each day, i.e., the process was rhythmic.

This finding presents a problem in interpretation: Do those members of the strain that display rhythms in both mating-type reversal and cell division have a separate clock for each, or does a single clock with two coupling units drive both rhythms? For the homozygous recessive clones, is a clock missing or simply a coupling mechanism? The answers are not known.

IN RODENTS

Cell-division rhythms have been found to be a common occurrence in many other unicellular and multicellular animals. The rodents have been especially popular subjects for these observations. Tiny punch biopsies are taken at hourly intervals from the pinna of their ears and examined microscopically for mitotic figures. The cell-division rhythms thus discovered peak at night when the animals are most active.

Cancer is a disease characterized by "runaway" cell division resulting in tumor growth. Thirty years ago, physicians first began to wonder what happens to the mitotic rhythm during this malignancy. Some very interesting results emerged from the pioneering work done at Western Reserve University on mice. A carcinogen, a substance which causes cells to become cancerous, was applied topically to a small region of mouse skin where, in time, it produced its insidious consequence. Then, at 4-hour intervals throughout a day and night, the mice were sacrificed and cell division rates in the normal epidermis of each mouse compared to that of the adjacent induced tumor. The results are shown in Figure 2-29 and are quite interesting for the following reasons. As expected, an epidermal cell-division rhythm was strongly expressed in the normal tissue. But this rhythm was lost from the tumor tissue which now reproduced at a constant, augmented rate (as is true of cancer tissue in general); but, at no time, did it divide faster than normal cells at the peak of their division rhythm.

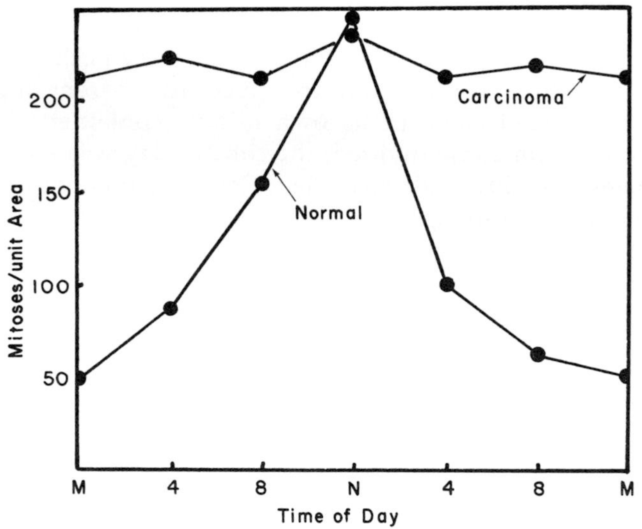

Figure 2-29 The loss of the cell division rhythm in mouse epithelium by the induction of cancer with topical application of 0.3% 3-methylcholanthrene. Both normal and malignant tissue samples were taken from individual mice at 4-hour intervals and the number of mitotic figures counted and compared. Cell division in the tumor has apparently "escaped" from the control of the biological clock. The mice were probably maintained in light-dark cycles, but the conditions were not stated. Modified and redrawn from C. M. Blumenfeld, *Arch. Pathol.* **35,** 667–673 (1943).

The interpretation of these results may run something like this: the biological clock produces a mitotic rhythm by inhibiting cell division at certain times of the day (at night in this case). If the clock is stopped, or becomes uncoupled from the process, the cells of a particular tissue are free to divide at their maximum rate (i.e., just as fast as they can grow, synthesize the necessary intercellular components and physically cleave in half), such as is seen in the tumor cells in Figure 2-29. Continuing this line of logic, cancer (in this case) might be the loss of clock control of cell division.

IN HUMAN CANCER

In studying the rates of cell division in several types of human cancers, either no sign of cell division rhythms are found, or in those rare cases where one is reported it is of such low amplitude and so erratic in period that only extensive statistical analysis can reveal its presence.

IN ROACHES

There is another case—not in man, but in a house guest of his, the cockroach—in which experimental disruption of the animals' rhythms produces cancer.

As those of you who live in metropolitan areas know all too well, our integration with roaches is complete. He is not seen during the daytime unless his hiding place is discovered, but at night he ventures forth into our kitchens to forage. This activity rhythm is controlled by the animal's clock and has been measured in the laboratory by simply tethering them to a recording pen with a fine thread. The rhythm persists for at least a week in the absence of light-dark cycles and kitchens.

Suspecting that hormones might be involved in the control of the locomotor rhythms, an ingenious experiment was performed. Roaches were maintained all their lives in constant bright light, a treatment that totally eliminated all rhythms (Chapter 1). A second batch of roaches was maintained in day-night cycles and they were therefore strongly rhythmic. To each arrhythmic roach was then attached a rhythmic one parabiotically, meaning that their circulatory systems were surgically connected. They were joined one on top of the other and the upper rhythmic one was immobilized by removing its legs—any rhythmic locomotor activity subsequently recorded must be that of the bottom, previously arrhythmic roach. In constant conditions, locomotion was found to be rhythmic, showing that hormonal material carrying information about rhythms had been passed between the conjoined insects.

The next experiment was equally flamboyant. The endocrine glands thought to be producing the hormone are located in the head, so rhythmic roaches were decapitated by callously drawing a tight thread garrote around their necks. This is not as dire to a roach as might be first thought, for a headless roach still locomotes, lives for up to 10 days, and can even copulate, but becomes arrhythmic. Into the abdomens of these animals were latter transplanted the endocrine glands from rhythmic animals. These glands, although removed from their rightful owners, continued to secrete rhythmically; consequently, the decapitated animals became rhythmic again. While the parabiotic experiment simply implicated hormonal material, this one identified the contributing gland and showed that it could function independently as a clock.

In a third series of experiments, and perhaps the most informative, half of the animals were maintained in reversed light-dark cycles for at least 3 weeks, a treatment that reversed their locomotory rhythms. Then for the next 2–4 days, or longer, one endocrine gland was transplanted each day from a reversed-cycle roach into the abdomens of a roach maintained previously in day-night cycles, while removing the previous day's transplant at the same time. This treatment presumed that the recipient roaches received hormone secretion from their own gland at the proper time each day and secretions from the implanted gland during the remainder of the day. The rhythmic secretion of the implant caused malignant tumors to form in the alimentary canals of most of the animals. These tumors could be transplanted to healthy roaches where, with time, they caused other tumors to form also, showing that they were truly cancerous. The experiment was then repeated in the same way, except that the rhythms of the new string of donors was not reversed and were, therefore, in phase with the recipients. With this design, the roaches again got double the daily dose of hormone, but all at the correct time of day, as dictated by their clocks. None of these animals formed tumors. Thus, it appears that it is the time of secretion which is important, rather than the amounts secreted, and that if an endocrine rhythm is disrupted—at least in the primitive cockroach—intestinal cancer is the consequence.*

* Two other laboratories have since attempted to repeat the last phase of the work and found no tumor production. Because of the obvious importance, it is hoped that the original investigator will come forth with further information. The subject is discussed further in Chapter 8.

Color-Change Rhythms

Unbeknown to many beachcombers and seashore naturalists, the fiddler crabs (Figure 1-1) and the green-shore crabs (Figure 3-5) they find on the beach change color each day. Located in the tissue just beneath the crabs' semitransparent exoskeletons are specialized cells, spiderlike in outline, that contain black, red, white, or yellow pigment granules. The pigment can be distributed in either of the two extreme phases: concentrated into a single pinpoint spot in the cell center, or evenly dispersed throughout the cell's body and extensions (Figure 2-30, top). In the latter state, that region of the animal—particularly the legs—takes on the color of the dispersed pigment. Crustaceans, such as the fiddler crab, are dark during the daytime (i.e., the back pigment is dispersed), but blanch out at night. This daily color-change rhythm will persist for weeks in constant darkness and unvarying temperature in the laboratory (Figure 2-30, bottom). The period is somewhat unusual in that it is precisely 24 hours long in constant conditions (rather than circadian) and is not altered when the cold-blooded crustaceans are studied in various constant temperatures between 6° and 26°C.

In originally describing this rhythm, large populations of crabs were studied as a whole, individuals were maintained together in a large tub of seawater, and each one was handled at regular intervals (often once an hour) when the pigment dispersion was assessed under a dissecting microscope. Only a population average response was calculated. This procedure was eventually questioned and an experiment designed to test for the possibility of mutual entrainment and the input of time clues associated with the observational procedures. This time the crabs were kept in individual paper Dixie cups, the color-change rhythms of individuals were followed, and the times for observations were randomly selected. In this way, it was discovered that the method of observations did not give timing information to the crabs and that most of the animals adopted their own circadian period in constant conditions; some were, at times, as much as 12 hours out of phase with the rest. But, if all the individual data were lumped together, the maxi periods canceled out the mini ones, and the population average period was indistinguishable from 24 hours.

MODIFICATION BY COSMIC RADIATION

As an extremely penetrating geophysical force, and one that undergoes a daily rhythm in intensity, cosmic radiation was suspected as

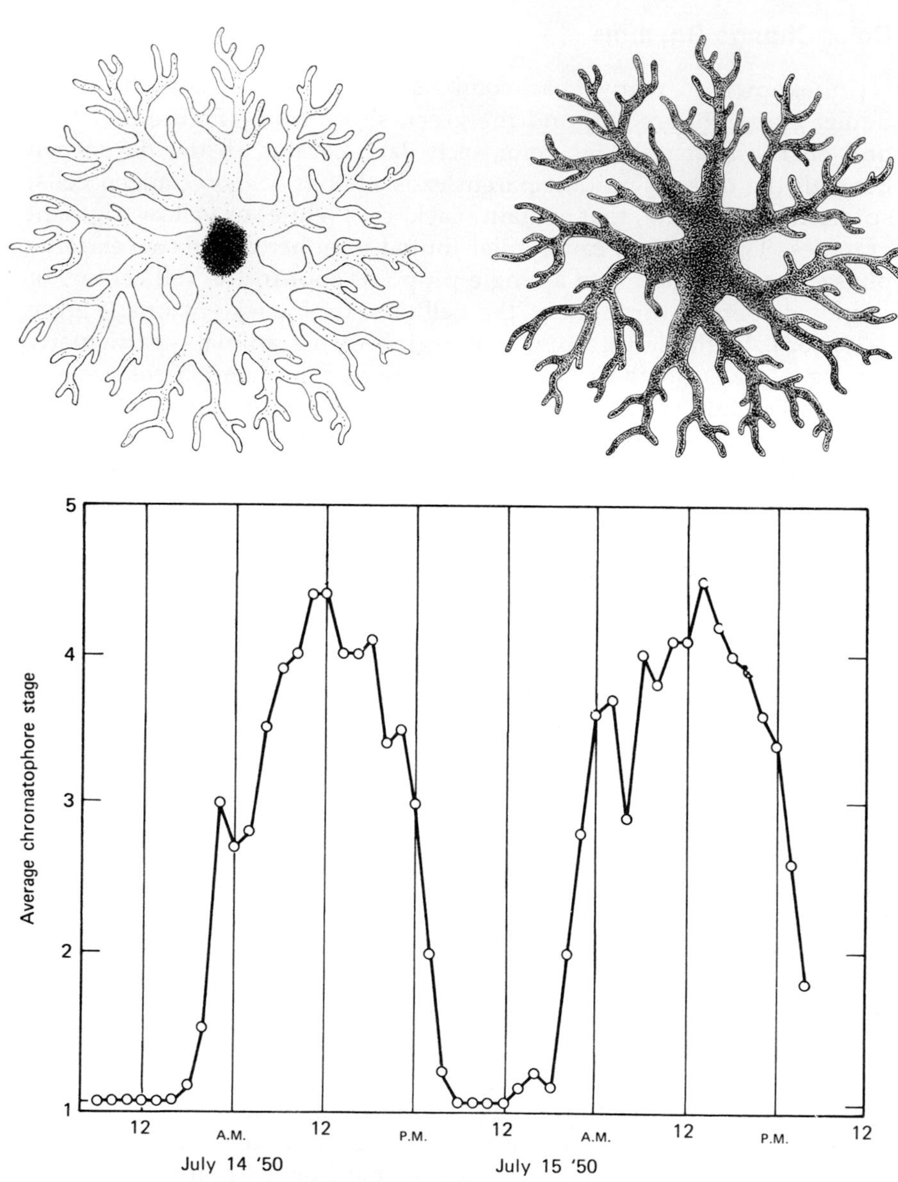

Figure 2-30 Top: The appearance of pigment cells in the fiddler crab as seen through the exoskeleton. On the left, the pigment granules are tightly clumped within the center of cell. On the right, the pigment is shown dispersed throughout the cytoplasm. In this configuration, the overlying part of the animal's shell takes on the general color of the dispersed pigment. Bottom: The average color-change pattern of 50 fiddler crabs during 2 days in constant darkness. The peaks signify maximum pigment dispersion. Modified from F. A. Brown, Jr. *et al., J. Exp. Zool.* **123,** 29–60 (1953); and J. D. Palmer, *Sci. Amer.* **232** (2), 70–79 (1975).

a possible exogenous, time-giving force for biological rhythms. A brilliantly simple experiment was designed to test the effect of this radiation on the fiddler crab color-change rhythm. When passing through lead of thicknesses up to 17 mm, the flux of cosmic radiation increases due to collision with this dense substance. The augmentation is known as cascade multiplication. Above thicknesses of 17 mm, absorption gains the upper hand, and lead becomes the protective shield expected. With this knowledge in mind, lead roofs varying between 3 and 21 mm thick were erected over washtubs of fiddler crabs and their color-change rhythm observed. A significant change in the rhythm was observed, but it was not exciting from a horological point of view. Only the amplitude was altered (it was increased 2–8%) and this, of course, really signified nothing about the *rate* at which the underlying bioclock was running. By way of analogy, if cosmic radiation had a similar influence on a grandfather clock, the chimes of the clock would simply ring louder under lead—the escapement mechanism would not be influenced.

In further experiments, the thickness of the lead roof was increased and decreased several times during the day so as to distort the normal daily intensity rhythm of cosmic radiation. This had no effect on the period of the rhythm.

TRANSLOCATION EXPERIMENTS

The color-change rhythm has also been the subject of a transcontinental displacement study in an experiment designed again to test for the role of a permeating geophysical force. A large population of fiddler crabs was divided randomly and packaged in two identical wooden buckets, and one was flown (surprisingly without airline delays) overnight from Massachusetts to California. The group remaining on the East Coast was even occasionally jostled to partially simulate the mishandling experiences of their flying brethren. At a predetermined time, both buckets were opened in identical constant conditions and the crabs' color-change rhythm followed for the next 6 days. The two laboratories were located 3.4 time zones apart, so that if a geophysical force was involved in the timing of rhythms, a 3.4-hour phase change might be expected to occur in the crabs translocated to California, since the potentially overriding effects of temperature and light cycles were precluded from the laboratory. Other than the flight itself, which seemed to have advanced the peak of the California crabs' rhythm by 22 minutes, no phase change occurred; thus, it was concluded that no geophysical force was involved.

Studies in Antarctica

In a different kind of attempt to examine the possible influence of an exogenous force in the timing of rhythms, a series of studies were conducted in a garage only 800 meters from the South Pole. This remote locality was chosen in an attempt to eliminate any 24-hour periodicity (caused by the rotation of the earth) in some impinging cosmic force. The reasoning was as follows: suppose that a cosmic influence of a constant intensity was emanating from a point source in space. Once a day, a particular longitude of the earth would face directly toward that distant source and 12 hours later, when the earth had revolved ½ turn, face directly away. In the latter position, organisms living at this longitude would be shielded from the empyreal force by the full mass of the intervening earth. Consequently, it is the earth's rotation that produces a daily rhythm in intensity of this otherwise constant force. Organisms perched on the South Pole would not be subjected to this periodic alternation in intensity and, if such a force was truly controlling rhythms, the latter should stop. As an added twist, turntables revolving opposite to the direction of the earth's rotation were used as experiment platforms.

Locomotion in the golden hamster, sleep movements in bean plants, zonation in bread mold, and fruit fly eclosion rhythms were studied in constant darkness and fairly constant temperatures. Persevering through all the tribulations encountered by working in rustic conditions and an outside summer temperature of 40°C below zero, the investigators performing these experiments found no alternation in their organisms' rhythms. One sidelight discovery was that the bean plants continued to twine in a counterclockwise direction as they normally did in the northern hemisphere, unlike the reversed maelstrom produced in the bathtub drain south of the equator.

Since the south geomagnetic pole does not coincide with the geographic one (it is about 1500 miles away near New Zealand), but revolves around the latter producing a low amplitude daily rhythm in several parameters of geomagnetism at the pole, a further experiment was conducted. On return to the California laboratory, three of the organisms were subjected to a man-made diurnal rhythm in magnetic flux alternating between 0.68 and 25 gauss (many times the strength of the earth's field). The rhythms were not influenced.

Rhythmic Susceptibility to Foreign Substances

As the medical profession became more aware of the rhythmic nature of physiological processes in man (as will be described in Chapter

4), the need to investigate the effect of medications, toxins, etc., over a time span of 24 hours became apparent. As indicated by many overt signs, such as the daily temperature curves (Figure 4-8) of feverish patients, it became obvious that people are sicker at certain times of the day than at others. It follows that the ill probably react differently to standardized doses of medication during the day. Furthermore, drugs should be metabolized and poisons detoxified at different rates at different times of day. One of the first studies concerned with this possibility involved the periodic administration to mice of identical doses of alcohol. They were injected intraperitoneally with a standardized amount of 25% ethyl alcohol at 4-hour intervals and the lethality observed thereafter. In a light-dark cycle with light on between 6 A.M. and 6 P.M., the greatest number of deaths was produced by injections given between 4 P.M. and 8 P.M. (Figure 2-31).

Work of this nature has now been extended to other agents with similar results. In experimental situations identical to the one just described, mice have been subjected to injections of *E. coli* endotoxin (a poison produced by a common intestinal bacterium of man), which

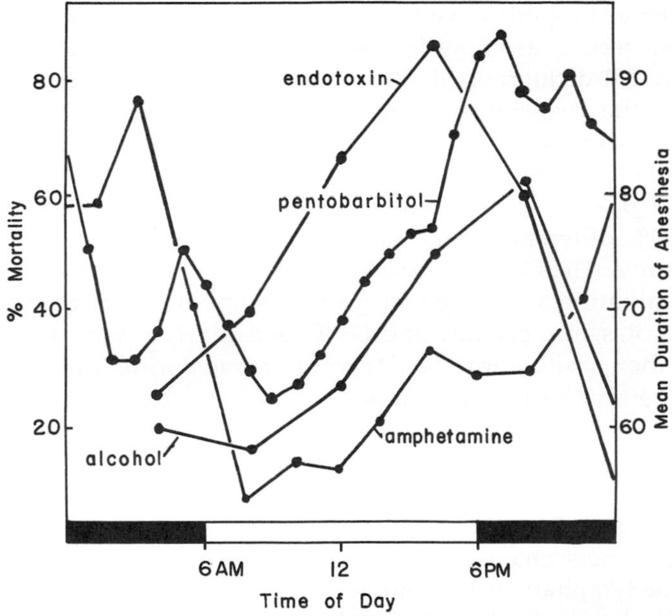

Figure 2-31 The rhythmic sensitivities of rodents, maintained in light-dark cycles, to toxic and narcotic agents. Drawn from the data of E. Haus and F. Halberg, *J. Appl. Physiol.* **14,** 878–880 (1959); F. Halberg *et al., Proc. Soc. Exp. Med.* **103,** 142–144 (1960); L. Scheving *et al., Anat. Rec.* **160,** 741–750 (1968); *Nature (London)* **219,** 621–622 (1968).

killed over 80% of them if injected at 6 P.M., but less than 20% when administered at midnight (Figure 2-31). When rats were subjected to amphetamine, identical doses became lethal overdoses for 78% of the animals injected at 3 A.M., but only 7% at 6 A.M. (Figure 2-31). Librium, a common tranquilizer taken by man, produced a lethality curve in mice similar to the one just described for amphetamine. A great many other substances, including strychnine, pneumococcal challenges, and x-irridation, have also been found to vary in lethality over the 24-hour day.

The temporal sensitivity of deer mice and rats to a standardized therapeutic dose of the barbiturate Nembutal (sodium pentobarbital) has also been tested. Subgroups of a rodent population living in natural light-dark cycles or constant conditions were injected periodically with this anesthesia and the average sleep time, i.e., the interval between injection and the time the animal was able to stand again, calculated. The greatest anesthetic effect occurred during the dark portion of the cycle (Figure 2-31), and the rhythm would persist in constant dim illumination.

As might be expected, rhythmic susceptibility is not restricted only to rodents: arthropods have been shown to have rather precise, time-dependent responses to pesticides. For example, the lethality of DDVP (dimethyl 2,2-dichlorovinyl phosphate), an acaricide fumigant, was tested on the two-spotted spider mite (*Tetranychus*). Female mites, otherwise maintained in a light-dark cycle, were affixed to microscope slides and dipped into 0.005% DDVP for 5 seconds at 2-hour intervals. The major peak of susceptibility occurred 2 hours after light off. There was an 80% difference in increase of kill between minimum and maximum values. The German cockroach (*Blattella*) maintained in constant dim illumination was killed in greater numbers by nighttime injections of potassium cyanide and DDT. Sensitivity rhythms to the narcotics, ether, chloroform, and carbon tetrachloride in crickets and mites have also been demonstrated.

Susceptibility Rhythms and Cancer Therapy

Practical use of susceptibility rhythms has been made. Leukemia is a form of cancer characterized by an overproduction of white blood cells in the lymphatic tissue and bone marrow. The hoards of cells produced circulate throughout the body, invade and take up residence in abnormal sites, and eventually kill the individual. Certain chemicals that inhibit cell division are especially effective in the treatment of leukemia. Mice, rather than men, are the initial subjects used in screening the effectiveness of potential therapeutic drugs.

To produce animals suitable for experimentation, healthy mice are injected with leukemia cells, which eventually spread throughout their bodies and kill them. If the inoculum of malignant cells is not too great, and treatment is begun early, the mice can be cured with medications such as arabinosyl cytosine, a drug that inhibits cell division and, with time, causes cell death. This drug is not, however, the universal elixer for leukemia therapy, for it attacks noncancerous cells also. For example, if the original inoculum of cancer cells injected into the mice is not large and small identical amounts of arabinosyl cytosine are injected at 3-hour intervals during a 24-hour period (the standard injection protocol), the mice are cured. As the size of the inoculum is increased, however, the amount of arabinosyl needed to eradicate the cancer increases proportionally to a point where a dose of say 240 mg/kg of body weight given over a 24-hour period causes the mice to die from the treatment rather than the disease. By taking into account the rhythm to be described next, this problem can be alleviated considerably.

At an earlier date, it had been found that by injecting mice with large doses (actually overdoses) of arabinosyl cytosine at different clock hours and then observing their survival rates the subsequent mortality differed as a function of the hour of injection. Significantly more mice were killed by injections given at night than during the day. Knowledge of this rhythm in sensitivity to the drug was then employed in the design of the following experiment.

As had been shown before, a total daily dosage of 240 mg of arabinosyl given in eight, 30-mg injections proved fatal. This grand total was used again, but doses were adjusted to compensate for the rhythmic sensitivity to the drug: only minute amounts were given to mice (who had previously received staggering inocula of leukemia cells) when they are most sensitive to the drug (middle of the night), and massive injections were given during midday when they are least troubled by its presence. The injection schedule (with doses in milligrams) is shown in the following tabulation.

	A.M.				P.M.				Total dose in 24 hours (mg/kg)
	2	5	8	11	2	5	8	11	
Dosage (mg/kg)									
Experimentals	7.5	15.0	30.0	67.5	67.5	30.0	15.0	7.5	240
Controls	30.0	30.0	30.0	30.0	30.0	30.0	30.0	30.0	240

This schedule was repeated for 4 days and the survival rates of the mice observed. The experimentals, which received the arabinosyl in a nonuniform regimen, survived about twice as long as the others.

Therefore, by taking into account the mouse's rhythmic sensitivity to the drug, more massive doses can be given with less harm to the mouse. Extrapolating from mouse to man, findings like these could be very important. Arabinosyl cytosine as presently used in the clinical treatment of malignancies must be given in borderline toxic doses to be effective. If man's tolerance to the substance is rhythmic—and while the experiment has not been done, I suggest that there is a good chance that it is—then total daily dosages might be safely increased with rhythm-based injections and the chemotherapy would be much more effective. Therefore, being aware of rhythms in physiology could save additional lives, especially in the administration of drugs that retain their activity for only a short time after administration.

SUMMARY AND CONCLUSIONS

1. Plant and animal rhythms in the most fundamental processes, such as activity, metabolism, photosynthesis, feeding, cell division, growth, and reproduction, were discussed. More exotic rhythmic processes such as luminescence, color change, and sensitivity to poisons and drugs were also described.

2. Because the experimental designs used in biorhythm work often differ greatly from those normally encountered in the laboratory, several are described here. An ideal design is one in which measurements can be automated (thus eliminating observational bias) and can be made without interrupting the otherwise constancy of the laboratory setting and the tranquility of the experimental subject. Suitable experimental organisms are those that take well to captivity and isolation, are easily cared for, and display *precise* overt rhythms.

3. There are individual differences in the precision of the periods of rhythms displayed in constant conditions.

4. Some rhythms, like locomotor activity in mammals and birds, and color change in crabs, will persist for months or years in constant conditions. Other rhythms damp out after just a few days. Persistence depends on the species studied, the conditions imposed, and individual tenacity.

5. The deep-seated nature of the 24-hour period has been demonstrated by (i) showing that bees could not be trained to periods that differed much from the interval of a day, (ii) that rhythms revert to a

period close to 24 hours after being entrained to artificial day lengths by light-dark cycles, and (iii) that organisms such as fruit flies do not live as long in artificial "days" longer or shorter than 24 hours.

6. Rhythms may appear spontaneously in organisms that have never been exposed to a day-night cycle (such as is found in the *Paramecium* sex-reversal rhythm and numerous other organisms not mentioned), or initially arrhythmic organisms can be caused to become rhythmic by a single, nonperiodic stimulus (as has been described for fruit flies).

7. The single-celled level of organization is sufficient for the expression of clock-controlled rhythms, as is shown by the mating-type reversal, and cell-division rhythms in *Paramecium;* and the several rhythms of *Gonyaulax.* The rhythms displayed by unicells show all the same properties as those of higher organisms, including alteration by D_2O, valinomycin, and alcohol.

8. Rhythms will persist even in enucleated cells such as *Acetabularia.* Dormant oat seeds, in which nuclear activities are suspended, are also rhythmic.

9. Segments isolated from multicellular plants and animals and maintained in culture continue to display the same rhythms of the intact organism.

10. Because organisms—even unicells—often display more than one rhythm, the question is raised as to whether each rhythm is goverened by its own clock, or if one clock controls all the rhythms, each via a separate coupling unit. There is some evidence suggesting the latter situation is the case.

11. Individuals have been shown to display simultaneously rhythms with different periods [such as the bean sleep-movement rhythms, the growth rhythms in the oat (and as will be described in the next chapter, many organisms possess both tidal and daily periods)] and different phases [e.g., the many rhythms of *Gonyaulax* and the two rhythms of *Paramecium* (and as will be described in Chapter 4, the 50 rhythms of man)]. This again raises the question of whether such organisms have separate clocks running at these different rates, or whether the frequency of a single clock may be transmuted into the different periods by frequency-transducing coupler units.

12. Deuterium oxide increases the period lengths of rhythms in beans, *Euglena, Gonyaulax,* mice, and birds. Also, the phase relationship between a light-dark cycle and these rhythms is delayed under the influence of heavy water.

13. Backcrosses between rhythmic mutants and nonrhythmic

members of a species (such as those done with the bread mold) indicate that the rhythmic trait is inherited as a single gene. As has been shown, however, the genes often do not affect the clock mechanism, but only the particular process controlled by it. However, inheritable alterations in the period length and phase (as described in the fruit fly) do suggest direct changes in the clock or coupling mechanism.

14. Translocation experiments across several time zones to identical constant conditions did not alter the feeding rhythm in bees or the color-change rhythm in crabs.

15. The intracellular clock continues to function accurately during cell division (when it too is presumably replicated). The clocks in the fruit fly also continue to run accurately throughout the extreme morphological changes and cellular disruptions accompanying development from the larva, to the pupa, to the adult.

16. The clock involvement in cell division appears to be lost in tissues that become cancerous.

Selected Readings

Aschoff, J. (1963). Comparative physiology: Diurnal rhythms. *Annu. Rev. Physiol.* **25,** 581–600.

Aschoff, J., ed. (1965). "Circadian Clocks." North-Holland Publ., Amsterdam.

Bennett, M. F. (1957). "Living Clocks in the Animal World." Thomas, Springfield, Illinois.

Biological Clocks. (1960). *Cold Spring Harbor Symp. Quant. Biol.* **25,** 1–524.

Bünning, E. (1957). Endogenous rhythms in plants. *Annu. Rev. Plant Physiol.* **7,** 71–90.

Bünning, E. (1973). The Physiological Clock," 3rd ed. Springer-Verlag, Berlin and New York.

Cumming, B. C., and Wagner, E. (1968). Rhythmic processes in plants. *Annu. Rev. Plant Physiol.* **19,** 381–416.

Enright, J. (1970). Ecological aspects of endogenous rhythmicity. *Annu. Rev. Ecol. Syst.* **1,** 221–237.

Harker, J. E. (1958). Diurnal rhythms in the animal kingdom. *Biol. Rev. Cambridge Philos. Soc.* **33,** 1–52.

Harker, J. E. (1964). "The Physiology of Diurnal Rhythms." Cambridge Univ. Press, London and New York.

Reinberg, A. (1973). Chronopharmacology. *In* "Biological Aspects of Circadian Rhythms" (J. N. Mills, ed.), pp. 121–152. Plenum, New York.

Sweeney, B. M. (1963). Biological clocks in plants. *Annu. Rev. Plant Physiol.* **14,** 411–440.

Sweeney, B. M. (1969). "Rhythmic Phenomena in Plants." Academic Press, New York.

Sweeney, B. M. (1972). Circadian rhythms in unicellular organisms. *In* "Circadian Rhythmicity," pp. 137–156. Center for Agricultural Publishing and Documents, Wageningen, Netherlands.

Webb, H. M., and Brown, F. A., Jr. (1959). Timing long-cycle physiological rhythms. *Physiol. Rev.* **39,** 127–161.

Wilkins, M. B. (1973). Circadian rhythms in plants. *In* "Biological Aspects of Circadian Rhythms." (J. N. Mills, ed.), pp. 235–280. Plenum, New York.

3

Tidal (Bimodal Lunar-Day) Rhythms

A SURVEY OF TIDAL RHYTHMS

The Activity Rhythm of the Fiddler Crab

That part of the shore called the intertidal zone is a region on most coastlines of the world which is alternately flooded and then uncovered by the sea twice each *lunar day* (an interval defined in Figure 3-1). The organisms living there often reflect this period in their behavior and physiology. For instance, as was described in the beginning of the first chapter, one of these shore dwellers, the fiddler crab, sits out high-tide inundations in the depths of its burrow; but, as soon as the tide recedes, he emerges to wander over the freshly exposed marsh surface. This behavior is studied in the laboratory using a technique so elementary in design that it is virtually foolproof.

The discarded plastic boxes in which fishing lures are purchased are used to house the crabs individually, without food, but with a supply of seawater. The boxes are balanced on knife-edged fulcrums so that as the incarcerated animals run between ends the container rocks back and forth like a teeter-totter. A tip in one direction closes a microswitch, while one in the opposite direction opens it (Figure 3-2); both movements cause a pen deflection on a moving chart which thus

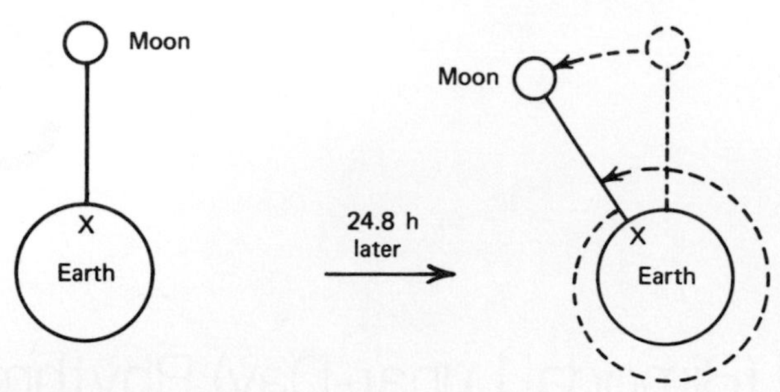

Figure 3-1 A view "from above" of the relative movements of the moon and earth, which produce the *lunar* day. The earth rotates on its axis in a counterclockwise fashion, and the moon circles the earth in a similar direction. As the earth completes one rotation (which takes 24 hours) relative to the sun, the moon's travels have placed it in a new location (therefore, it is not a stationary reference point). Thus for a given longitude (signified above as X) to face the moon again, the earth must "catch up" by rotating an additional 13°. Consequently, the interval between successive moonrises at a given longitude is 24 hours and 51 minutes. (By comparison, a solar day is the interval between successive sunrises and, of course, totals 24 hours.) It is the moon's gravitational attraction that causes the tides on earth, which produces 2 inundations each lunar day. From J. D. Palmer, "Biological Clocks in Marine Organisms." Wiley (Interscience), New York, 1974.

records the event. The whole uncomplicated, homemade device is elevated to scientific respectability by calling it an actograph.

Sixty or more crabs, each in its own actograph, are then placed in constant conditions in a standard-sized incubator and their spontaneous activity recorded. The persisting rhythmic pattern of an especially precise animal—named Supercrab because of his punctuality—is seen in Figure 1-1. Just as the period of solar-day rhythms usually lengthen or shorten in constant conditions and are thus referred to as being circadian, the periods of persistent bimodal lunar-day rhythms also often alter and, when they do, are called *circalunadian*. Such a change can be seen in Figure 1-1 by drawing a straight line between the points on

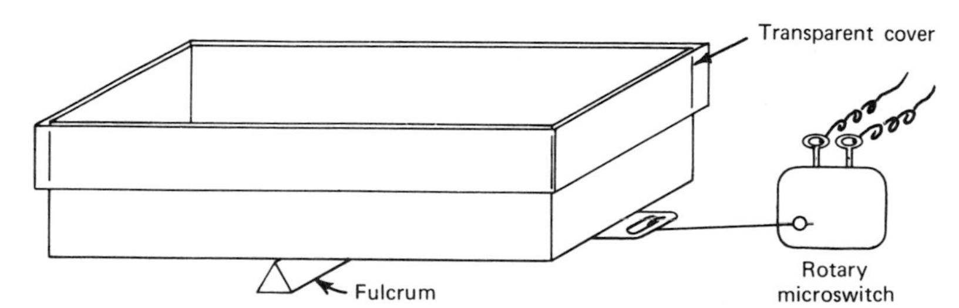

Figure 3-2 A teetering-box actograph. When the crab within the box moves from end to end, the container tips in either direction, depressing or lifting the arm on the rotary microswitch. The switch in turn causes a pen mark on a moving chart. From J. D. Palmer, "Biological Clocks in Marine Organisms." Wiley (Interscience), New York, 1974.

the left-hand sides of the top and bottom figure frame, and then joining the same two points on the right-hand sides of the same frames, so that two parallel diagonal lines are superimposed over the figure; the points at which they intersect each daily curve represent the approximate midpoints of low tides on that day. Thus done, it will be quickly seen that the circalunadian period of Supercrab is longer than 24 hours and 51 minutes—it is, in fact, about 25 hours and 40 minutes in length.

The basic metabolism of the fiddler crab also reflects the state of the tide. This discovery was made by placing the animals in the reaction vessels depicted in Figure 2-7 and sealing them away in constant conditions. The rhythm displayed mimicks in form and phase the one for activity (Figure 1-1) and is caused, in fact, by the increased oxygen consumption required by the crabs "running in place" on the slick concave walls of the glass vessel at the times of low tide.

Dual Rhythmic Components in the Penultimate-Hour Crab

Intertidal creatures like the fiddler crab are subjected to two major dominating periodicities: the 24-hour interval of day-night light cycles, and the twice per lunar day tidal inundation. Many of these organisms display both these frequencies in their persistent rhythms. As an example, the laboratory behavior of the penultimate-hour crab (*Sesarma*), an intertidal crustacean with a dominating nocturnal activity rhythm that peaks at 11 P.M., will be discussed. When sample populations of these animals are maintained in constant darkness and their combined daily activity pattern displayed using the array analysis technique (page 61), a figure such as 3-3 is produced. Here the strong nocturnal component is seen as a bold band of activity occurring between 7 P.M. and 4 A.M. each night. Additional major bursts of activity are seen to appear around the midpoints of high tide (indicated by the diagonal lines on Figure 3-3). Therefore, here is a crab with dual rhythmic components (both unimodal solar-day and bimodal lunar-day frequencies) in a single physiological manifestation—locomotion (these components are summarized diagrammatically in Figure 3-4). Notice that the periods of these components are indistinguishable from 24 and 24.8 hours, rather than becoming circadian and circalunadian. This again is an effect of lumping the data of many individuals together and showing only the average response of the population. Individual animals have circa periods, both longer and shorter than their geophysical counterparts, but these tend to cancel each other out when lumped together and expressed as an average response.

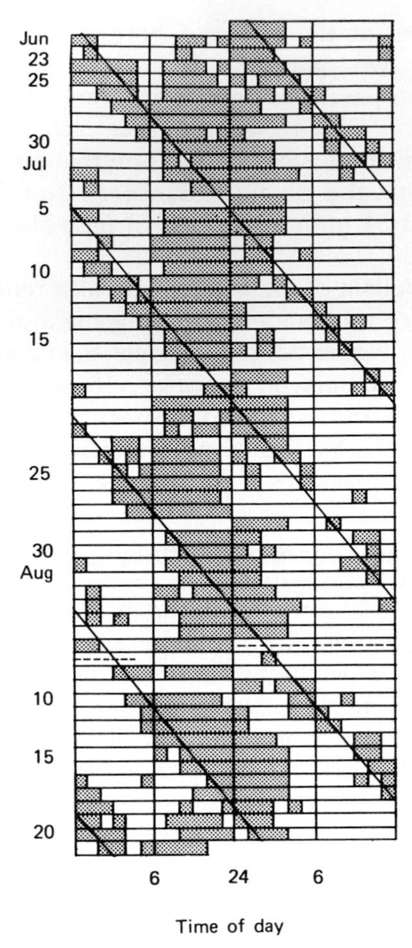

Figure 3-3 The mean solar daily and lunar daily components in the rhythmic locomotor behavior of the penultimate-hour crab, *Sesarma.* The darkened areas signify the hours of each day when the population average activity was higher than the daily mean. The solar-day component is signified by the blocks between 7 P.M. and 4 A.M. The bimodal lunar-day component is indicated by the activity bursts falling on the parallel oblique lines which are drawn through the midpoints of daily high tides. The dashed lines signify mechanical failure of the recording system. This study includes over 12,000 "crab hours" of data. Modified from J. D. Palmer, *Nature* (*London*) **215,** 64–66 (1967).

Dual Rhythmic Components in the Green Shore Crab

The green shore crab (*Carcinus*) (Figure 3-5) is also a common inhabitant of intertidal regions. It lives under rocks and in burrows whose entrances are often centered in a chimney of excavated mud. Like the

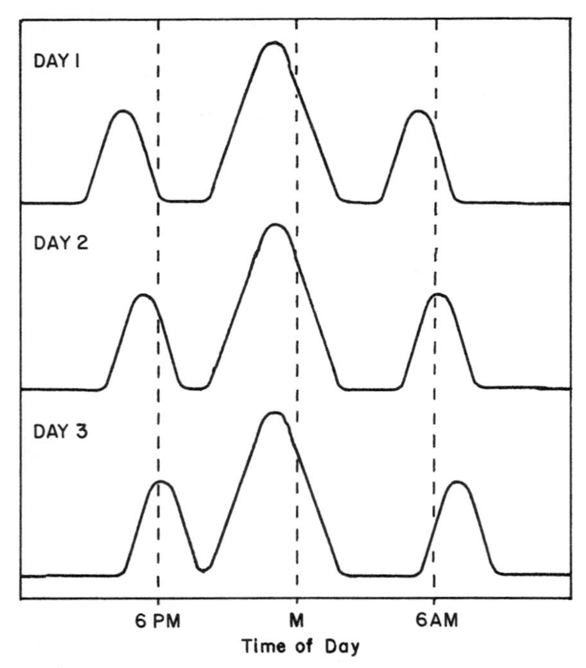

Figure 3-4 A diagrammatic representation of the dual rhythmic components in the activity of the penultimate-hour crab on 3 successive days. Note that the peaks of the bimodal lunar-day component (the low amplitude peaks) proceed to a point 50 minutes later each day relative to the large solar-day peaks. This is, of course, the same rate of progression made by the tides across the solar day.

penultimate-hour crab, they emerge from their place of repose during high tide, thus avoiding the dangers of being eaten by sea gulls and marsh-inhabiting mammals. They also display a daily, as well as a bimodal lunar-daily component in their persistent activity rhythm, but it is expressed differently than in the penultimate-hour crab. It makes itself manifest mainly by augmenting the nighttime peak of the tidal rhythm (Figure 3-6).

The tidal component of this rhythm will persist in light-dark cycles in the laboratory for only a week or two and then disappears, being replaced by a nocturnal solar-day rhythm. However, while performing a variety of experiments with temperature (where it was found, incidently, that the period of the tidal rhythm was unaltered by different constant temperatures), it was discovered that the tidal component could be reinstilled by subjecting the crabs to one brief cold spell (8 hours at 4°C).

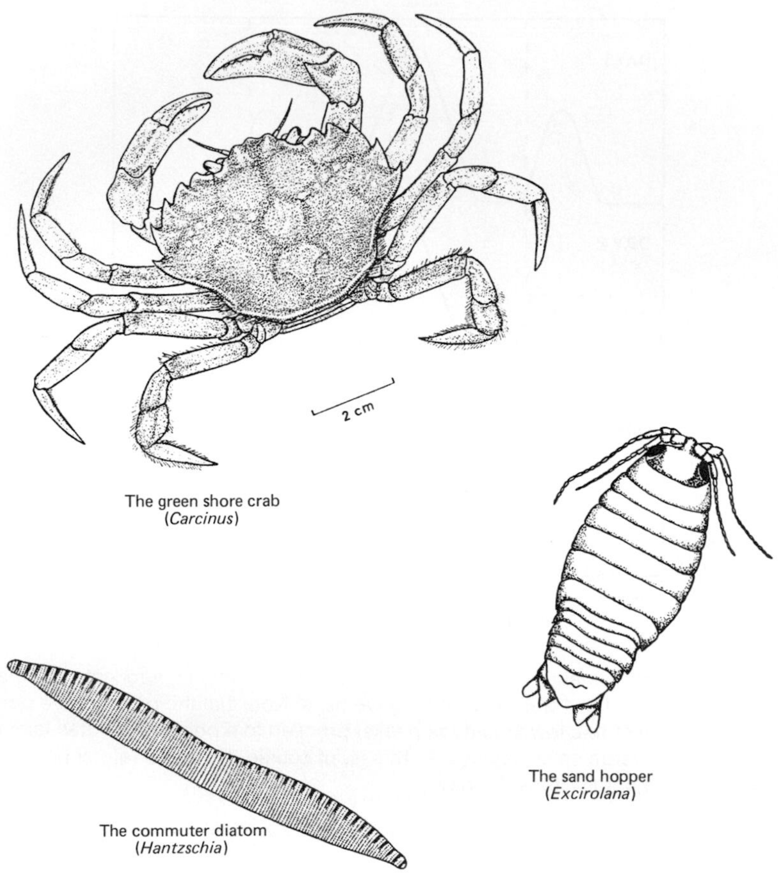

Figure 3-5 (A) The green shore crab (*Carcinus*); (B) the sand hopper (*Excirolana*); (C) the commuter diatom (*Hantzschia*).

INNATE NATURE OF THE TIDAL RHYTHM

The cold-spell treatment has been further used to demonstrate the innateness of the tidal rhythm in this crab. Green crabs were raised from eggs in the laboratory under alternating light-dark cycles and when they finally reached a size and weight sufficient for study, placed in actographs. Their spontaneous locomotor activity was found to be rhythmic, but displayed only a daily period (Figure 3-7A). The crabs were then subjected to a single 15-hour exposure at 4°C and returned to the actographs where it was found that the tidal component was now also present (Figure 3-7B). Because the crabs could not be expected to have learned about repeating 12.4-hour intervals from a single 15-hour cold

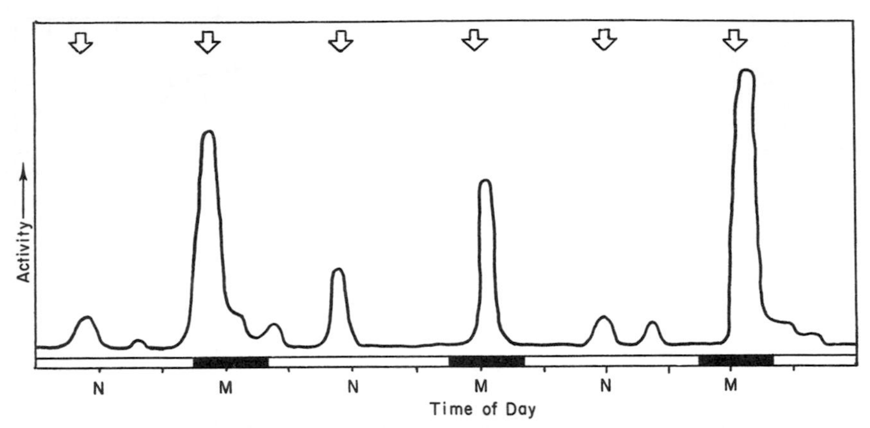

Figure 3-6 The spontaneous locomotory activity of the green shore crab (*Carcinus*) in constant conditions. The midpoints of high tide are signified by the falling arrows. The darkened portions of the abscissa signify the hours of darkness *outside* the laboratory. Note that the diurnal component of the activity rhythm is made manifest by an augmentation in amplitude of the "nighttime" tidal peak. Drawn from the data of E. Naylor, *J. Exp. Biol.* **35,** 602–610 (1958).

pulse, it is safe to conclude that this rhythmic frequency is innate. It is not known whether the cold pulse caused a tidal clock to start running, or whether it was already running but uncoupled from locomotory behavior and the temperature treatment just served to initiate coupling. The latter alternative would be the expected one if a situation such as portrayed in Figure 3-14 actually obtains.

The Activity Rhythm of *Excirolana*

An inhabitant of open beaches (which, unlike the crab's habitat, are periodically subjected to the direct onslaught of the pounding surf) also displays a tidal activity pattern, but one that differs somewhat in form from those previously described. This tiny animal (Figure 3-5), known as a beach flea, or sand hopper (an isopod crustacean of the genus *Excirolana*), lives buried high up on the shore in a zone that is flooded only at the very peak of each high tide. When the first waves wash over its habitat, it wriggles up out of the fine sands to tread water and feed in the wave wash. For about 2 hours it continues this activity until the tide begins to recede and it reburrows back into the sand.

This behavior pattern is studied in the laboratory by placing about 100 of the animals in a beaker of seawater with a sand-covered bottom. In here, the beach fleas spend their time alternately buried in

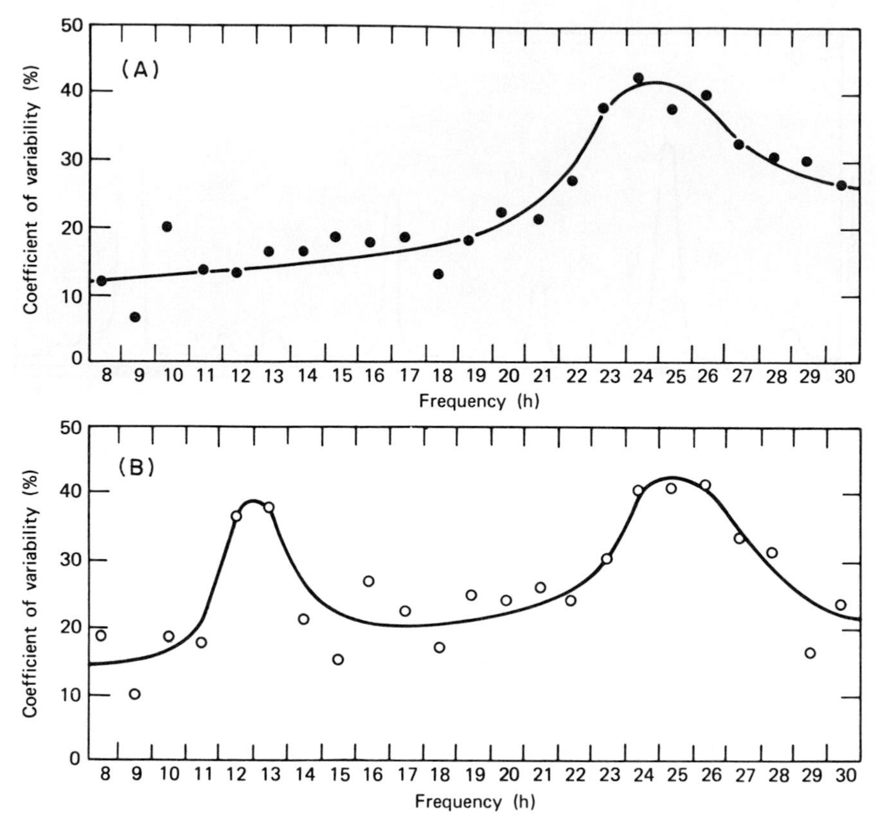

Figure 3-7 The induction by a cold pulse of a tidal rhythm in a green shore crab (*Carcinus*) raised from an egg in the nontidal setting of the laboratory. On reaching an adequate size for actograph studies, the crab's locomotory behavior was studied for several days and then analyzed by a statistical technique that scans the data for repeating periods. As seen in (A), there were a preponderance of peaks repeating at, or near, a 24-hour interval, as might be expected since the crab had been raised in natural light-dark cycles. The crab was then exposed to 4°C for 15 hours and its subsequent activity reanalyzed. As seen in (B), the treatment initiated a tidal periodicity with peaks coming at 12.4-hour intervals. From B. G. Williams, and E. Naylor, *J. Exp. Biol.* **47,** 229–234 (1967).

the sand or swimming in the overlying water. Their activity, or lack of it, in the beaker is recorded by time-lapse photography and, in this way, it was discovered that the tidal rhythm would persist in constant conditions for several days.

The form of this rhythm varies, depending on the pattern of the tide displayed at the time when the animals were collected. The tidal configuration on the beach where the animals were collected is called a "mixed" one. On some days of the month, both peaks of high tide are of equal height; at other times, the morning peak is higher than the afternoon one, or vice versa. Additionally, during a few days each month, there is but a single tidal crest each lunar day. Freshly collected animals examined in the laboratory reflect, in the form of their rhythms, the pattern of the tide on the day previous to their capture. For example, if they were collected on a day when there was only one interval of high water, that was the frequency they tend to describe in constant conditions (Figure 3-8A). Animals subjected to two tidal crests/day, each of equal amplitude, approximate this in the laboratory (Figure 3-8B), while those taken during times of unequal tidal amplitudes expressed this form in the laboratory (Figure 3-8C). Just how the modulation in amplitude is impressed on the tiny crustaceans will be described at the end of the chapter.

THE ROLES OF ALCOHOL AND D_2O IN PERIOD AUGMENTATION

The roles of alcohol and deuterium oxide in modifying the activity rhythm of *Excirolana* were tested by simply adding them to the seawater in the animals' container. Just as had been found with the circadian rhythms in plants and other animals, the period of this circalunadian rhythm was also lengthened by both substances.

The Vertical Migration Rhythm in *Hantzschia*

The last rhythm that will be described is a rather bizarre one, but is carried out by a unicell and therefore demonstrates that the single celled level of organization is sufficient for the expression of a tidal rhythm—just as it is for a solar daily one. The rhythm is one in vertical migration and is a recurring event in the life of a diatom called *Hantzschia* (Figure 3-5).

This microscopic alga lives buried in intertidal sands during high tides and at night, but ascends up onto the sediment surface during daytime low tides. On the surface, they accumulate in such enormous numbers that the exposed sand assumes the color of the algal pigment—visible through their transparent body covering—and thus becomes golden brown. Just prior to the return of the flood tide, the diatoms burrow back into the substratum and avoid being washed away. In fact, there is strong environmental pressure selecting for their timely reburrowing since the returning tide washes away any nonconforming stragglers. These ups and downs in the daily lives of the diatoms are appropriately described as vertical migrations and are particularly fascinating because the commutations will persist in sand samples removed to the constancy of the laboratory (Figure 3-9).

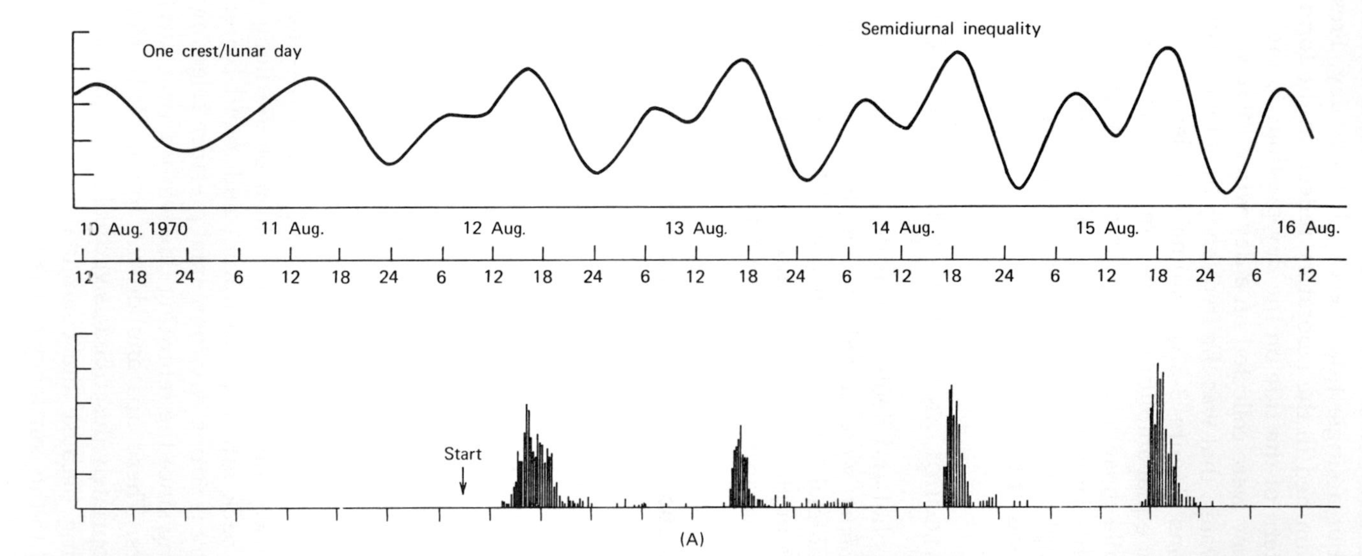

Figure 3-8 (A) The form of the persistent tidal activity rhythm in the sand hopper (*Excirolana*) in the laboratory after being collected at a time of the month when the tide consisted of but a single peak per day (upper curve). Note that this is the frequency displayed in the lab (while the tide in nature began to become bimodal and is thus referred to as displaying a "semidiurnal inequality" character and is labeled as such).

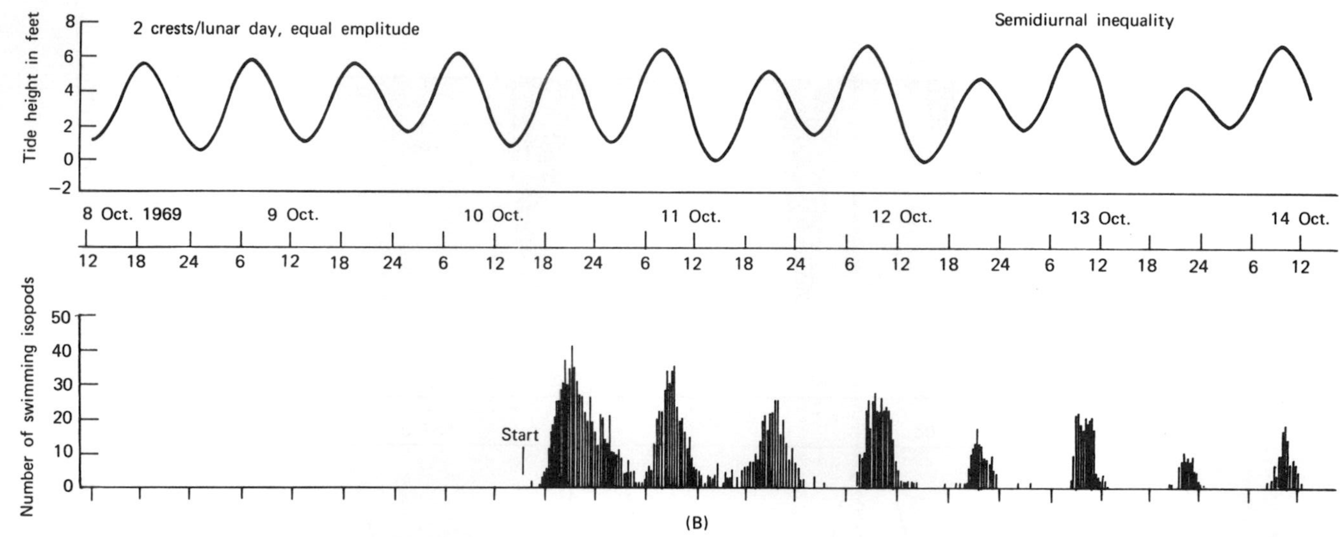

Figure 3-8 (B) These animals were collected from a twice/lunar day, equal amplitude, tidal situation, and this form is—at least at first—displayed in constant conditions. Then, however, the animals appear to spontaneously adopt the newly expressed unequal form of the natural tide without ever being exposed to it.

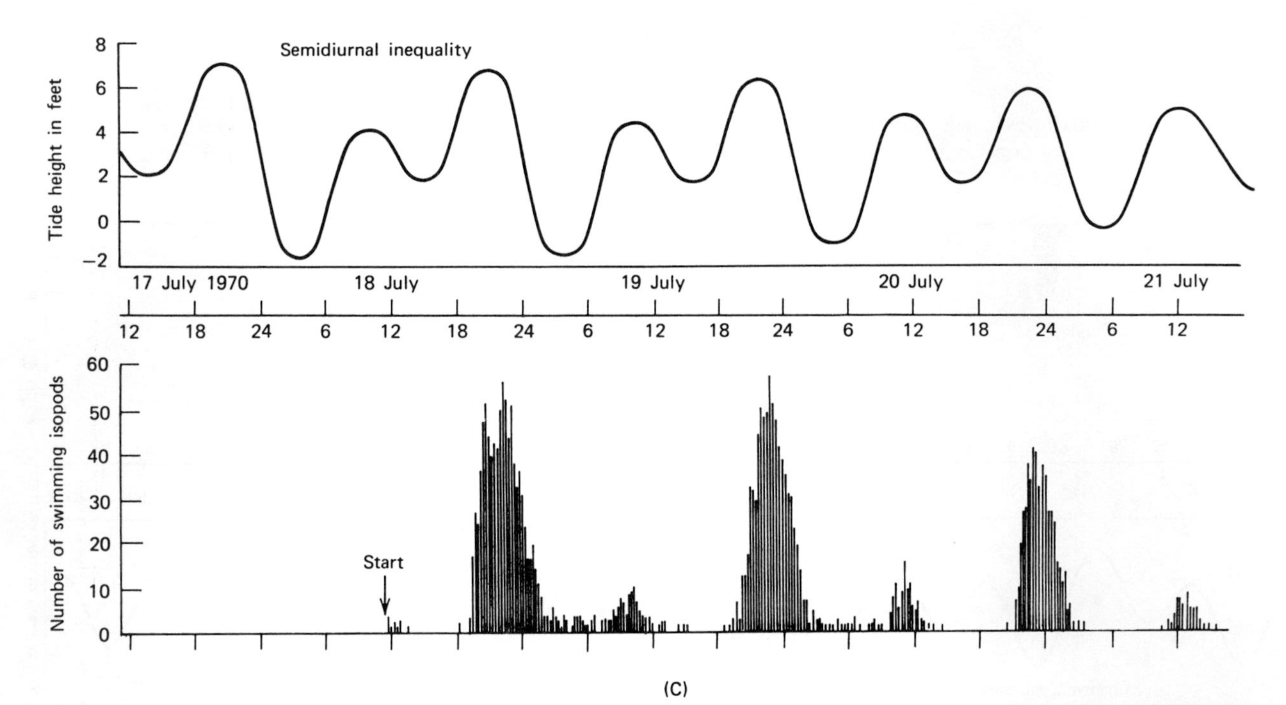

(C)

Figure 3-8 (C) This shows the laboratory expressed pattern mimicking the inequality of external tidal crests. Modified from L. A. Klapow, *J. Comp. Physiol.* **79,** 233–258 (1972).

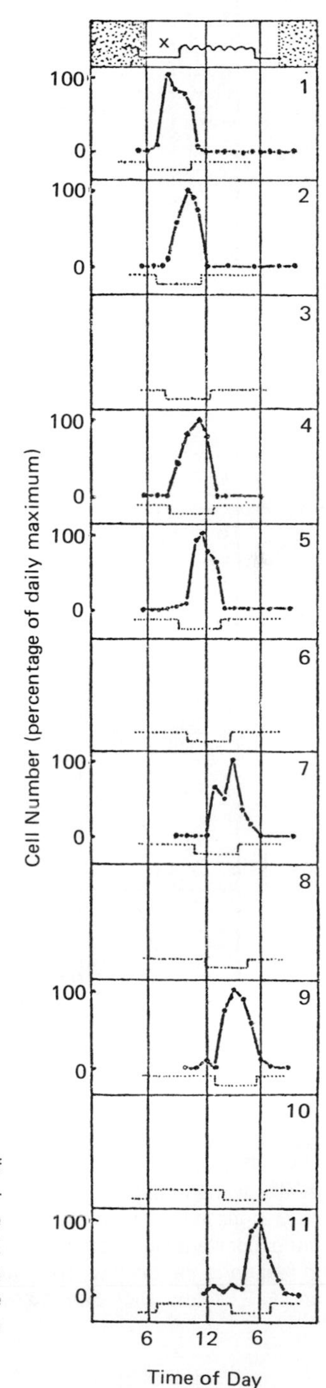

Figure 3-9 The persistent vertical-migration rhythm of the commuter diatom (*Hantzschia*) in constant conditions. Consecutive days are aligned one beneath the other; the stippling indicates the hours of darkness; X marks the time of collection of the sample; the intervals of low tide in the natural habitat are symbolized by the depressed segments of the dotted lines subtending the daily curves. Modified from J. D. Palmer and F. E. Round, *Biol. Bull.* **132,** 44–55 (1967).

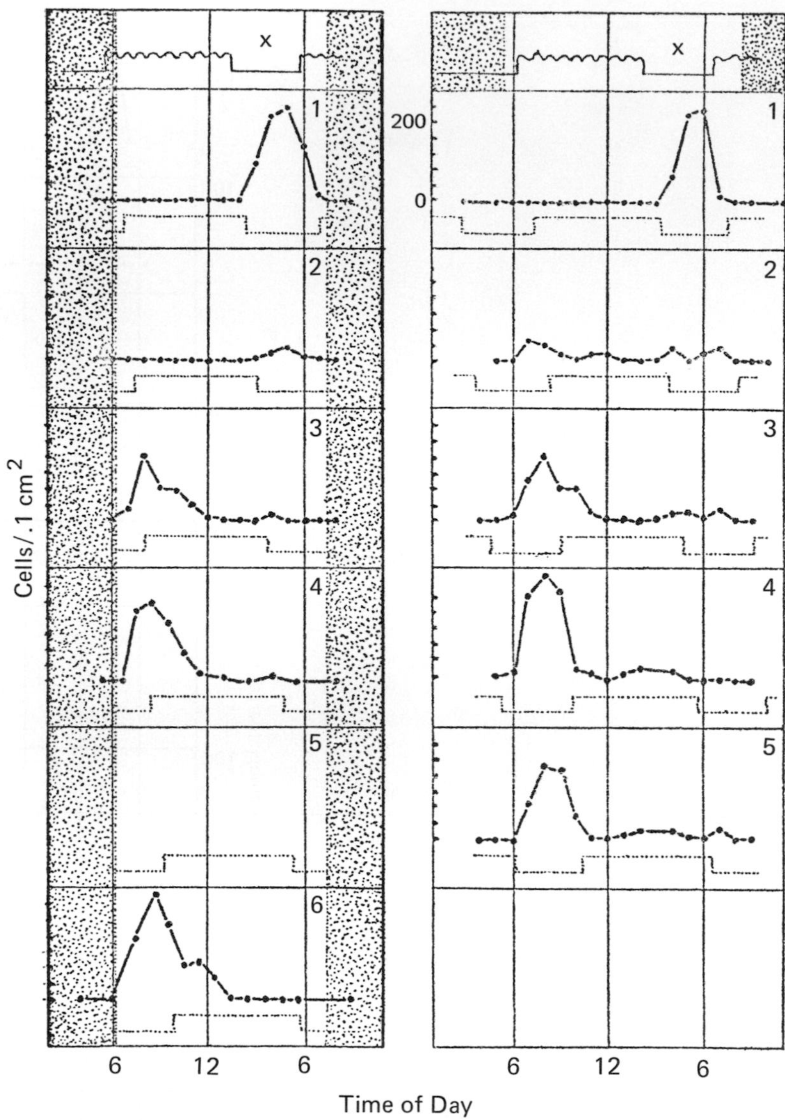

Figure 3-10 The clock programmed phase change in the commuter diatom (*Hantzschia*). The rephase from the late afternoon to the early morning hours of the day, as seen in alternating light-dark cycles on the left, suggests that interation of the rhythm with darkness is the causative stimulus for the rephase. However, the same rephase takes place under constant illumination, as shown on the right, indicating that the response is a built-in feature of the display. The symbols are the same as in Figure 3-8. From J. D. Palmer and F. E. Round, *Biol. Bull.* **132,** 44–55 (1967).

This rhythm is identical to all other tidal rhythms in that the daytime peaks come approximately 50 minutes later each day in approximate synchrony with the times of low tide on the alga's home beach. It differs radically in that it is a unimodal rhythm, i.e., there is only one peak per lunar day, with none ever being expressed at night. In nature and in the laboratory in natural light-dark cycles, the unimodality is expressed as follows. The single peak works its way across the solar day at a rate of 50 minutes/day until it begins to coincide with sunset (or light off in the laboratory). The peak then collapses, but quickly builds up during the hours just after sunrise (or lights on in the laboratory) and, once established, again scans the day at a tidal progression rate (Figure 3-10). This pattern gives the impression that the dark boundaries around the interval of light produce the radical and rapid rephase to the early hour. As obvious as this might seem, it is not the case, for the rephase "backwards" is displayed in the laboratory even when the lights are left on continuously (Figure 3-10). Therefore, the late afternoon collapse and postsunrise reconstruction are a programmed part of this organisms temporal physiology.

Rather than simply accepting this display as a unique quasitidal rhythm, an explanation has been derived using some of the known characteristics of other tidal rhythms. The behavior is explained as the final expression (Figure 3-11C) of the combined action of a normal 2-peaks per lunar day tidal rhythm (Figure 3-11A) and a 24-hour rhythm which functions to inhibit the nighttime tidal peak (Figure 3-11B). As a consequence of the presence of the solar-day rhythm, the overt behavior of the diatom is a single daytime commutation which occurs 50 minutes later each day. And, when the late afternoon tidal peak falls under the influence of the repression phase of the daily rhythm, it collapses; but simultaneously, the other tidal peak escapes to the permission "window" of the daily rhythm and is thus expressed. The overt manifestation of this activity appears to the observer as a radical rephase backwards to the hours of sunrise.

DETERMINATION OF PHASE

Light-Dark Cycles

As described in Chapters 1 and 2, the phase of a solar-day rhythm is determined primarily by day-night cycles of light and temperature. Twenty-four cycles in these two environmental parameters cannot be expected to influence a 12.4-hour tidal rhythm—entrainment would, of course, destroy the tidal nature of the rhythm. However, 12.4-hour

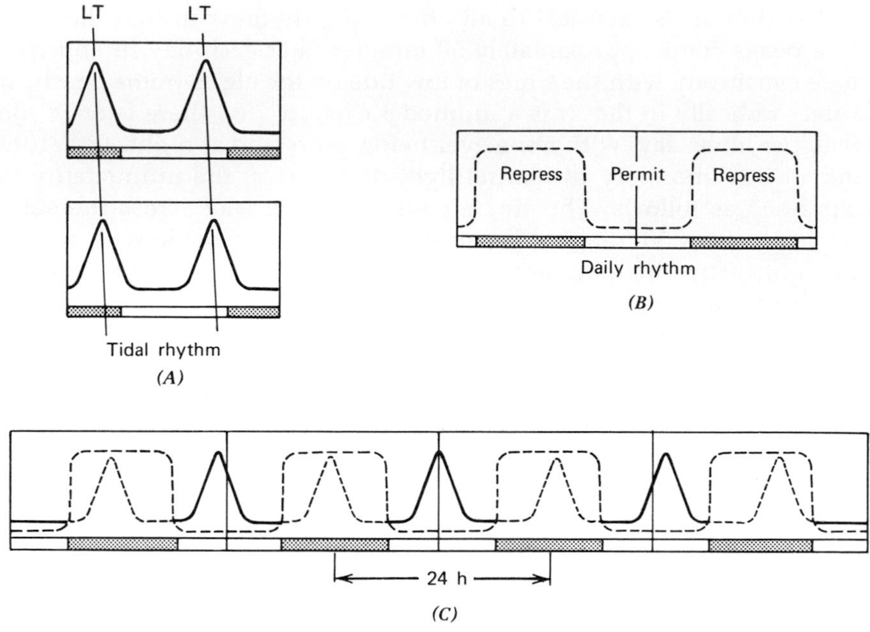

Figure 3-11 A diagrammatic representation of the interaction of a bimodal lunar-day rhythm in vertical migration (A) and a 24-hour rhythm that either permits or inhibits, depending on its phase, the expression of one peak of the tidal rhythm (B). When the two interact, those portions of the tidal rhythm indicated by the solid lines in (C) are expressed to an observer. LT, midpoints of low tide; shaded bars, hours of darkness. From J. D. Palmer, "Biological Clocks in Marine Organisms." Wiley (Interscience), New York, 1974.

light-dark cycles might be expected, *a priori,* to be effective for some tidal organisms—those that live in dark burrows during high tides and emerge out into the sunlight and moonlight at low tides. Therefore, intertidal animals such as the penultimate-hour crab and the sand hopper have been exposed to cycles of 6.2 hours of light, alternating with 6.2 hours of dark, offered so as to conflict with the phase of their persistent activity rhythms. No phase entrainment resulted. Tidal rhythms also persist as such, when subjected to natural day-night light cycles in the laboratory.

However, a measurable effect of solar-day light-dark cycles (12 hours of light alternating with 12 hours of darkness) can be demonstrated in one category of marine animals: those that display both solar-day and tidal components in a single rhythmic function. In these organisms, as might be expected, the solar-day component should respond in the usual way and, as will be described, changes in this component cause alterations in the tidal one also. The first experi-

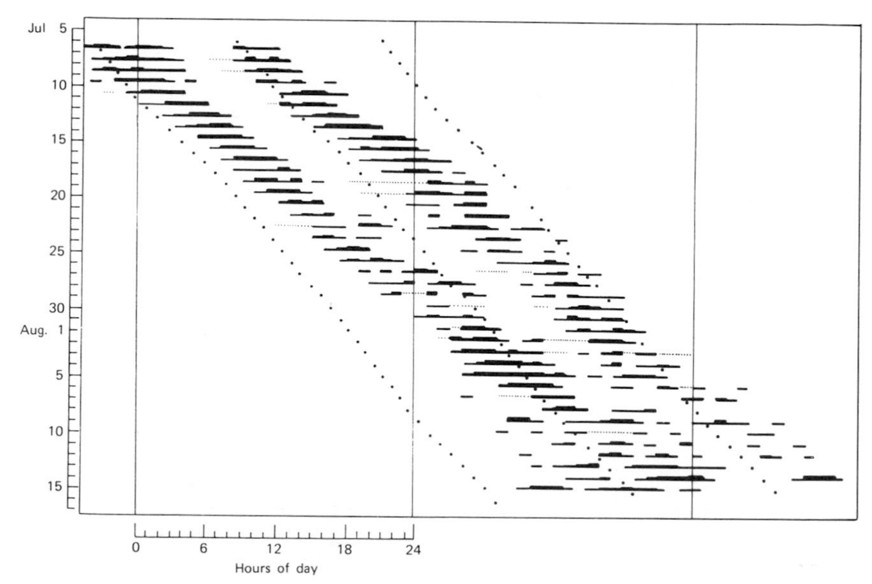

Figure 3-12 The circalunadian rhythm of a fiddler crab (*Uca*) in constant conditions. The height of the blocks indicate relative activity, the diagonal chain of bold dots indicate the midpoints of high tides on the crab's home beach, and the horizontal dotted lines signify mechanical failures of the recording system. Note that the period of this predominately bimodal lunar-day rhythm is decidedly longer than 24.8 hours. Drawn from the data of F. H. Barnwell, *Biol. Bull.* **130,** 1–17, (1966) and reproduced from J. D. Palmer, "Biological Clocks in Marine Organisms." Wiley (Interscience), New York, 1974.

ments demonstrating this were performed on the fiddler crab activity rhythm, which, though not mentioned previously, also contains a low amplitude solar-day component merged with the predominantly tidal frequency. In the laboratory, it was found that when the crabs were maintained in constant conditions, the tidal period increased as expected to a circalunadian one (Figure 3-12), but when the crabs were maintained in alternating day-night conditions, the period of the manifest tidal rhythm was a strict, bimodal 24.8-hour one (Figure 3-13). It appears then that if the daily component is prevented from becoming circadian the tidal component is also prohibited from becoming circalunadian.

Even further modification of the tidal rhythm in this organism can be produced with light-dark cycles. For example, by placing freshly collected fiddler crabs in an altered day-night cycle in which "dawn" suddenly was made to occur at midnight and "sunset" at 6 A.M. and by maintaining them in this cycle for 3 days, it was found that the diurnal component of their activity rhythm was advanced by 4.9 hours

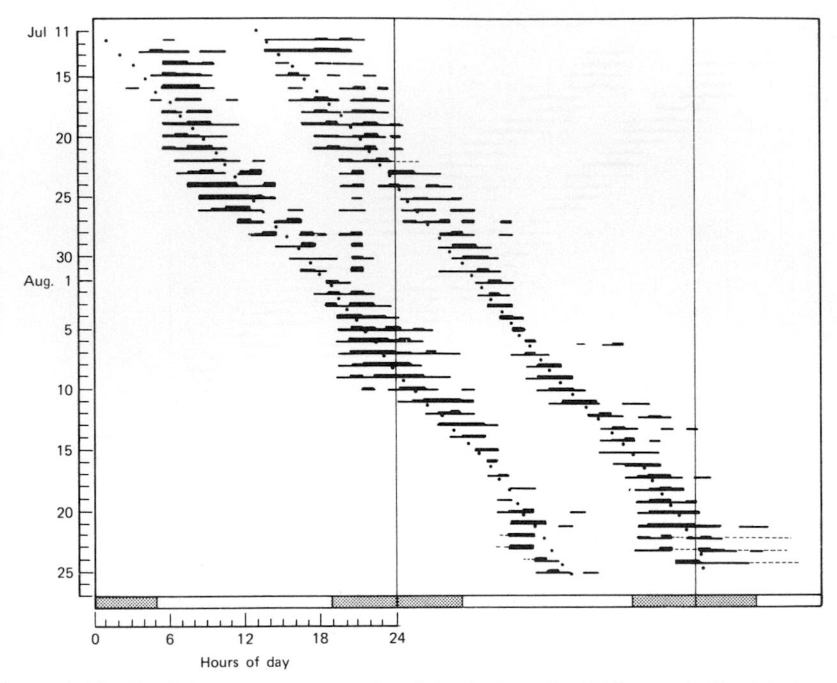

Figure 3-13 The laboratory expressed activity rhythm of a fiddler crab (*Uca*) in natural light-dark illumination. The symbols have the same meanings as those in Figure 3-12. Note that under this daily light-dark cycle the tidal rhythm is indistinguishable in length from the bimodal natural tidal interval. Compare this response with Figure 3-12. Drawn from the data of F. H. Barnwell, *Biol. Bull.* **130,** 1–17 (1966), and reproduced from J. D. Palmer, "Biological Clocks in Marine Organisms." Wiley (Interscience), New York, 1974.

and the tidal component by about 4.6 hours, i.e., both by essentially the same amount. These results also indicate that the daily and tidal components are intimately associated.

Therefore, light-dark cycles can play roles in the determination of period length and phase setting of a tidal rhythm, but, in comparison to the major role they play in daily rhythms, the effect is very small. However minuscule a role, evidence of this type indicates that daily and tidal rhythms can be very intimately related, so close in fact, that the possibility exists that they may both be under the simultaneous control of a single clock. This can be explained as follows: Interpolated between the clock and each of the processes that it causes to be rhythmic is a coupling mechanism (Chapter 1). And, if at least one of these couplers has the additional property of being able to transform the basic frequency of the clock into another one, say a tidal frequency, then the single clock would be sufficient to drive all of an intertidal organism's rhythms (Figure 3-14). The situation would be

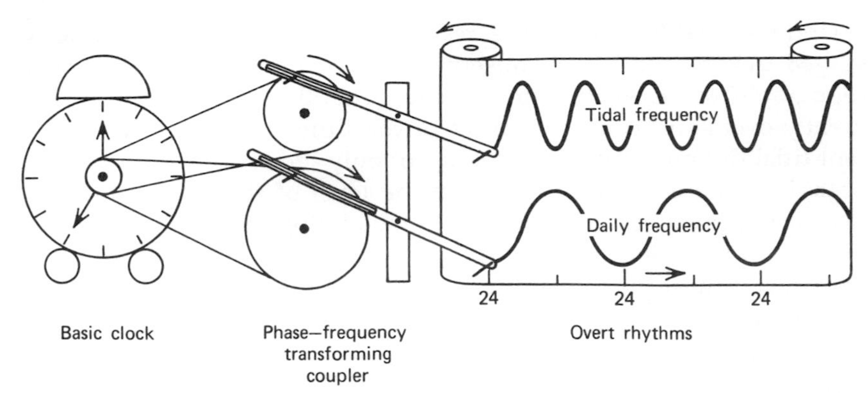

Figure 3-14 A physical analogue representation of phase-frequency transforming couplers interposed between a "biological" clock and the "rhythms" they drive. In this scheme, the size of the belt-driven pulleys determines the period length of the individual rhythms. An organism equipped with a transducer for tidal frequencies would be especially well adapted for life on the shoreline. From J. D. Palmer, *Biol. Rev. Cambridge Philos. Soc.* **48,** 377–418 (1973).

directly analogous to the functioning of the wristwatches often worn by surf fishermen, the escapement of which produces only one frequency. But, via multiple gear combinations (the "couplers" between escapement and dial), the hands of the clock are made to signal both time of day and time of tide.

The suggestion that one clock might drive both rhythms is, of course, highly speculative, but probably no more so than daring to consider that some sort of timing mechanism actually does underly overt rhythms. With this speculative nature in mind, here are some reasons for believing that only one clock may be involved. Both alcohol and deuterium oxide have been shown to alter circalunadian rhythms in a way exactly identical to their effect on daily rhythms, which indicates that daily and tidal clocks are identical in makeup- . . . or, are one and the same. Additionally, it would seem unnecessarily redundant for nature to have evolved two separate clocks whose periods differed by only about 3% and yet could drive rhythms with circa periods that easily overlap both basic frequencies. Whatever the actual situation may be, certainly tidal and daily rhythms can be coupled together in some way, as was especially indicated before in the discussion of the role of light-dark cycles on the fiddler crab rhythm.

Returning to the topic of phase determination, the prime suspect is, of course, the tide itself. A series of simple experiments have shown this elementary supposition to be correct, but they have also revealed a surprising paradox.

Unquestionably, the tide is responsible for the phase setting assumed by a rhythm. Three observations prove this: (i) Because of the topography of the sea bottom and shoreline, the times of the tide vary greatly over short lateral distances of coastline. The phase of a persistent tidal rhythm is initially that of not only the beach from which the organisms were collected, but also the level of their habitat on that beach. For example, the burrows of crabs living high up on the beach are uncovered first by the receding tide and, consequently, the rhythms of these crabs peak sooner than those living lower down on the beach incline. (ii) Crabs translocated from one tidal situation to another quickly adopt the new tidal regimen. (iii) Normally intertidal crabs that have somehow become established in a nontidal habitat display only solar-day rhythms when studied in the laboratory. When exposed to a tidal situation, this rhythm also is quickly established in these animals.

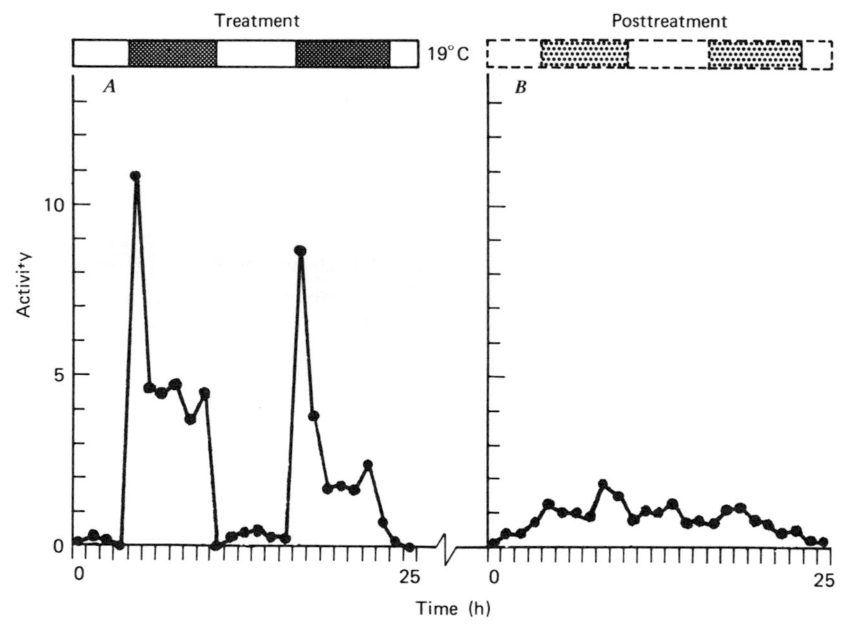

Figure 3-15 The ineffectiveness of periodic inundation as an entraining agent of the green shore crab (*Carcinus*) activity rhythm. (A) The mean lunar-daily activity pattern for crabs during 5 days exposure to cycles of 6.2 hours of immersion in water (signified by cross-hatching above the curves) alternating with 6.2 hours of exposure to air. Both air and water temperatures were 19°C. Note that peak activity took place during the hours of inundation just as it does in the natural setting. (B) The average curve for 3 days posttreatment while the crabs were maintained in moist air (the stippled portions of the overhead bars signify the times of "expected" immersion). There was no significant carryover from this treatment with immersion cycles. Modified from B. G. Williams and E. Naylor, *J. Exp. Biol.* **51,** 715–725 (1969).

Inundation and Temperature

The question then arises as to just what aspects of the tide produce the actual entrainment. Many possibilities suggest themselves, with periodic physical inundation being the number one candidate. Paradoxically, however, water per se does not seem to be involved. This was neatly demonstrated by exposing arrhythmic green shore crabs to 12.4 hour inundation cycles (6.2 hours in air alternating with 6.2 hours under water) for 5 days. Both the air and water temperature were held constant at 19°C. During the 5 days of treatment, the crabs were most active during the hours of inundation; but, on the sixth day, when they were placed in actographs and maintained in moist air, no carry-over from the treatment was seen (Figure 3-15).

The experiment was repeated again, but this time the water and air temperatures were not identical. After subjecting arrhythmic crabs to 5 days in 12.4 hour cycles of 6.2 hours immersion in 13°C water alternating with 6.2 hours in air warmed to 24°C, it was found that when they were moved to constant conditions a tidal rhythm had been established in the animals. The peaks of the activity rhythm were centered on the times that had been intervals of cool immersion.

The final version of the experiment eliminated the periodic immersion: 6.2 hours of 13°C alternated with an equal interval at 24°C. Five days of this induced a tidal rhythm in the crabs that would then persist under constant conditions (Figure 3-16). Therefore, quite sur-

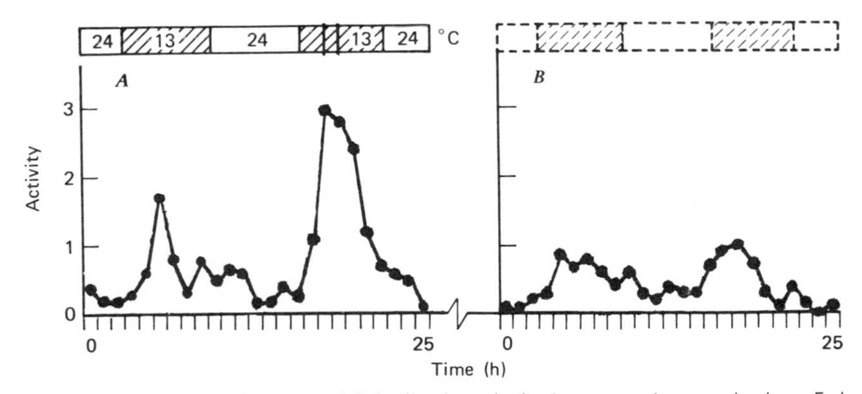

Figure 3-16 The induction of a tidal rhythm in arrhythmic, green shore crabs by a 5-day treatment with temperature cycles. (A) The average response of 5 crabs to 6.2 hours at 13°C alternating with 6.2-hour intervals at 24°C. (B) The average for 3 days post-treatment at a constant temperature of 13°C, showing that the rhythm instilled by this temperature cycle would then persist in its absence. The hatching in the superscribed bar indicates the "expected" times of low temperature. Modified from B. G. Williams and E. Naylor, *J. Exp. Biol.* **51,** 715–725 (1969).

prisingly, it is not inundation that establishes the rhythm or sets the phase, it is, instead, the temperature change delivered by the tide that is the effective stimulus.

Pressure

Temperature is not the only important phase-setting stimulus in the habitat of the green shore crab; pressure also has a well-defined role. To demonstrate this, arrhythmic crabs were subjected to pressure cycles consisting of 6 hours of ambient atmospheric pressure alternating with an equal interval at ambient plus 0.6 atmospheres of pressure. As seen in Figure 3-17, the treatment instilled a tidal cycle that would then persist in constant conditions, with the peaks corresponding to what had been intervals of high pressure. While each high pressure interval was identical, the persisting rhythm produced is characterized by alternating high and low peaks—the form displayed in nature (Figure 3-6).

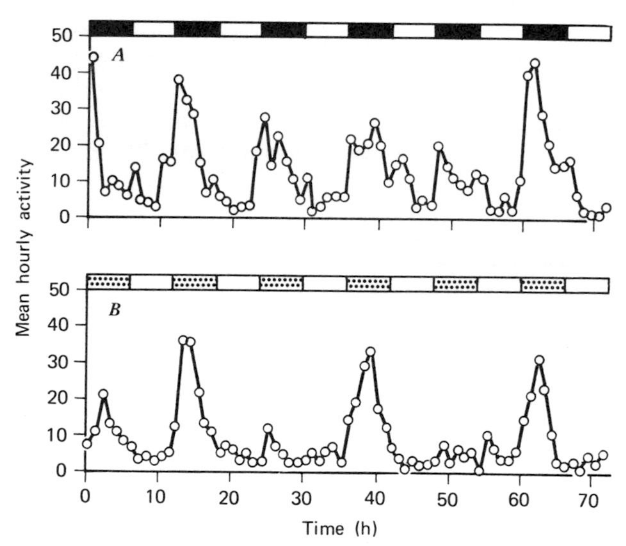

Figure 3-17 The induction of a tidal activity rhythm by pressure cycles in the green shore crab (*Carcinus*). (A) The average hourly activity of six previously arrhythmic crabs subjected to pressure cycles of 6 hours at ambient atmospheric pressure, alternating with 6 hours of ambient plus 0.6 atmospheres (signified by overhead horizontal dark bars). (B) The instilled persistent rhythm at constant atmospheric pressure (stippling signifies the times of "expected" pressure increase). Note in (B) the alternating high and low peaks—the pattern that is typically shown in the natural habitat. Modified from E. Naylor and R. J. Atkinson, *Symp. Int. Soc. Cell Biol.* **26,** 395–415 (1972).

Mechanical Agitation

The prime entraining stimulus of the activity rhythm in the beach flea (*Excirolana*) is not temperature or pressure, but mechanical agitation. It will be remembered that this animal, unlike the fiddler and green shore crabs, lives on the open beach and is thus subjected twice per lunar day to the violent pounding of the high tide surf. It has been found that by simulating the action of the surf in the laboratory the phase of this organism's rhythm can be set to any time of the day and the form of the rhythm can also be molded into various patterns.

The "wave simulator" used in these observations consists of nothing more than a jar of seawater containing the animals and a mechanism to stir the water with sufficient vigor to swirl the animals up into suspension. A few repetitions of 6 hours of stirring alternating with 6 hours of calm are sufficient to entrain the animals to the cycle.

As previously described (Figure 3-8A,B,C), the form of the rhythm displayed by this animal in constant conditions is determined by the form of the tide to which it was exposed just prior to capture. The rhythm can therefore take on an unimodal or bimodal pattern, and the peaks of the latter may be of equal or unequal amplitude. In the laboratory, the wave simulator can be used to create any of these patterns. For example, a collection of sand hoppers was brought into the laboratory and divided into two groups. One population was subjected to 7 days of periodic agitation consisting of 120 minutes of swirling each morning and 30 minute intervals of the same in the afternoon; the two treatments were separated from one another by 12.5 hours (Figure 3-18A). The other population received the same cycles except that the 120-minute intervals of swirling were offered in the afternoon (Figure 3-18B). Obviously, these agitation cycles were designed to roughly imitate the bimodal, unequal amplitude form of the natural tide. As vividly shown in Figure 3-18, the treatments did instill these patterns.

PHASE-RESPONSE CURVES

As was described in Chapter 1 for daily rhythms, entrainment to a particular environmental cycle is produced by one or more directed phase shifts. The direction of change (either advance or delay) and the magnitude of each are described by the organism's phase-response curves; the best known of which are for light and temperature.

Because tidal rhythms are controlled by a clock that is very similar to the one that controls daily rhythms and since they can be entrained by pressure and agitation cycles, it was reasoned that it should be pos-

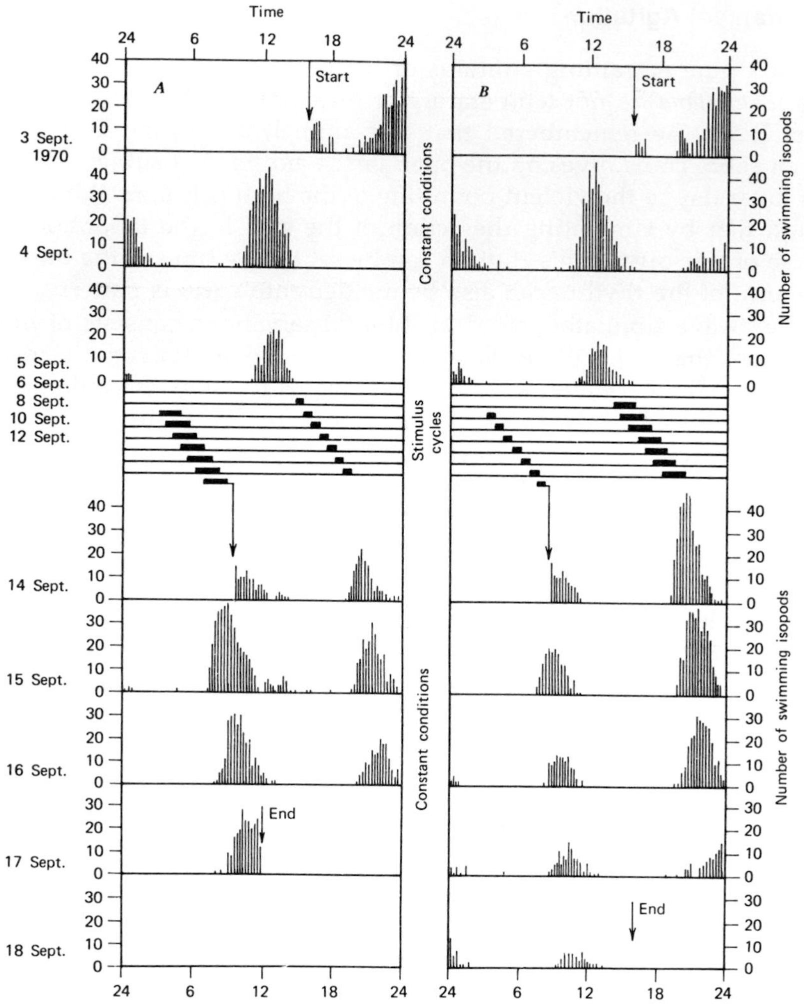

Figure 3-18 The entrainment and control of rhythm form in the sand hopper (*Excirolana*) by alternating short- and long-agitation stimuli. Alternating at 12.4-hour intervals, 30- or 120-minute intervals of swirling (solid horizontal bars) were offered (in reverse order in A and B) to two different populations of animals. Seven days of treatment set the phase of the rhythms of both populations and also established the inequality in the amplitude of successive peaks; both changes persisted in constant conditions (from 14 September on down). From L. A. Klapow, *J. Comp. Physiol.* **79,** 233–258 (1972).

sible to demonstrate phase-response curves for these entraining stimuli also. The one for pressure has not yet been sought after, but a phase-response curve for agitation has now been defined. The experimental subject was, of course, the beach flea. Laboratory populations

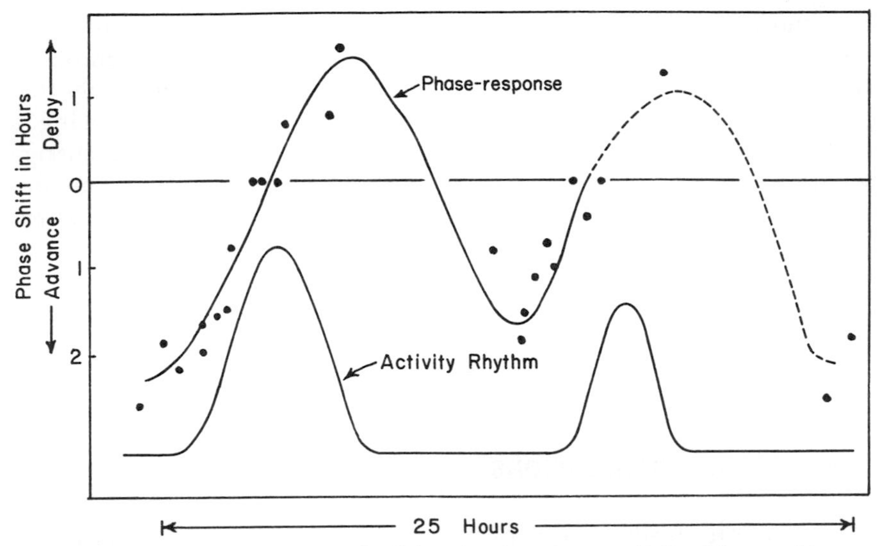

Figure 3-19 The response curve signifying the changing sensitivity of the sand hopper activity rhythm to mechanical agitation. The dashed portion of the upper curve is surmized, since the observations had not yet been made at the time of this publication. As can be seen, stimuli given at the times of the activity maxima produce no phase change; stimulation during the ascending phase of the cycle produces phase advances; while mostly delays are produced during the descending portion of the activity curve. This phase-response curve for tidal rhythms is uniquely different from the one for solar-day rhythms (Figure 1-7) in that it is bimodal and stimulation at identical portions of either peak produce the same phase change in the rhythm it governs. Plotted from the data of J. T. Enright, *In* "Biological Rhythms in the Marine Environment" (P. J. DeCoursey, ed.), pp. 103–114. Univ. of South Carolina Press, Columbia, 1976.

of this creature were given 2-hour agitation treatments with the wave simulator during most of the hours of the day (observations for all times of the day had not been completed at the time of this writing) and the direction and degree of phase change plotted. The response curve (or rhythm in sensitivity to agitation) thus delineated is depicted in Figure 3-19 where it is seen that, in general, agitation offered during the ascending phase of each peak of the activity rhythm produced phase advances; those given at the peaks of the rhythm produced no phase change; those imposed during the descending phase caused phase delays; and at other times advances resulted. This rhythm in sensitivity to agitation, while producing the same end results as are produced by its counterpart for solar-day rhythms, is quite different. It is bimodal and stimulation of identical points on *either* peak will produce the same phase change in the activity rhythm.

The form of this rhythm is just what would have been predicted before the experiment was performed, because a scheme like this one would serve to tune precisely the rhythm to the tides. For example, if the activity rhythm was in perfect synchrony with the tides, only that portion of the response curve that was ineffective in producing a phase change would be exposed to the surf and no phase change would result; if for some reason the activity peaks should become delayed, the advance portion of the sensitivity rhythm would then be exposed to wave action and the overt rhythm would be advanced in phase; and if the peaks accidently occurred too early, the delay portion of the response curve would be exposed to the tides, again adjusting the rhythm in the proper direction. The phasing mechanism is therefore a perfect adaptation to the open beach environment.

SUMMARY AND CONCLUSIONS

1. Shore-dwelling organisms of the intertidal zone often display cyclic behavioral and physiological patterns which are in synchrony with the tides in their habitat.

2. The phase relationships of tidal rhythms are species specific, e.g., the penultimate-hour crab centers its activity on high tides, while the maxima of the fiddler crab activity rhythm occur at the time of low tide.

3. These tidal rhythms will often persist in constant conditions in the laboratory; because the period is usually altered by the isolation from the natural habitat, the rhythms are referred to as being *circalunadian*.

4. Persistent tidal rhythms are innate.

5. The unicellular level of organization is sufficient for the expression of tidal rhythms.

6. In intertidal organisms, often a single process will contain both lunar- and solar-day rhythmic components. The solar-day component manifests itself either as a distinct peak, as in the penultimate-hour crab; by augmenting one of the tidal peaks, as in the green shore, and fiddler crabs; or by inhibiting one peak of the tidal rhythm as in the commuter diatom.

7. The period of a tidal rhythm, like that of a solar-day one, is virtually independent of temperature.

8. The form of the tidal rhythm of the beach flea can be determined by the pattern of the tides on the animal's home beach, or by exposure to a wave simulator in the laboratory.

9. Alcohol and deuterium oxide alter tidal rhythms in the same

manner as they do daily rhythms: they increase the period length in a dose-dependent fashion.

10. It is possible that via a frequency transforming coupler, a single clock frequency is used to drive both tidal and daily rhythms.

11. The phase of the tidal rhythms is, paradoxically, not set by the periodic wetting of high tides. Temperature and pressure changes, delivered by the tides, are the most important phase setters for the intertidal organisms whose habitat is not exposed to the direct pounding of the surf. Mechanical agitation is an effective phase setter in organisms of the exposed beach, such as the sand hopper.

12. A phase-response curve, representing a changing sensitivity to mechanical agitation, has been described for the sand hopper. It is uniquely different from the one for solar-day rhythms.

Selected Readings

Bennett, M. F. (1974). "Living Clocks in the Animal World." Thomas, Springfield, Illinois.

Brown, F. A., Jr. (1954). Biological clocks and the fiddler crab. *Sci. Am.* **60,** 34–37.

DeCoursey, P. J., ed. (1976). "Biological Rhythms in the Marine Environment." Univ. of South Carolina Press, Columbia.

Enright, J. T. (1965). Endogenous tidal and lunar rhythms. *Proc. Int. Congr. Zool., 16th, 1963* Vol. 4, pp. 355–359.

Enright, J. T. (1976). Resetting a tidal clock: The phase response curve. *In* "Biological Rhythms in the Marine Environment" (P. J. DeCoursey, ed.), pp. 103–114. Univ. of South Carolina Press, Columbia.

Fingerman, M. (1960). Tidal rhythmicity in marine organisms. *Cold Spring Harbor. Symp. Quant. Biol.* **25,** 481–489.

Palmer, J. D. (1973). Tidal rhythms: The clock control of the rhythmic physiology of marine organisms. *Biol. Rev. Cambridge Philos. Soc.* **48,** 377–418.

Palmer, J. D. (1974). "Biological Clocks in Marine Organisms: The Control of Physiological and Behavioral Tidal Rhythms." Wiley (Interscience), New York.

Palmer, J. D. (1975). Biological clocks in the intertidal zone. *Sci. Am.* **232,** 70–79.

Palmer, J. D. (1976). Clock-controlled vertical migration rhythms in benthic organisms. *In* "Biological Rhythms in the Marine Environment" (P. J. DeCoursey, ed.), pp. 239–256. Univ. of South Carolina Press, Columbia.

4

Human Rhythms

After the two previous chapters, it should come as no great surprise that many aspects of human physiology are also rhythmic. In broaching the subject, two seldom discussed examples will be introduced.

RHYTHMS IN ALCOHOL METABOLISM

When alcohol is consumed, it is absorbed unchanged into the bloodstream and circulated throughout our bodies. On reaching the brain cells, it produces the "desired" effects which may range from pleasant relaxation, to intoxication, and eventually to unconsciousness. As long as the alcohol remains in the circulatory and body fluids, and depending on its concentration therein, one continues to experience some stage in this spectrum of symptoms. The influence of the alcohol wears off as it is metabolized, excreted, expired, or in those maladroit extremes, lost through retching. Experimentation has shown that the length of time that alcohol remains circulating in the blood is rhythmic.

In the elucidation of this fact, five eager volunteers consumed identical amounts of whiskey at 1-hour intervals throughout the day, except that the 1 and 2 A.M. doses were combined with the midnight dollop, and the 4 and 5 A.M. alcohol combined with the 3 A.M. cocktail, so that the period of sleep was not too drastically interrupted. The standard ration consumed hourly by each participant was adjusted so that none became severely intoxicated. Just prior to each imbibition, the blood or saliva alcohol content of each subject was determined. The resulting composite curve (Figure 4-1) portrays the circulating level of alcohol in the body at any one time. It may be interpreted in a practical way as follows, if one is willing to assume that these five thirsty volunteers are representative of the population as a whole. Those imbibers concerned with getting the most for their money should confine their drinking to the hours between about 2 A.M. and noon, because during this time of day alcohol is cleared from the blood most slowly and has longer to act on the brain cells. During the rest of the day—which of course includes the cocktail hour—it is burned more rapidly (often 25% faster), which means its effects wear

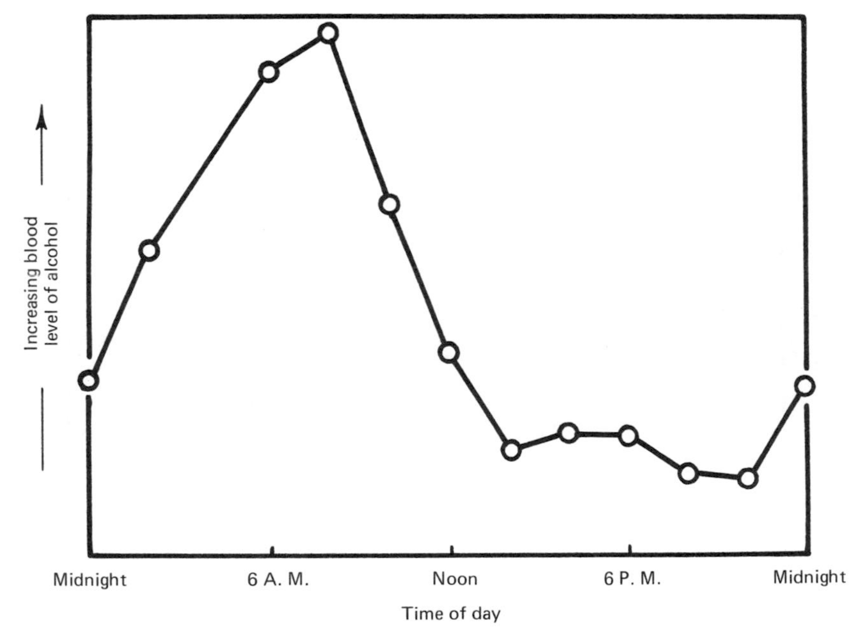

Figure 4-1 The daily rhythm in the metabolism of alcohol. Several imbibers consumed equal proportions of whiskey at fairly regular intervals throughout the day and night. At similar intervals the amount present in their body fluids was determined and used to construct this curve. Between 2 P.M. and midnight, alcohol is burned more quickly than during the other hours of the day. Consumption at these times of day should therefore have the lowest intoxicating effect. Drawn from the data of R. Wilson, E. Newman, and H. Newman, *J. Appl. Physiol.* **8,** 556–558 (1956).

off that much faster. It might also be worth mentioning that the last drink of a party—"the one for the road" after the bewitching hour—is metabolized relatively more slowly than the preceding ones and will produce a more lasting rise in blood alcohol—a feature that could prove embarrassing and even expensive should one be challenged by a traffic policeman.

RHYTHMS IN PAIN TOLERANCE

Man's tolerance to pain, at least as far as skin and teeth are concerned, has also been found to change over the day in a rhythmic manner. As seen in Figure 4-2, when the teeth of volunteer subjects were periodically tested with an identical painful stimulus, the extent of the discomfort they sensed varied with the time of day. Unfortunately for cavity-prone people, it is during the usual working hours of dentistry that the teeth are most sensitive to picking and drilling.

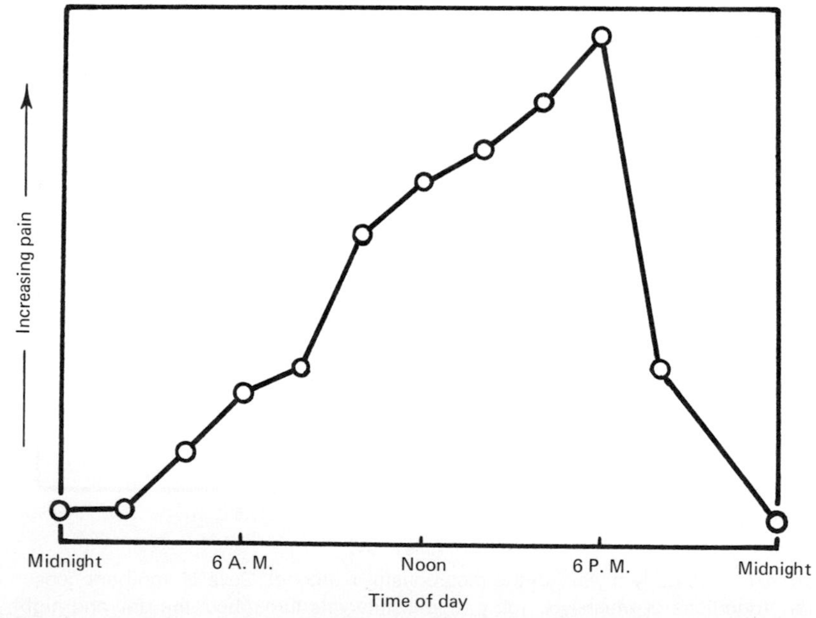

Figure 4-2 The daily rhythm in the changing sensitivity of the teeth to the same painful stimulus. Between the hours of 8 P.M. and 8 A.M. the teeth are much less sensitive than during the daily office hours of the dentist. Redrawn from A. Jores and J. Frees, *Dtsch. Med. Wochenchr.* **63,** 962–963 (1937).

Again, turning to the practical side of things, perhaps it would be worthwhile for the "chicken hearted" to search out a "moonlighting" dental student for treatment at night.

Just as painful responses are rhythmic, so are pleasurable ones—at least in the rat. To demonstrate this fact, electrodes were implanted in particular parts of the rodent brain, so that small amounts of electrical current could be pulsed into it. The rats quickly learned that by pressing a lever in their cages, the current would flow. Stimulation of these brain centers apparently caused a pleasurable sensation since the levers were pressed again and again, sometimes as many as 122,000 times a day. However, the effect emanating from the cerebral electrode was not constant over a 24-hour period. The sensation, as measured by the number of times the bar was pressed, was much more pleasurable during the nighttime than at day. This ecstasy rhythm was repeated daily for the length of the study, one month.

The rhythms just discussed could simply be induced by physical or social changes in the environment, or they could be under the control of man's biological clock. As a first step in distinguishing between the two possibilities, it is necessary to isolate the subject from environmental day-night cycles. If the rhythm persists in isolation, it suggests that it is under the control of the living clock. Studies in nonperiodic conditions have been done; but, before discussing them, a brief description of man as a subject for rhythm experiments will be presented.

MAN AS AN EXPERIMENTAL SUBJECT

Man is probably the most difficult subject upon which to perform long-term physiological experiments. His psychological makeup is such that it is hard, if not impossible, to keep him in isolation as is required in the exploration of biological rhythms (solitary confinement is, in fact, considered a form of torture). His emotional vagaries easily and routinely disrupt normal physiological patterns; and it can be especially perplexing to an experimenter if a psychosomatic upset occurs in the middle of a long set of observations. Humans are also very expensive to keep: they eat a lot, are fastidious in choice of food, and will not reside in quarters lacking the standard creature comforts. In addition, some experiments require hard work, which some human subjects might find distasteful. Therefore, it is very difficult to get an adequate number of subjects to submit to experimentation. Luckily, there are two categories of persons who can be procured: (i) those whose price is low enough to fall within the animal feed and care budget of federal research grants, and (ii) the eternally popular graduate student, whose traditional penury makes him eager to trade periodic blood letting, and urine sampling, in return for free room and board, the quiet of constant conditions in which to study, and a chance to temporarily escape from the pressures of graduate school. Once obtained, the human subject provides a benefit over lower animals: at the end of the experiment, he can relate all his experiences (colored, of course, by his own prejudices) to his paymaster. Thanks to the captive willingness of these graduate students, a great deal about human rhythms has been learned.

Another mass of data comes from the hospital ward. These data are usually less informative since experimental conditions can only be poorly controlled at best and because many medical researchers and experimental psychologists are apparently sated simply by the discovery that another human process is found to be rhythmic. Few of these investigators seem to be concerned with why a process is rhythmic, which, because rhythms are a fundamental property of life, is one of the truly important scientific questions to be answered today.

Man in Prolonged Isolation

To screen man from his cyclic environment, the first studies were carried out in caves and then in World War II bunkers. The results of these studies proved so rewarding that specially designed underground living quarters were constructed at the Max-Planck Institute in Germany. They contain a kitchen so that one can make his own meals, a shower, and the other standard exigencies for routine survival. Except for a few psychological tests and the restriction of no napping after lunch, each subject is allowed to do whatever he pleases. Most subjects are students who spend much of their time cramming for exams. Watches are verboten during the 2- to 6-week isolation period. Entrance to the bunker is through an antechamber guarded by doors at either end, one leading to the outside and the other to the inner sanctum. The locking system is such that both doors cannot be opened at the same time. In the antechamber is the refrigerator in which fresh foodstuffs (and a bottle of Andechs, the local superlative beer) are replenished on a random schedule, and where, in scientific matter-of-factness, urine samples, on their way out for analysis, are stored side by side with the comestible items. The only way of communicating with the outside support crew is by sending and receiving letters in the refrigerator. No ill effects of the isolation have ever been reported.

SLEEP-WAKEFULNESS RHYTHMS

Cave Studies

Over a decade ago, a young speleologist became obsessed with the idea that it was scientifically important for him to live alone, underground, *sans* clocks, for two months. He chose an inhospitable cave in the French Alps where he lived at the 375-foot level. Here the temperature hovered at 32°F, the relative humidity remained unchanged at 100%, and the darkness was complete save for a small battery-powered light in his tent. Each time he awoke, ate, or prepared to retire he called over a field telephone to a surface camp, where his lonely words and the times of his calls were recorded. He claims that the inexorable cold and dampness reduced his body temperature to less than 97°F, and he was constantly threatened by avalanches and cave-ins—still, he held out for the sake of science and whiled away his time writing a best-selling novel about his subterranean adventures, anxiety, and building libido.

Throughout his underground stay, he tried mentally to keep track of the passage of time on the surface. When the men in the surface camp informed him on September 14 that his experiment was over, he thought it was only August 20. His subjective judgment of the passage of time had been exceedingly sluggish; mentally, he had lost 25 days! A major problem in his estimation of the passage of time was the fact that he commonly imagined that he was taking a short siesta after his midday meal, where in reality he had been awake for 16+ hours and his "naps" lasted approximately 8 hours. However, his living clock, as evaluated by the times of his retiring and awakening phone calls to the surface, had ignored his mental confusion and guided his body functions all the while, measuring off periods of activity and sleep that totaled just longer than a day: 24 hours and 30 minutes on the average (Figure 4-3).

As seen in Figure 4-3, because the period of the cave dweller's rhythm was slightly longer than 24 hours, his sleep-wakefulness cycle quickly fell out of phase with the actual day-night cycle in the French Alps above him. Only once again during the experiment did it come into phase with the light cycle outside, which led to a rather interesting result. "Daily," during his underground sojourn, he had entered limited scientific observations and numerous complaints in a log. Save for one entry, the diary is a hodgepodge of chronicled discomforts, misadventures, perpetual intestinal uprisings, cave-ins, and real and imagined terrors. In this particular entry, however, the diary tells us that "for the last few days I have felt very optimistic, I suffer less from

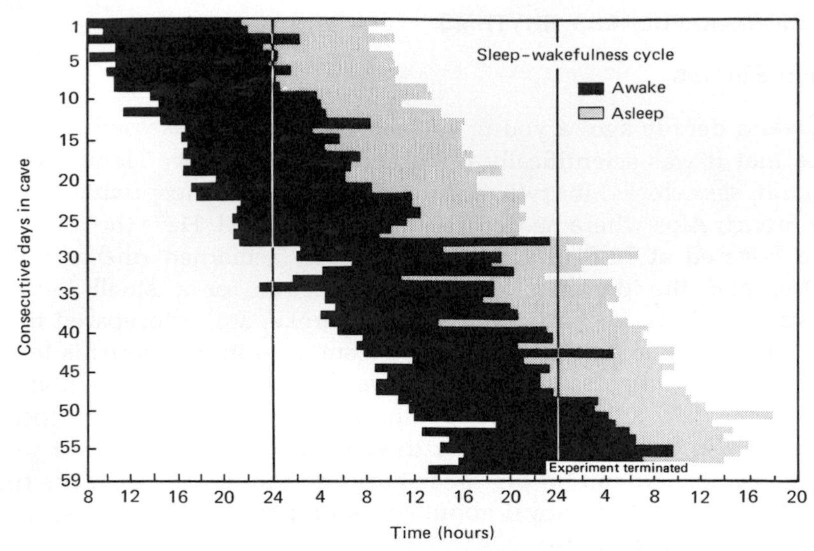

Figure 4-3 The sleep-wakefulness rhythm of a human subject isolated in a cave for 59 days without a watch. His rhythm had an average period of 24 hours and 30 minutes. Drawn from the data of M. Siffre, "Beyond Time." McGraw-Hill, New York, 1964, and appearing in J. D. Palmer, *Nat. Hist.* **79**(4), 53–59 (1970).

the cold; I am better adapted to conditions." During this optimistic period (days 36–39 on Figure 4-3), his sleep-wakefulness rhythm was again in phase with normal day-night cycles outside the cave.

Another troglodyte, an English potholer named Geoffrey Workman, vowed to live alone for 100 days in a cave, and had the fortitude to even surpass this goal. The cave temperature was constant at 44°F, and the atmosphere nearly saturated. Light was provided only by candles and a miner's forehead lamp. Workman was allowed to keep his wristwatch and he spoke via a field phone to a surface vigil about once a day.

On first entering the cave, Workman had intended to maintain his suprasurface sleeping habits, but soon discovered that he could not fall asleep at his usual time and that he had a compelling tendency to oversleep in the morning. Finally, after three weeks, he abandoned his plan and retired when he felt sleepy, which turned out to be about 40 minutes later each day. The period of his sleep-wakefulness rhythm stabilized at 24.7 hours and when he ascended to the surface, 105 days after his descent, he had lost just over 2.5 real days.

Of the six best-studied isolationist spelunkers, who have remained alone in caves for between 8 and 25 weeks, all adopted circadian sleep-wakefulness periods of longer than 24 hours, with the average being 24 hours and 42 minutes.

Laboratory Studies

In Germany, in the specially designed bunkers described before, further studies of sleep-wakefulness rhythms have been conducted. These chambers represent a less extreme way of excluding clues to the passage of day and night, and therefore attract a less daring coterie of volunteers who, perhaps, more closely represent the average man. At any rate, before entering the bunkers, the subjects often asked what they should do, and were told, in essence, "anything you want." Invariably, they would respond, "First I'll catch up on all the sleep I've missed lately." Ironically, most commonly, they remained awake for unusually long intervals for the first couple of days before settling down to a fairly regular routine. Also, in the initial days of their isolation, most of the subjects were greatly preoccupied—as documented in their diaries—with the question of real time. However, after a few days, this obsession is forgotten and many agreed that their behavior would probably not have been much different with a clock present (just as was true in Workman's case). All found that the decision to rise was more difficult than to retire, as they could not help but question whether the residual sleepiness they suffered on waking was normal or signified a short nap rather than a full night's sleep. Paradoxically, the waking times of most subjects were more regular than the times of repose.

Once isolated in the bunker, the times of wakefulness and movements in the bed were recorded automatically. (Other physiological parameters were also measured; these will be subjects of later chapters.) The incarcerated subjects were, depending on the experiment, either made to live in constant darkness, or with the light burning day and night, or were allowed to turn them off at bedtime. A representative sleep-wakefulness rhythm (along with other rhythms) is shown in Figure 4-4. The subject here portrayed was released on day 10 after his last period of sleep which ended at 3 P.M.; his rhythm had slowed to the point where he had "lost" 0.5 real days. He reported that he had had difficulty in deciding whether he had slept long enough; on day 8, he had gotten up after only 3 hours of sleep and prepared breakfast, but then feeling drowsy slept for 3 more hours. His clock had obviously "corrected" his mistaken effort of will. Note that on days 8, 9, and 10 there is a slight blip in the temperature curves at the times corresponding to the "mistaken" awakening.

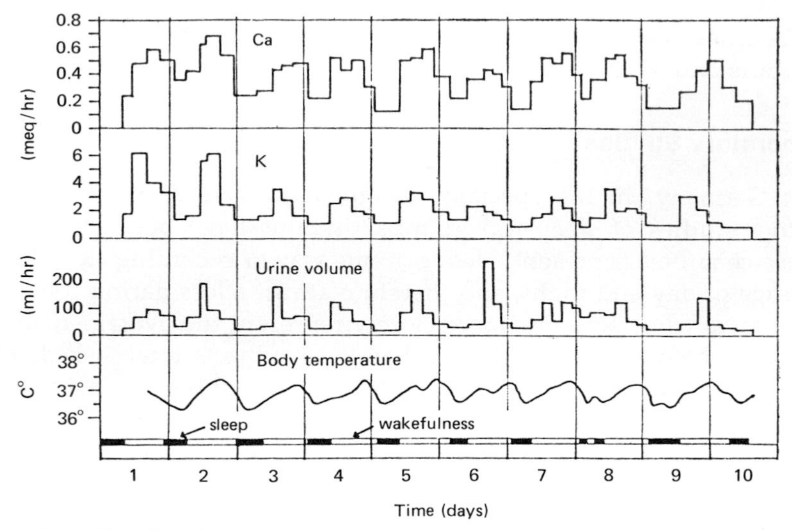

Figure 4-4 Circadian rhythms of one subject isolated in an underground bunker. The upper three curves represent rhythms in two excretory products, calcium and potassium, and total urine volume. Below these are deep body-temperature and sleep-wakefulness rhythms (solid bars signify intervals of sleep). Modified from J. Aschoff, *Science* **148,** 1427–1432 (1965).

As characteristic of the caveman studies, the bunker dwellers also described rhythms with periods slightly longer than 24 hours—the average for just over 100 cases is 25.1 hours. Only a few of these subjects have ever displayed periods shorter than 24 hours, and one of these, a potential transcendentalist, could modify his timing at will: during his first 10 days under a self-controlled lighting schedule, he displayed a rhythm with a period of 19 hours, but then willed an increase which resulted in a 25.8-hour period. The conscious manipulation of period by at least some subjects must interject a substantial degree of unwanted variability into some experiments.

To test the role of social stimuli on the entrainment of sleep-wakefulness rhythms, four subjects were placed together in the same bunker. The lights in the chamber were kept constantly on. Right from the beginning, it became apparent that one member of the team was an early riser (Figure 4-5), but all managed to keep pretty well in phase until the thirteenth day, when the others began taking catnaps and finally "broke away" from his pattern by the seventeenth day. A few days later, another began to shorten his period so that amusingly, when the experiment was terminated and the investigators broke into the bunker to inform the subjects of the fact, they found all four seated

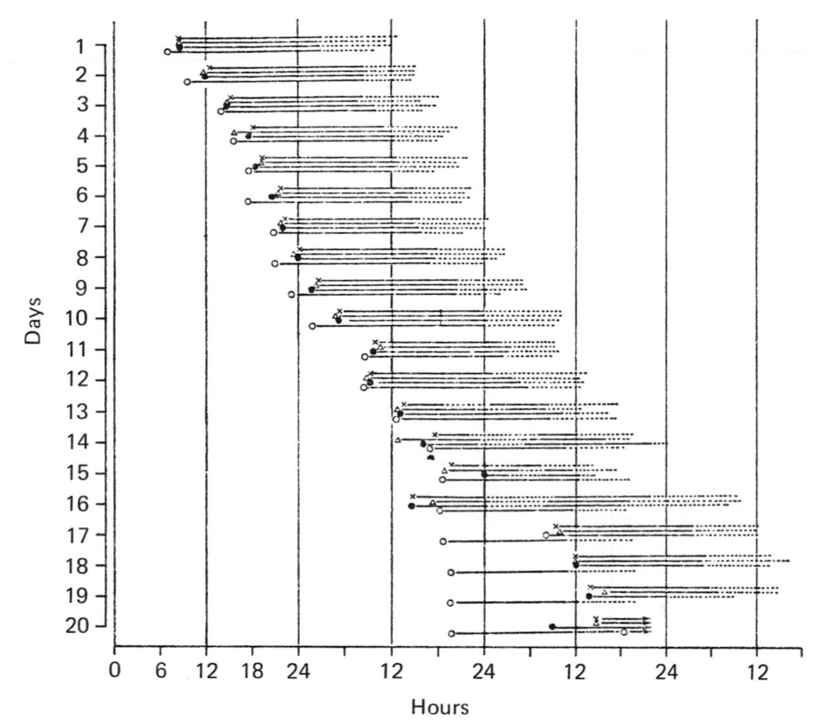

Figure 4-5 The persistent sleep-wakefulness rhythms of four male subjects isolated together in constant conditions. During the first 12 days, they remained entrained to one another with the slight exception of one subject who consistently awakened earlier than the others. They then began to display different sleep patterns and by the seventeenth day no longer remained in synchrony. The early riser assumed a period of 24.1 hours while the others extended their rhythms to 27.2 hours. By the last day of the experiment, a second subject also had begun to shorten his period. The solid horizontal lines represent wakefulness; the dotted lines sleep. Modified from E. Pöppel, *Pflüegers Arch. Gesamte Physiol. Menschen Tiere* **299,** 364–370 (1968).

at the table eating; however, one was having breakfast, one lunch, and the other two dinner.

Development of Sleep Rhythms in Infants

The sleep-wakefulness rhythm is not present at birth, but develops soon after. The sleeping habits of 19 newborn children were studied by requiring cooperative parents to record on tally cards the times of sleep and wakefulness of their infants as accurately as possible. The study encompassed an interval of time beginning when the mother and offspring left the hospital, up to the time that the baby would no

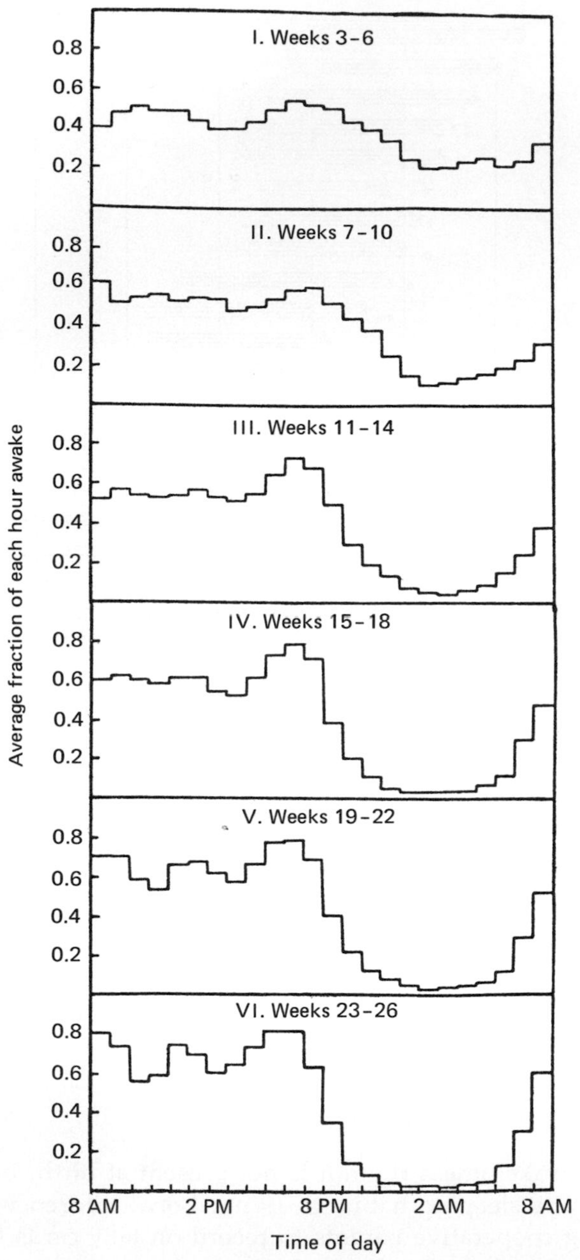

Figure 4-6 Average hourly wakefulness of 19 children plotted as a function of week after birth. The day-night asymmetry is already apparent in the top curve. The mean daily sleeping time drops from 14.7 hours in the top curve to 13.9 in the bottom. (From N. Kleitman and T. Englemann, *J. Appl. Physiol.* **6,** 269–282 (1953).

longer "stay put" in the crib. The results were lumped together and portrayed in Figure 4-6. Contrary to popular belief, the newborn does not sleep 21–22 hours a day, but is instead awake 8–9 hours out of each 24. As early as the first week of postuterine life, the sleep rhythm had already become apparent; at this age, the average accumulation of sleep time between 8 P.M. and 8 A.M. was 8.5 hours; while between 8 A.M. and 8 P.M., it was 6.5 hours. Night sleep gradually rose to 10 hours by the tenth week and remained constant at this level through the twenty-fifth week. During the same interval, the neonate's day sleep decreased a little over 3.5 hours.

One baby girl in the study, a first child, had parents indulgent enough to allow their offspring to determine its own sleep-wakefulness schedule, rather than forcing it to conform for convenience to their pattern of living. At first, sleep was haphazard; but, starting about the fourth week, a period of about 25 hours was developed (approximately the same as reported for the cave and bunker adult isolates). Not until about the eighteenth week was the 24-hour periodicity assumed, with major sleep confined to the nighttime.

TEMPERATURE AND PERFORMANCE RHYTHMS

That the human body does not maintain an even 98.6°F temperature throughout the day, but instead undergoes a regular 24-hour variation, has been known since the middle of the last century. John Davy, a British army physician, was one of the first to describe such a rhythm after using himself as an experimental subject. As a way of comparing 1845 temperature taking with the present, I have excerpted the following from his paper given before the Royal Society in London:

> The thermometer I have employed is a bent one, about 12½ inches long, its bulb about an inch long, and, where widest, half an inch thick; its curvature is about 3½ inches from the bulb, and its stem, to which the scale is attached, nearly at right angles to the bulb, so that when inserted under the tongue, the observer has no difficulty in distinguishing accurately the degrees himself, whether near-sighted or the contrary; in the latter instance using merely a common magnifying glass.
>
> . . . it is necessary that the thermometer remain in the mouth many minutes . . . a shorter time being required . . . if the mouth has been kept closed for a quarter of an hour previously. . . .

He made observations on himself nearly daily for 8 months and found that his temperature vascillated normally from a low in the small hours of the morning to a high, usually in the late afternoon and early evening.

Repetitive studies during the next century verified Davy's early finding many times (Figure 4-7). Now it is common knowledge that the rhythm becomes first noticeable at an age of 4–5 weeks and, depending on age, describes a curve vascillating approximately between 97.2° and 98.35°F during a 24-hour period. Consequently, an afternoon

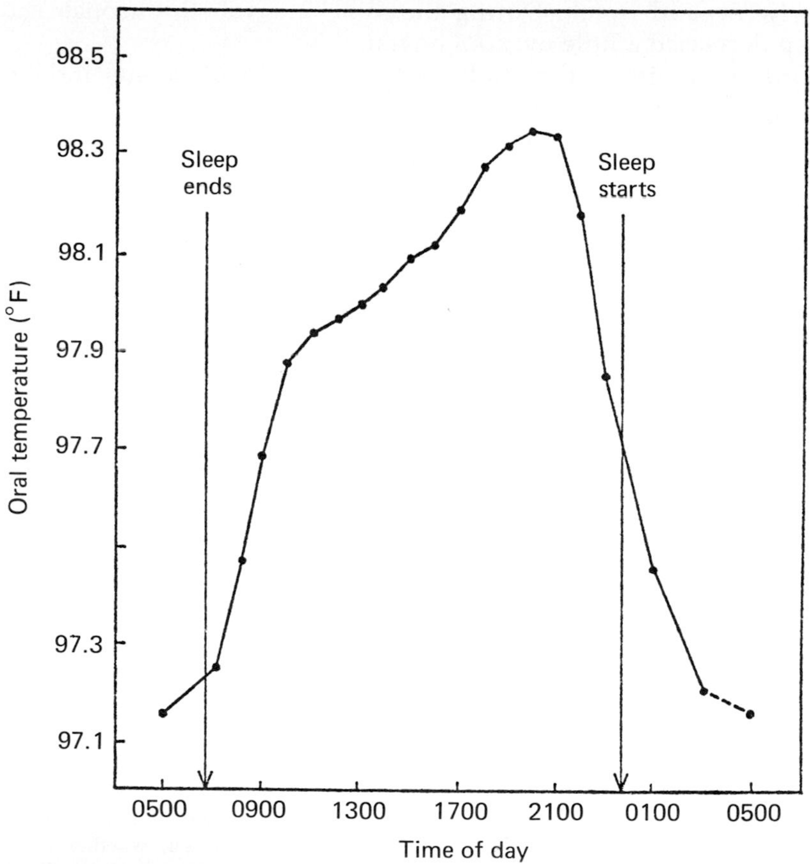

Figure 4-7 The average daily rhythm in the oral temperature of 70 English seamen. Temperatures were recorded at 1-hour intervals during the waking hours of the day and every second hour during sleep (a three-point moving average was used to smooth the curve). The temperature rhythm is characterized by a rapid rise on waking, followed by a more gradual ascent to a peak in the evening and a precipitous, almost linear, drop after the maximum. The average temperature difference over the day was 1.19°F. W. P. Colquhoun, *in* "Biological Rhythms and Human Performance," (W. P. Colquhoun, ed., pp. 39–107. Academic Press, New York, 1971.

temperature of 99°F does not necessarily mean that one is slightly feverish, though it does if recorded on waking.

Body temperature depends on the balance between heat production in internal tissues and heat loss from the skin. Therefore, it was only logical for many early human physiologists to explain the rhythmic change they observed—without recourse to experimental verification—as an axiomatic consequence of muscular activity, digestion of foodstuffs, and maintenance metabolism (all heat-producing processes) during the waking hours, alternating with a lack of these and bed rest during the night. As sometimes happens to the chagrin of scientists, their inductive jumps prove to be incorrect, as was subsequently shown by a variety of different observations. For example, subjects confined to bed and fed identical meals at regularly spaced intervals throughout the day, or fasting during the measurements, still displayed the rhythm with no decrease in amplitude. Even a young man tragically paralyzed with poliomyelitis for 16 months was found to display a normal temperature rhythm throughout his incapacitation. In illness, an accompanying fever simply increases the amplitude of the rhythm, so that febrility is most intense in the late afternoon (Figure 4-8). Even sleep deprivation will not disrupt the temperature rhythm, although it does decrease the mean body temperature (Figure 4-9). [An interesting sidelight of deprivation studies is the revelation of a daily rhythm in fatigue. Subjects were asked to evaluate their feelings of fatigue according to some quasi-objective ranking method such as (1) not tired; (2) somewhat tired; (3) pretty tired; or (4) dead tired. The results of such a study of 63 soldiers kept awake for 3 days are superimposed over the temperature data in Figure 3-3. It is clear that their feelings of fatigue were rhythmic and

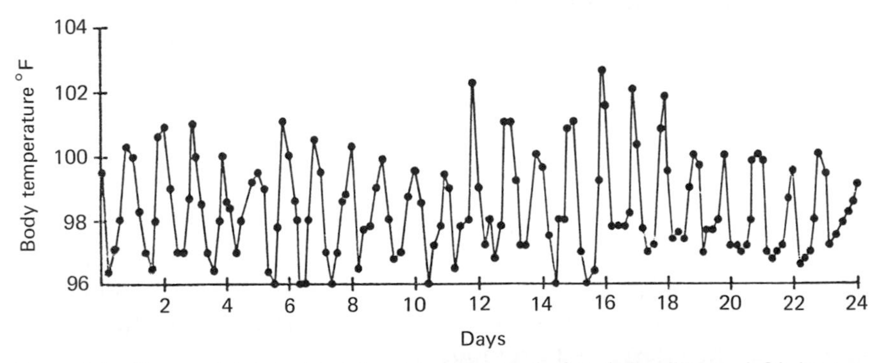

Figure 4-8 The persistence of the daily temperature rhythm during illness. A 24-day temperature record for an 18-year-old girl suffering from an allergy to milk. Drawn from the data of H. Rowe, *Ann. Allergy* **6,** 252–260 (1948).

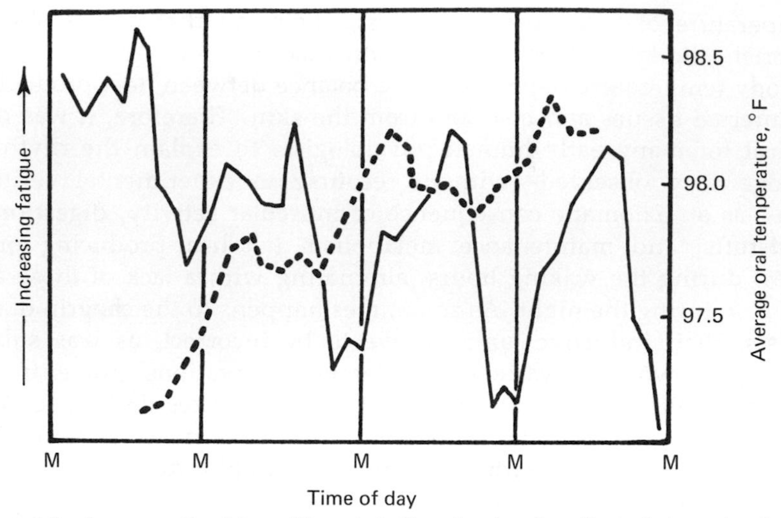

Figure 4-9 A composite of two different studies showing the effect of sleep deprivation on the temperature rhythm and subjective fatigue. The descending curve shows the persistence of the daily temperature rhythm of 15 men during 98 consecutive hours without sleep. The ascending curve represents the self-assessed fatigue for 63 soldiers kept awake and active for 3 days. Note that fatigue is also rhythmic and tends to mirror image the temperature curve. M stands for midnight. Drawn from the data of E. Murray, H. Williams, and A. Lubin, *J. Exp. Psychol.* **56,** 271–273 (1958), and J Fröberg and L. Levi, as cited by L. Levi, *Sartr. Föersuarsmed.* **2,** 3–8 (1966).

peaked in the wee hours of the morning. This rhythm has been undoubtedly experienced by most of us as students staying up all night cramming for exams, or at marathon parties—we become sleepy and less lively sometime after midnight, but then tend to perk up again around dawn. As pointed out by David Hubbard (a psychiatrist and expert on political terrorists), during the 1972 violence at the Olympic games, since the Arab guerillas did not sleep, German police should have stalled until the predawn hours and only then attempted to either negotiate or overpower them, to take advantage of the fact that these barbarians were at their daily physiological and psychological low point.]

The conclusion from these observations is that the temperature rhythm appears to be independent of other body oscillatory physiology, and probably also environmental cueing. To demonstrate the latter, subjects were isolated in the constant conditions of the experimental bunkers described previously.

In these studies, the body temperature of volunteers was measured continuously by a sensor probe inserted in the rectum. Most commonly, the thus impaled subjects were made to live for weeks at a time with the house lights left on continuously and held at some constant intensity, although in one set of experiments they were compelled to live for 4 days in complete darkness. In spite of the absence of external clues as to time of day, their temperature rhythms persisted, usually with a circadian period ranging closely to 25 hours (Figure 4-4). Often, under these unnatural conditions, the temperature rhythm was seen to dissociate from the sleep-wakefulness rhythm by adopting a new period (see Figure 4-20). Therefore, the sum of all the experiments indicates that the temperature rhythm is definitely under the control of a living horologe.

Because of the ease in measuring body temperature, this rhythm has been subjected to numerous experimental manipulations. In one study, two men regulated their living schedules to 21-hour "days" (which produce 8 "day weeks") and later 28-hour "days" (6 "day weeks"), each for six weeks, while trying to participate in university activities when the times of sleep and activity permitted. Between these two experiments, they lived normal 24-hour days for six weeks. One subject, the younger of the two, easily adapted to both the 21- and 28-hour routines within a weeks' time. Curiously, as his day length increased, the mean amplitude of the temperature curve also increased from 1.21°, to 1.66°, to 1.75°F for the 21-, 24-, and 28-hour days respectively. However, the daily temperature curve of the older man remained at 24-hours in spite of the diligently followed routine. Surmising that the interruptive social stimuli associated with this study may have influenced the latter's ability to adjust to unnatural "days," a follow-up experiment was performed in which the same subject, plus a new companion, housed themselves in a dark chamber of Mammouth Cave in Kentucky. In there, darkness was absolute, the temperature varied less than a degree in a year, and the humidity was constant at just under saturation level. Artificial lighting, beds, and a table and chairs were installed. The subjects forced themselves to live on a 28-hour day, with 19 hours of wakefulness alternating with 9 hours of sleep. As before, the older man was not able to adjust well, having difficulty falling asleep unless the time of retiring happened to coincide with his customary sleeping habits aboveground. His temperature rhythm always maintained a 24-hour period. The rhythm of his new companion in this adventure adapted completely to the 28-hour day within one week of subterranean existence. In a subsequent experiment, lasting 30 real days, the 2 men forced themselves to stay awake for 40 hours and to sleep only 8 hours; but in this case, their temperature rhythms "ignored" the self-imposed cycle of existence and continued to display one peak every 24 hours.

The conclusion, based on this work in particular and a great deal of similar experimental designs, is that there is considerable variation in the adaptability of individuals. When adjustment can be accomplished, the test cycle cannot differ by more than a few hours from the interval of a natural day.

There are two other categories of human rhythms—time estimation and psychomotor performance—that are intimately associated with (if not directly caused by) the temperature rhythm.

Time-Perception Rhythms

It has been found that man's subjective time perception varies rhythmically on a 24-hour basis. For example, if one is asked to estimate the passage of a 60-second interval at different times of the day, he tends to overestimate—indicate periods longer than 60 seconds—in the morning and evening and underestimate them during the day. The *form* of the curve is roughly bell-shaped and tends to mimic the form of an individual's daily change in body temperature (Figure 4-10). This correlation suggests that possibly the hourly differences in time estimation may be just a secondary consequence of the changes

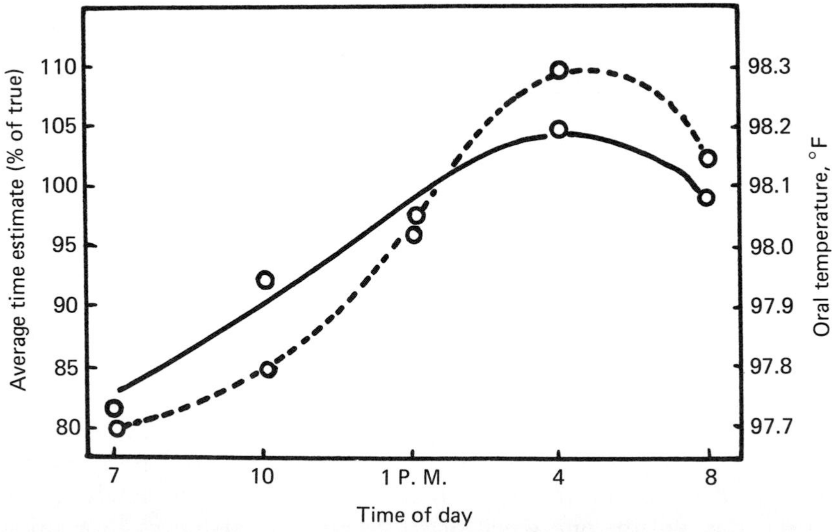

Figure 4-10 Diurnal variations in time estimation and body temperature. The average body temperature of 10 subjects, taken at 5 different times throughout the day, plotted (broken line) against the group average error of estimations of intervals of time ranging between 10 and 60 seconds. Redrawn and modified from D. Pfaff, *J. Exp. Psychol.* **76,** 419–422 (1968).

in body temperature. Work such as that to be described next showed that this was the case.

An eminent Harvard physiologist was sent to the drugstore for medicine by his wife who was sick with influenza. When he returned fifteen minutes later he was berated by his spouse for having been gone so long. Rather than cowering in typical noncelibate fashion, he took her temperature and made her estimate intervals of sixty seconds by counting at a rate of one numeral per second. Being an accomplished musician and priding herself on her good sense of duration and rhythm, she indulged him by performing this chore forty times during her illness. Each time he measured the actual duration of her count with a stopwatch and plotted the results against her oral temperature—the latter varying between 97.4 and 103°F. He discovered that she unknowingly counted faster at higher temperatures than at lower ones (Figure 4-11). Therefore, at higher body temperatures she subjectively surmised that time passed rapidly; but, since this judgment was erro-

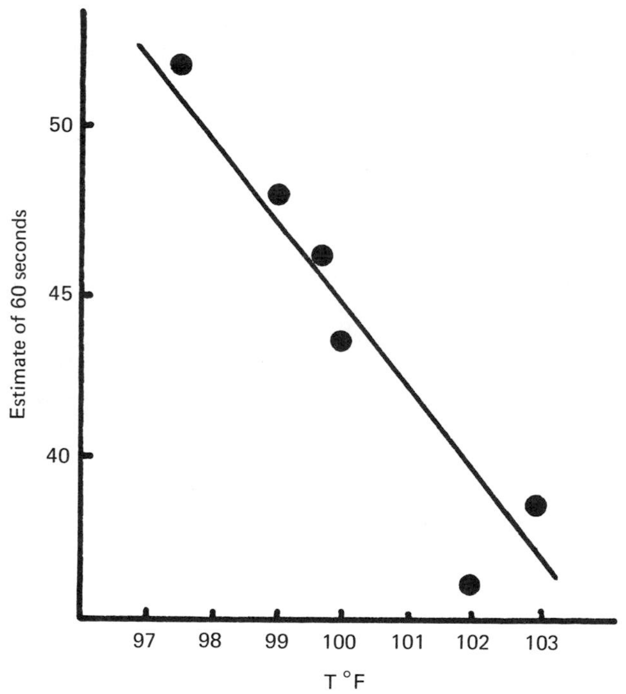

Figure 4-11 The accuracy of one woman—suffering from the flu—in estimating an interval of 60 seconds, plotted as a function of her body temperature. Each point is the average of about six estimates. Drawn from the data of H. Hoagland, *J. Gen. Psychol.* **9,** 267–287 (1933).

neous, the time seemed to drag. With this study the dom not only vindicated himself of dawdling on the way to the drugstore, he also augmented his publication record in the true fashion of a Harvard Man.

Being concerned by the fact that his studies included only one infirmed subject suffering from the flu, he extended his investigations to several other volunteers who were wrapped in blankets for insulation and subjected to artificial febrility created by diathermy. The results obtained with these men were in every way compatible with his original finding, as were those done in another laboratory in which subjects were cooled. The general relationship between body temperature and time sense approximates a 10% speeding up for each 1°F rise in temperature.

Therefore, the subjective perception of time passage is a secondary consequence of the body's daily temperature rhythm, rather than being under the direct control of a biological clock. This conclusion is further strengthened by the finding that this form of time sense can be modified by various chemicals, e.g., cannabis preparations and hallucinogens such as LSD and psilocybin cause time to pass very slowly or even stop; while alcohol, amphetamines, and narcotics such as opiates cause it to pass more quickly. As has been described previously, true biological rhythms are not generally influenced by external temperature change or by exposure to most chemicals.

Daily Rhythms in Psychomotor Performance

Recognizing the fact that so many human physiological processes are rhythmic, it should be obvious that one is not precisely the same person from one hour to the next (but at the same time each day, people are much like they were the day before and much like they will be tomorrow). Accepting this, it should not be too *avant garde* to suspect that the speed and accuracy with which people perform daily tasks might also vary in a recognizably rhythmic fashion. Studies of this type fall under two headings: on the job field studies, which are concerned with practical performance throughout the normal working day, and lab studies, which sometimes additionally span the hours of nighttime. The lab situation is truly the better, since many of the unwanted variables, such as interruptions in the flow of raw material, machinery breakdowns, and increased motivation because of the temporary presence of the boss, are eliminated. Of course, in the lab, there is a substantial loss of realism, but the novelty of the contrived task often provides sufficient motivation to replace the socioeconomic incentive of factory performance.

Several studies on the time-dependent nature of speed and accuracy of simple digital manipulations and mental activities in general have been made. For example, in one study, subjects were given tests that could easily be performed at home 5 to 10 times per day. These tests included such tasks as determining the time necessary to deal a "deck" of 156 cards into four hands, or sorting the same pack according to denomination, or multiplying 8-digit numbers by one another. The subjects determined their oral temperatures after each test. As seen in Figure 4-12, all the resulting scores varied in a regular pattern over the day, and the forms of the curves resembled the subjects' personal mean daily temperature curve. A great many other daily endeavors have been found to be rhythmic also, such as the response time to a light signal while driving, calculation speed in addition, steadiness of the hand, vigilance in monitoring a radar screen and sonar, and memorization ability. The forms of almost all of these rhythms mimics that of the tested individual's daily temperature curve.

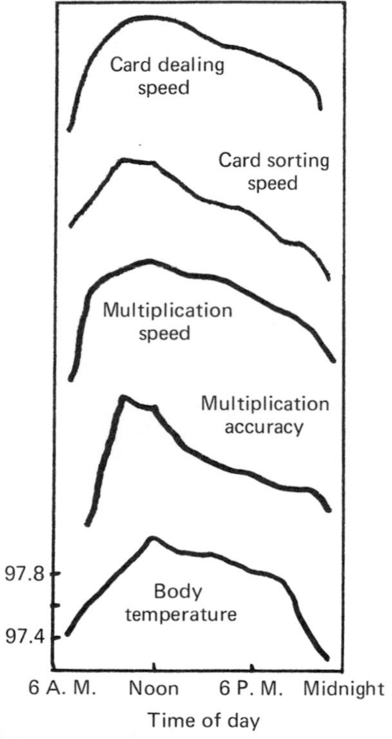

Figure 4-12 Average daily rhythms in speed and accuracy of performance of simple tasks by one subject over 20 days. Temperature taken orally and recorded in °F. Redrawn and greatly modified from N. Kleitman, *Amer. J. Physiol.* **104,** 449–456 (1933).

"Early Birds" and "Night Owls"

Commonly bandied about by the laity is the thought that some people are "early birds," meaning that they wake up with the sun, eagerly jump out of bed to begin the day's toil, and unsurprisingly, retire early each night. Their counterparts have difficulty getting up each morning, do not get into the "swing of things" until relatively later in the day, and remain active well into the night; they represent the "night owl" segment of the population. (In addition to these extremes, a considerable portion of the population do not fit clearly into either of these avian categories.) Experimental psychologists have verified the existence of these types.

With the use of a Heron Personality Inventory, a test that purports to measure sociability, 47 relatively extreme introverts and extroverts were culled out of the group of English seamen used in the study portrayed in Figure 3-1. Oral temperatures were taken periodically over a span of 2 days and used to produce the two curves seen in Figure 4-13. Quite clearly, the average daily temperature curve of those sailors defined as introverts (which, as will be described later, are early birds) differ from their more outgoing companions: their body temperatures

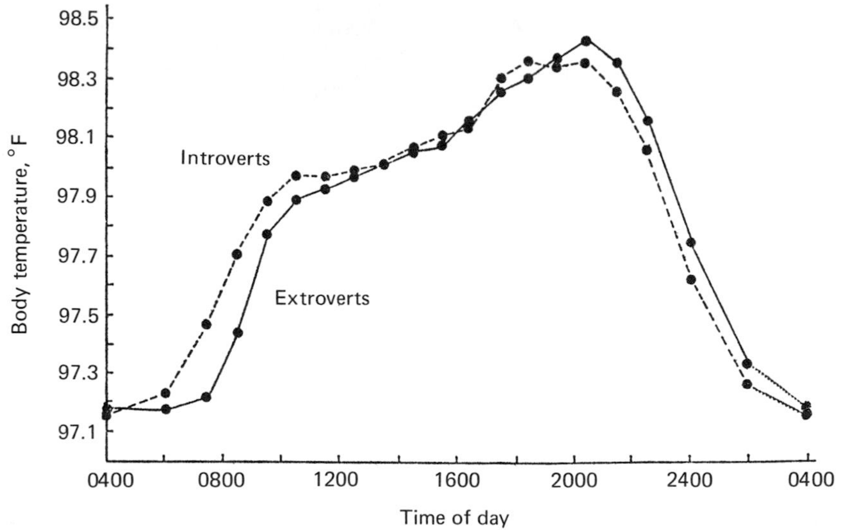

Figure 4-13 The average body-temperature rhythms of 25 introverts and 22 extroverts during two 24-hour periods separated by 1 week. It is seen that the temperature cycle of the introverts rose earlier in the morning, peaked sooner, and fell earlier at night than that of the extrovert group. From M. J. F. Blake, *Nature* (*London*) **215,** 896–897 (1967).

rise, peak, and fall earlier in the day than those of their night owl, extrovert shipmates.

The peaks of hourly efficiency of these two groups also differ in the expected way. Several simple experiments, such as the one that follows, have demonstrated this. Five times during their waking hours, 12 introverts and 10 extroverts were asked to cross out all the letter "e"s—as quickly and accurately as possible—in a story taken from *Punch* magazine. Each subject was scored on the number of letters he cancelled in 30 minutes. Obviously, this "letter cancellation" task involves no other skill than being able to read and scribble. Their temperatures were taken at each measurement. Both the temperature and the output curves differed in the expected way, with the introverts warming up sooner and performing better in the morning and the extroverts peaking later in the day (Figure 4-14). The performance

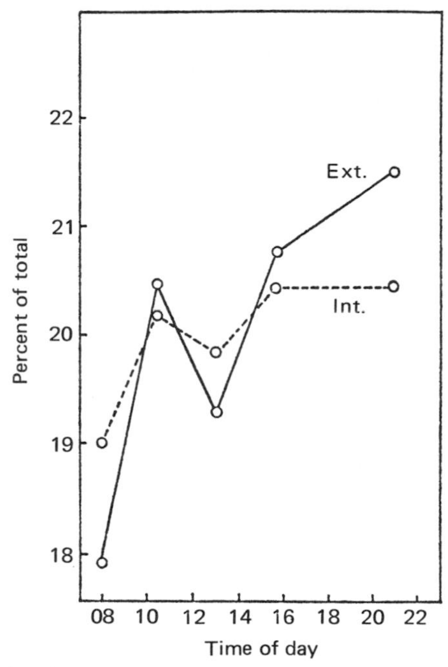

Figure 4-14 The number in percentages of letter "e"s cancelled out of a single magazine article in 30 minutes at five different times during the day. Note that the average performance of 12 introverts was greater at 8 A.M. than that of 10 extroverts but then fell below the latter for the remainder of the day when the extroverts, as predicted, finally "got going" for the day. See also the postprandial dip which is often recorded in time-dependent performance studies. M. Blake, *in* "Biological Rhythms and Human Performance" (W. P. Colquhoun, ed.), pp. 109–148. Academic Press, New York, 1971.

of both groups corresponded to the form of their group temperature curves with one blatant exception, a drop in efficiency after lunchtime. This postprandial dip is a reported characteristic of about one-half of all published performance studies.

The "dip" was first discovered in 1916, in a group study conducted on an introductory psychology class at the University of California. All 165 students in the course participated by first stating their sleep patterns and preferred hours of study, and then submitting to five repetitive memory tasks at 1-hour intervals over three consecutive days. The tests were as follows: *auditory memory,* the accuracy with which students could remember and write down a series of digits, 4–12 numbers long, after hearing the numbers read; *visual memory,* the same procedure except the numbers to be remembered were shown, rather than read, to the subjects; *substitution test,* the rapidity of learning what nonsense symbol stood for a particular number; *recognition test,* the accuracy with which geometrical figures were remembered; and *logical memory,* a simple test for the memory of ideas.

The results were threefold. The first, and to the modern day student probably the most surprising, is the fact that the majority of the stu-

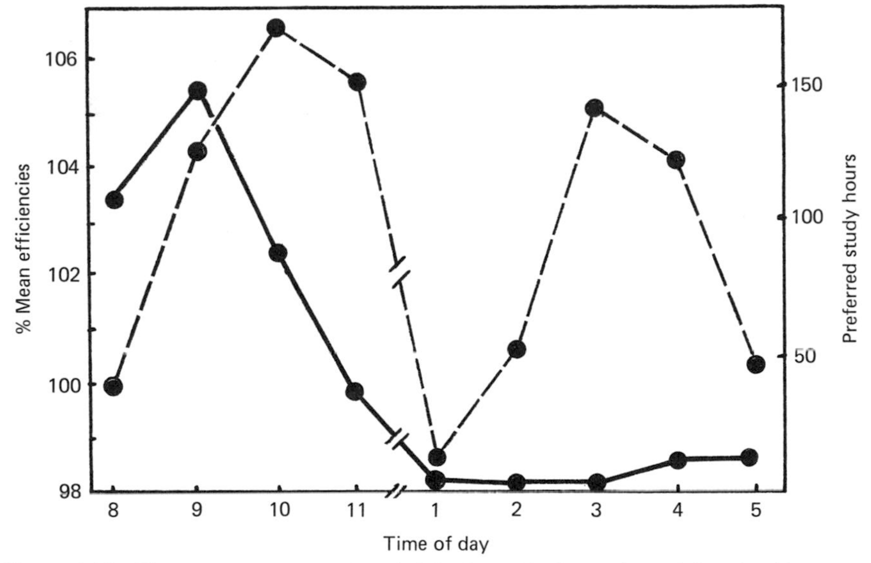

Figure 4-15 The preferred hours of study (single-peaked curve), as determined by questionnaire, for 165 college students, and their combined performance on five simple tests of memory. Prominent in the data is the postprandial dip in efficiency and the fact that the students did not study at the hours when the tests indicated that they were most capable. Drawn from the data of A. I. Gates, *Univ. Calif., Berkeley, Publ. Psychol.* **1,** 323–344 (1916).

dents awakened at 7 A.M. and retired at 10:30 P.M. Second, as shown in Figure 4-15, they preferred to study between 8 and 10 A.M. (and also after dinner, but this information has been omitted from the figure since testing was not carried on after 5 P.M.). Third, all five of the resulting average performance curves, while differing in amplitude, were otherwise virtually identical. Therefore, a single composite curve is portrayed in Figure 4-15. The data show that the student's subjective interpretation of their times of greatest efficiency was not at all accurate—in particular, the entire afternoon (when body temperatures would have been highest) was unused. The dip described by the curve emphasizes the modern student's wisdom of spending the postlunch interval playing Frisbee on the college mall.

To date, there is no good explanation for the postlunch decline. The usual reason given by students, that the school cafeteria lunch deadens their senses, is incorrect; for the depression occurs even if the noon meal is skipped.

RHYTHMIC MOODS

Using an "adjective check list"—a list of words each of which suggest an attitude or condition—psychologists can obtain what they feel is a rather objective measure of a person's mood at the time the test is administered. In the study reported here, a list of 58 words or short phrases, such as "alert, depressed, grouchy, and full of pep," were presented several times during a day to a group of 36 college students who were told to assign a score between 0, "not at all," and 4, "extremely," to each word to describe how closely it corresponded to their present mood.

The outcome of the study revealed that moods changed over the day in a rhythmic fashion. Anxiety and depression were highest in the morning and dropped continuously during the day. Cheerfulness and friendliness were just the opposite, while alertness and friendliness peaked at midday.

CELL DIVISION RHYTHMS

As early as 1851, it was discovered that in flowering plants the rate at which root and stem cells divided was rhythmic. The tips of onion roots were sectioned and stained in essentially the same way it is done in many freshman biology courses today, and the number of mitotic figures counted and compared with the total number of nondividing cells. The rhythms were found to persist even when the plants were maintained in constant light and temperature in the greenhouses.

In the years intervening since this initial discovery, many other mitotic rhythms have been described in a variety of organisms including algae, protozoa, and even up to a closer relative of man, the rodent (Figure 2-29). Scientists then began to question whether similar rhythms might also be present in humans. However, any approach to the answer required the periodic removal of living tissue from a volunteer. Eventually, in 1939, the problem of suitable subjects was overcome by an ingenious lady physician who capitalized on the custom of circumcision. Working in New York City where this is especially common, she collected tiny excised foreskins at most clock hours of the day. Her final collection was especially homogeneous, since all the donors were 6 to 11 days old and lived together in the same environment of the nursery. The data gathered from 57 penes are portrayed in Figure 4-16, which is kind of a classic as the first representation of a daily rhythm in phallus growth; obviously, cell division was greatest during the afternoon and early nighttime.

In the years following these observations, the dignity and financial status of at least some medical students had declined to a point where they could be "coerced" to serve side-by-side with laboratory mouse

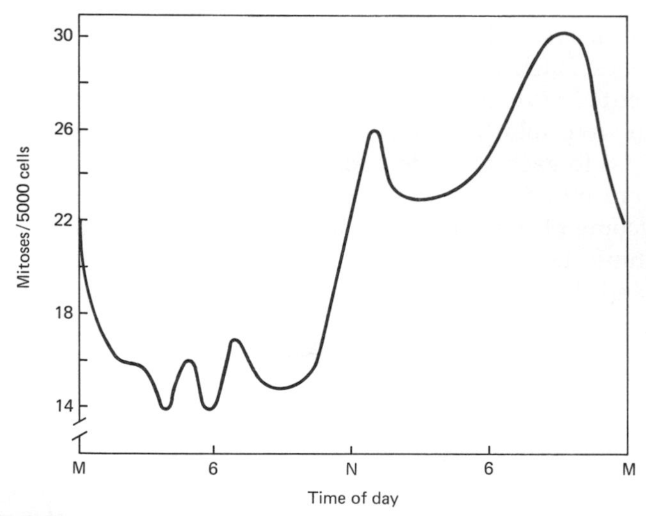

Figure 4-16 Daily rhythm of cell division in the foreskins of the human penis. The study included data from 57 neonates, 6–11 days old, living in standard hospital conditions. Curve drawn from the data of Z. Cooper, *J. Invest. Dermat.* **2,** 289–300 (1939); using a 3-hour moving average.

populations in cell-division studies. Twelve students donated 193 fragments of skin from their shoulders, each being obtained with a biopsy punch (an implement similar to a standard cork borer). Their loyalty to science resulted in a confirmation that the human epidermis does indeed replace lost cells in a rhythmic fashion, and a completed thesis for their unpunctured project director.

HEART-RATE RHYTHMS

The average rate of heart beat for a young adult is slightly over 70 beats/minute when measured during a physician's working hours (and, naturally, in his office). But, when similar determinations are carried out over a span of 24 hours—at, for instance, 2-hour intervals—it is found that a diurnal rhythm is described in the changing rate of pulsation (Figure 4-17), with the heart beating faster during the daytime than at night. What is the cause of this rhythm?

Early in the study of human physiology, a relationship was noted between heart rate and body temperature. In fact, in the years preceding the introduction of the oral thermometer to medicine, the doctor, after diagnosing a fever by placing his hand on a patient's forehead, commonly quantified the degree of febrility by determining his cus-

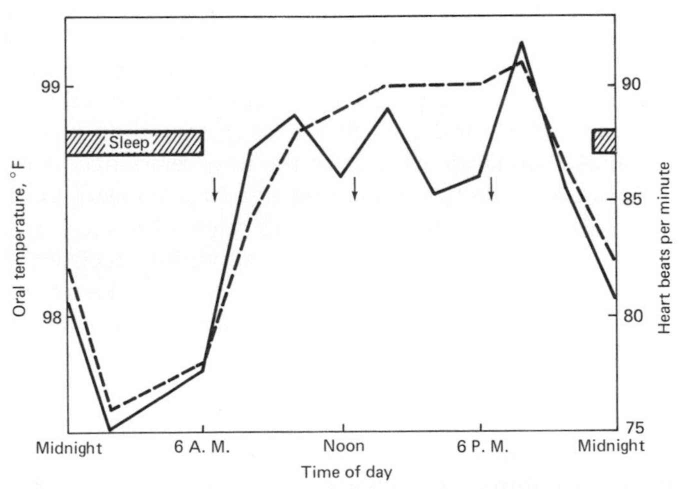

Figure 4-17 The average daily rhythm in heart rate (solid line) and temperature (dashed line) for a young woman during 168 days of measurement. In this subject, each 1°F rise in body temperature increased her heart rate by 10 beats/minute. Falling arrows signify usual mealtimes. Redrawn and modified from N. Kleitman and A. Ramsaroop, *Endocrinology* **43,** 1–20 (1948).

tomer's pulse rate: for every 1°F rise in temperature, the heart rate increases by 10–15 beats/minute. This relationship is especially striking when an individual's heart rate and temperature rhythms are plotted side by side (Figure 4-17). Because of the association, it was surmised that the parallelism must be created by the daily temperature oscillation molding the heart rate into a rhythm. It has since been found that while body temperature does play a significant role in the genesis of the heart-rate rhythm there is an intrinsic basis as well, as will be described in the next section.

In Transplanted Hearts

Within the heart is an area of specialized muscle tissue called the pacemaker, which periodically sends out electrochemical stimuli that cause the heart cells to beat in unison. It is this command unit that produces the standard 71 beats/minute. Here also is where the nerves from the brain and spinal cord join the heart and tell it to speed up its beat during activity and to slow down when the body is at rest. The electrical signals emanating from the pacemaker also spread to the body surface where they may be recorded as an electrocardiogram.

In the preparation of a patient for a heart transplant, all but a small flap of tissue on the back side of the diseased heart is removed and discarded. The remaining piece of heart is used as a mooring to which the new heart is sutured; it also contains the patient's pacemaker which continues to transmit contraction stimuli, even though the heart cells it controlled are no longer present. The new recycled heart has its own intact pacemaker, and it is this unit (rather than the recipient's pacemaker) that commands its beat in the new residence. A new nerve supply does not grow to the implanted heart for weeks (if ever).

In the one observation done so far, 32 days after receiving a heart transplant, the patient lay quietly in bed while electrocardiograms were periodically made. The authors of this work make no mention of the ambient lighting conditions in the hospital room, which lead to the assumption that some form of daily schedule prevailed. Interestingly, rhythms in electrical activity were found for both the transplant and recipient pacemaker tissue and both had periods of 23.4 hours. This shortened period is very unusual, because under a daily lighting regimen, rhythms are always entrained to a strict 24-hour period—they only deviate when light-dark cycles are eliminated. If there were enough nighttime interruptions in the hospital so the patient's sickly body interpreted the lighting as constant, then a circadian frequency would be expected, but as described previously, very

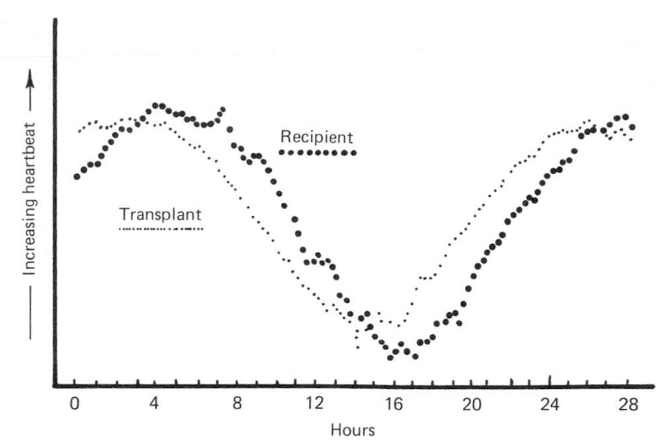

Figure 4-18 Computer printout of the heart-beat rhythms of a transplanted heart and the remaining pacemaker of the recipient of the new heart. Both rhythms have periods of 23.4 hours and the two are 135 minutes out of phase with one another. Redrawn and modified from I. Kraft, S. Alexander, R. Leachman, and H. Lipscomb, *Science* **169,** 694–695 (1970).

few human persistent rhythms are ever shorter than 24 hours. Even more intriguing, the two pacemaker rhythms were 2 hours and 15 minutes out of phase (Figure 4-18), which showed their autonomy from one another and proved that the patient's daily rhythm in temperature change does not cause the rhythm. In addition, since the implanted heart had no connection with the host's nervous system, it eliminates that as the governing factor in the rhythm. It suggests that the clock must be contained within the heart cells themselves.

This fact has been demonstrated in hamsters. It is possible to remove their hearts and keep them alive and beating for several days outside the animal. Even when thus isolated, the daily heart-rate rhythm persisted. If the mutilation is carried one step farther, the intercellular cement that glues the heart cells together can be dissolved and the living cells separated. Individually, in nutrient medium, they will survive for very long periods of time, where they not only continue to beat, but still undergo a daily rhythm in the rate they pulsate! They truly do have their own personal clocks.

Rhythms in Space

The persistence of the human heart-rate rhythm can be observed during orbital space flight when astronauts are subjected to very short external "days." (When orbiting at a radiation-safe altitude, 125 to 490 miles below the Van Allen Belt, each journey around the earth lasts

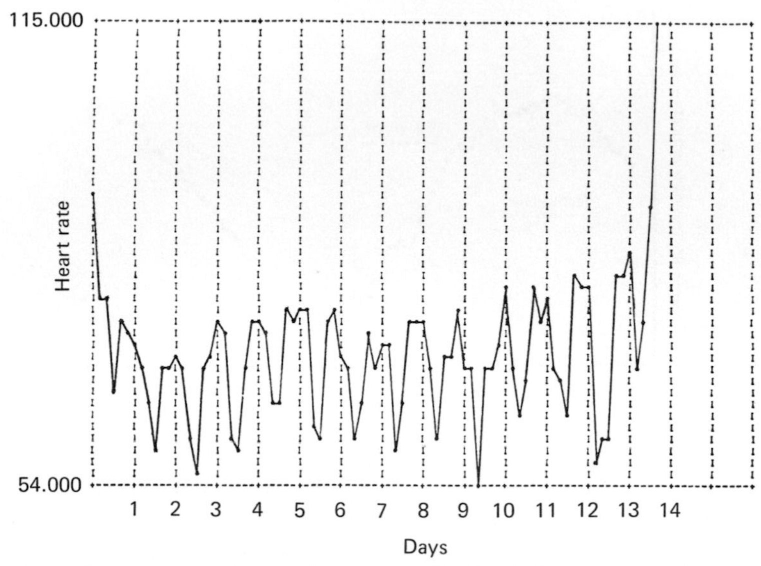

Figure 4-19 The heart-rate rhythm of command pilot Frank Borman during the Gemini VII flight. The curve starts at 2:30 P.M. at "lift off" and ends 14 days later at "splash down." Each point is a 4-hour average. During the flight the astronauts lived on 23.5 hour "days"; computer analysis of the heart-rate data produced a period of 23.7 hours. From J. Rummel, E. Sallin, and H. Lipscomb, *Rass. Neurol. Veg.* **21,** 41–56 (1967).

from 80 to 130 minutes, depending on the shape and altitude of the orbit. Not more than 30% of each orbit is spent in the earth's shadow.) During the 14-day Gemini VII flight, the heart rate of the command pilot, Frank Borman, was continuously monitored by an on-board biomedical tape recorder via electrocardiographic telemetry. Figure 4-19 shows that his heart-rate rhythm persisted throughout the flight, with the unsurprising exceptions seen at "lift off" and "splash down."

The Russian cosmonauts on Vostok flights 3 and 4 also displayed distinct 24-hour rhythms.

Development of the Rhythm in Infants

The tiny fetal heart begins to beat about four weeks after conception when it is no larger than the head of a pin. By the seventh month of gestation, the obstetrician can easily hear the fetal heart sounds with his stethoscope; a study was undertaken to see if by this stage of development, the heart rhythm had begun. It was found that while the mothers-to-be in the study all had well-defined heart-rate and temperature rhythms, their unborn children did not. During the seventh month of intrauterine life, the beating rate was a nonoscillating 133 beats/minute; throughout the ninth month, it averaged 129. In fact, it is not until 4 to 5 weeks after birth that the rhythm first makes its appearance, when the nighttime rate begins to decrease.

RHYTHMS IN RENAL PROCESSES

Rhythmic Desynchronization

The substances dissolved in the body fluids and carried in the blood must be maintained within a rather narrow range of concentrations for good health; this job is carefully done by the kidney. For example, after a very salty meal is consumed, the kidney filters out and excretes sufficient quantities of sodium and chloride so as to maintain precisely the required amounts circulating in the blood. If, on the other hand, only meager amounts of these substances are ingested, they are conserved by the kidneys and not passed from the bloodstream into the urine. The regulating process is such a very exacting and demanding job that, on an ounce for ounce basis, the kidney consumes more oxygen in its labors than the heart.

In 1843, it was observed that, with the possible exception of bed wetters, the amount of urine passed was reduced at night. No profundity was attached to the "discovery" of this rhythm, since sleep and the abstinence from food and fluid intake at night was sufficient explanation. However, by the end of the first quarter of the twentieth century, the rhythm was shown to have an intrinsic basis as well. This fact became apparent after finding that the rhythm persisted in fasting subjects or those fed identical meals at regular intervals; in those remaining recumbent or continuously active throughout a 24-hour period; and in those deprived of sleep or made to live in constant conditions.

In addition to the rhythm in the total volume of urine excreted each day, various components also oscillate in concentration on a 24-hour basis. Substances such as sodium, potassium, chloride, and phosphate (collectively called electrolytes, i.e., substances which dissolve in water causing it to be a conductor of electricity). All these substances, plus certain hormones, are excreted in lesser amounts at night than during the daytime (Figures 4-4 and 4-20). These rhythms will persist under the same controlled conditions as were used in the study of the water volume rhythms. The phosphate-excretion rhythm was even found to persist for 14 days in a fasting subject (the spontaneous

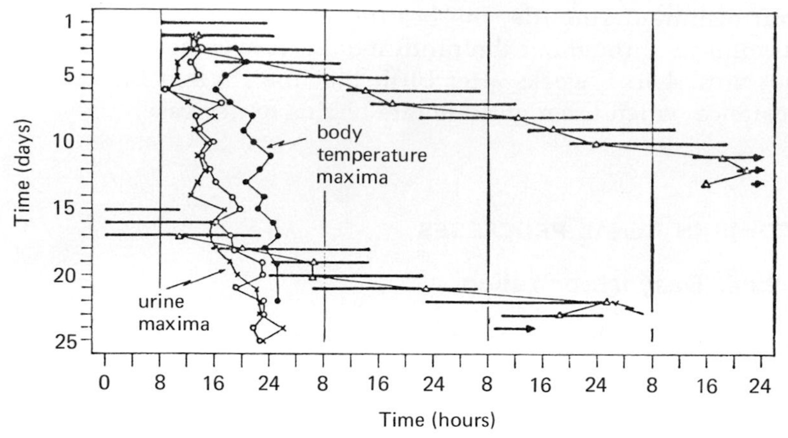

Figure 4-20 Desynchronization of various rhythms in a man living in constant conditions. Black bars stand for times of wakefulness; open triangles, maxima of calcium excretion rhythm; open circles and crosses, maxima of water and potassium (respectively) excretion rhythms; solid circles, maxima of the temperature rhythm. For graphic clarity of the sleep-wakefulness rhythm, the base of the graph has been stretched to a length totaling 4 days; if all data were plotted against a single 24-hour day, it would be seen that the sleep-wakefulness and calcium rhythms coincide with the other rhythms every 3 to 4 days. From J. Aschoff, *Science* **148,** 1427–1432 (1965).

experiment was terminated when the lost welfare checks were finally delivered to the graduate-student subject).

These rhythms are not present immediately after birth. The urine flow rhythm first appears in the fourth week postpartem, while the sodium and potassium excretion does not become rhythmic until about the fifteenth week.

In one very sophisticated study of a single individual, by using an enormous amount of data and intricate statistical analysis, a low amplitude *weekly* rhythm in excretion was identified. It was subsequently discovered that the "rhythm" was simply the result of this subject's regular custom of drinking a bottle of beer each Sunday after church.

Studies examining the effect of constant conditions on human excretory rhythms have been carried out in the experimental bunkers at the Max-Planck Institute in Germany. As expected, all the rhythms persisted in constant conditions and all assumed circadian periods. Several of the subjects tested have displayed very interesting period alterations and phase desynchronization between various other body and excretory rhythms. In one such case (Figure 4-20), the sleep-wakefulness and calcium-excretion rhythms assumed a maxiperiod of about 32.6 hours, while the water volume, potassium content, and temperature rhythms adopted more expected periods of about 24.7 hours. If one considers the sleep-wakefulness rhythm, this subject "lost" over 5 real days during the stay in the bunker, while two of the kidney rhythms and the daily temperature cycle lost only ½ of a real day over the same time period. The way in which Figure 4-20 has been drawn does not make obvious the fact that every 3 to 4 days the phase of the calcium and sleep rhythms come into the accustomed synchrony with the volume, potassium excretion, and temperature rhythms. The peaks of the calcium and sleep rhythms continuously race ahead and repeatedly "overtake" the peaks of the latter rhythms because of the difference in their period lengths. In perusing the intimacies of this subject's diary after the experiment, it was found that he occasionally felt particularly well and fit, and these occasions of mild euphoria corresponded exactly to the times when all the rhythms temporarily came back into normal phase relationship.

Entrainment to Artificial Days

An interesting set of experiments was carried out to discover whether renal rhythms could be entrained to unnatural day lengths. The experiments were conducted on the Spitsbergen Islands (north of Norway and within the Arctic Circle) during the summer months when there is little difference between day and night in either light or temperature. A total of 19 subjects lived for 6 weeks in three isolated camps in a desolate portion of this region. Before and after the Spitsbergen sojourn, their excretion and temperature rhythms were studied under normal day-night cycles in England. On arrival in Spitsbergen, camping gear was issued which included sham watches rigged to signal 12 hours in either 10½ or 13 real hours, which produced for the unsuspecting clock-watchers, 21- or 27-hour "days," respectively. The subjects, who were unaware of the intentional deceitfulness of their watches, were separated by watch type into three camps (one group had orthodox watches), and were required to carry out their daily activities within the framework of the time indicated by their watches. The skies were overcast during most of the summer, so that no regular clues about the real time of day could be obtained by studying the position of the sun. At 2- to 5-hour intervals, urine samples were collected and analyzed for sodium, potassium, and chloride content; the volume passed was also measured. Oral temperatures were taken at each voiding. The subjects' intake of food and fluid was kept as constant as possible. Under these abnormal time schedules, every eight experimental 21-hour days were equivalent to seven real days, and eight experimental 27-hour days totaled nine real days.

For those subjects with conventional watches, the daily fluctuations in the temperature and excretory rate of urine components retained their 24-hour periods, in spite of the perpetual daylight of the Arctic summer. The renal rhythms of some of the campers living under abnormal days immediately adjusted to the 21- or 27-hour days, while, in others, it took a considerable number of exposures to the artificial days for entrainment to take place. The potassium rhythm was especially recalcitrant to change and, in most cases, maintained its old 24-hour periodicity. The sleep-wakefulness and temperature rhythms of all but one subject adjusted quickly to the 21- and 27-hour days. Clearly then, in the absence of other time cues, the periods of these rhythms can be molded into artificial lengths merely by looking at the hand of one's watch. On return to the British Isles, all the rhythms, and as expected the behavior of the subjects, returned to the traditional English way of life.

In Kidney Transplants

The kidney receives regulatory stimuli from the central nervous system via nerves emanating from the spinal cord. It was therefore reasonable to suspect that the clock controlling the renal rhythms may reside in the central nervous system and exert its influence on the kidney through these nerves. This, however, is not the case as was shown in the studies of patients who have received kidney transplants and survived long enough to demonstrate their excretory rhythms.

In each of the 25 transplant studies completed to date, in which excretory rhythms were looked for, they were found to have persisted. In 7 of these, the rhythms persisted unchanged. In the 18 other cases, which included a transfer from an identical twin, the kidney rhythms were reversed. In other words, the rhythms were not the same as those of the host or donor, but were rephased, apparently as a result of the operation. Peak filtration rates and electrolyte disemboguement were highest at night, rather than during the day, and retained this phase for the remainder of the studies which extended for 3 months after transplant in some cases.

A DAILY RHYTHM IN TRANSPLANT SUCCESS

To learn whether a rhythm might affect the success of transplanting kidneys, two different strains of rats were used. (Man, of course, cannot be the subject for this kind of experimentation.) No histocompatibility is known to exist between the two, so that the transplanted kidneys should rather quickly be rejected by the recipient, who would signal the fact by dying. The rats were kept under natural alternating light-dark conditions and an equal number of transplants made at 4 A.M., 8 A.M., noon, 4 P.M., 8 P.M., and midnight. The average survival time of all the recipients was 7–8 days, but those operated on during most of the daylight hours rejected their new kidneys in less than 4 days, while 50% of the 8 P.M. recipients survived longer than 3 weeks. If one was bold enough to extrapolate from these data to man and if one takes into account that rats are noctural while man is diurnal, the most successful kidney transplants would be expected to occur from operations made at 8 A.M.

RHYTHMS IN BIRTH AND DEATH RATES

The clock functions in mysterious ways. For example, natality and mortality for the individual are unique one-time events; yet, the clock appears to be active (the maternal one in the former and the personal one in the latter case) in these processes too, which permits them to occur only at times that it deems acceptable. Elucidating these facts is a matter of scrutinizing hundreds of years of old hospital and governmental records, an endeavor first completed in 1829, with the discovery of a daily population rhythm in the time of birth. This fact of life has been rediscovered many times since then; the study to be reported here, which was accomplished just recently, agrees quite closely with the results of earlier works. This study encompasses the onset of 207,918 spontaneous labors and 2,082,453 natural births, which are presented in Figure 4-21. By onset of labor is meant the initiation of painful contractions and/or the rupture of the fetal membranes (signified by the release of the "waters"); this begins most commonly at night. The peak in birth rate occurs between 1 and 7 A.M. If pregnancy terminates in stillbirth, the rhythm is greatly altered (Figure 4-22).

The month-to-month birth rate differs also, so that a prominent annual rhythm can be defined. Many factors—the use of contraceptives being a major one—modify this rhythm, which makes it very difficult to get an undistorted picture of the true biological pattern. For example, in a study of 39,000,000 births, in which no form of birth control was practiced, the annual peak rate (in the Northern Hemisphere) was January 7. This date, however, was probably produced as a result of religious guidelines for living, rather than being a natural biological phenomenon. This is because a large portion of the above

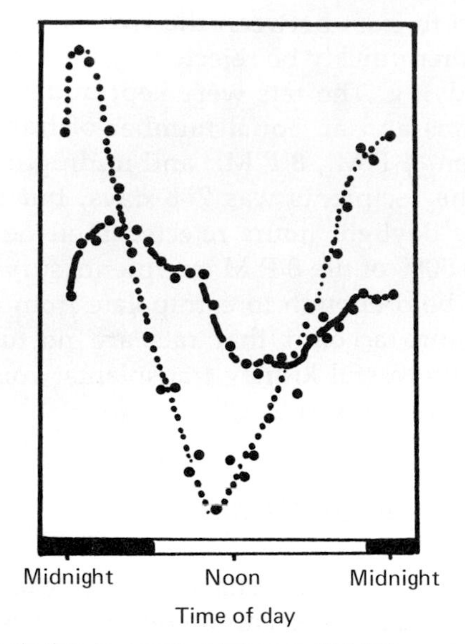

Figure 4-21 The daily rhythm in onset of labor in 207,918 women (dashed curve) and birth times of 2,082,453 babies (solid curve). The shaded region of the bar subtending the curves indicates the probable hours of normal sleep. Redrawn and modified from M. Smolensky, F. Halberg, and F. Sargent, in "Advances in Climatic Physiology" (S. Itch, K. Ogata, and H. Yoshimura, eds.), pp. 281–318. Springer-Verlag, Berlin and New York, 1972.

sample consisted of Roman Catholics, for which marriage is discouraged, and sometimes even sexual abstinence practiced, during Lent. If one counts backwards 266 days (the average human gestation interval) from January 7, he arrives at April 15 as the usual date of conception—just a few short days after Easter.

The living clock also affects the time of day at which people die! This was first learned in 1814, and has been verified in a modern study of 432,892 death records, the results of which are summarized in Figure 4-22. As can be seen, the most common time of death is 6 A.M., with a secondary peak at 4 P.M.

There is also an annual rhythm in human mortality rate. The two most common causes of death, cardiovascular and pulmonary mortality, were selected for study; it was found that the death rates for both were highest in January in the Northern Hemisphere and in July in the Southern Hemisphere (Figure 4-23), Obviously then, natural fatality is greater in wintertime the world round. It is not known whether climate is responsible for this rhythm, and it will be interesting to watch for signs of a flattening of these curves should the luxuries of winter heating and summer air conditioning ever spread over larger portions

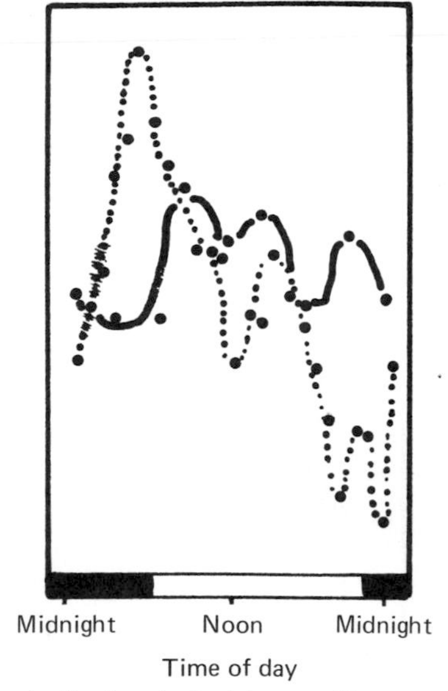

Figure 4-22 Daily rhythm in the hourly incidence of human mortality. The large amplitude curve (dashed) represents the temporal risk of dying as based on a study of 432,000 adults. The solid curve represents 12,081 stillbirths (i.e., full-term pregnancies producing nonviable neonates). (Compare this curve with Figure 4-21.) Drawn from the data of M. Smolensky, F. Halberg, and F. Sargent, in "Advances in Climatic Physiology" (S. Itoh, K. Ogata, and H. Yoshimura, eds.), pp. 281–318. Springer-Verlag, Berlin and New York, 1972.

of the globe. However, the extremes of winter may not be the cause of the annual mortality rhythm. In two other studies, the blood content of cholesterol in prisoners and traffic policemen were studied at monthly intervals for a year or more. An annual rhythm was found in the amounts circulating in both groups, and the phase was identical to that of the annual cardiovascular mortality rhythm (which goes along with the fact that high levels of blood cholesterol and heart disease correlate positively). Because the prisoners were confined to the indoors and were forced to eat a constant daily level of dietary fat, it suggests that this rhythm is independent of exogenous seasonal input. Possibly, the annual rhythm in susceptibility to death is similarly intrinsic.

RHYTHMIC SENSITIVITY TO ALLERGENS AND DRUGS

Two studies will be described here, one on the reactivity of the skin to an antigenic substance, and the other on the rate of aspirin retention in the body. Before making any of the measurements described

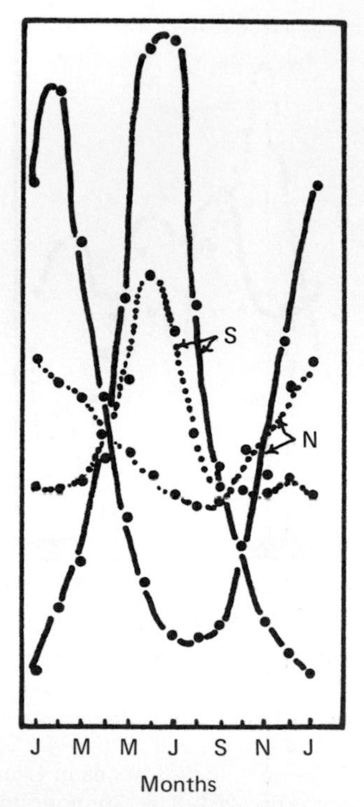

Figure 4-23 Annual rhythms in death rate in the northern (N) and southern (S) hemispheres. The curves for cardiovascular-disease deaths (solid lines) are based on a sample of 420,000 people, while those representing respiratory associated deaths (dashed curves), 140,000 people. Drawn from the data of M. Smolensky, F. Halberg, and F. Sargent, *in* "Advances in Climatic Physiology" (S. Itoh, K. Ogata, and H. Yoshimura, eds.), pp. 281–318. Springer-Verlag, Berlin and New York, 1972.

below, all the subjects in the experimental populations were put on the same routine (sleep from 11 P.M. to 7 A.M.) for at least a week, so that all their rhythms would be in synchrony.

Six adults, who were unfortunately allergic to ordinary house dust, were given interdermal injections of a standard house-dust extract at 4-hour intervals throughout the day and night. The effect of the injection was measured by evaluating the intensities of erythema (inflammatory reddening of the skin around the puncture wound) and wheal (skin welt, like that caused by a mosquito bite) 15–20 minutes after the administration of the allergen. Both response patterns were found to be rhythmic with the most violent reactions occurring after the 11 P.M. injection and the mildest responses coming 12 hours later. (An identical test, using subjects sensitized to penicillin, produced an almost identical rhythm in sensitivity to standarized scratch tests.)

These allergic responses (erythema and wheal) are brought about by the liberation of histamine from the affected tissues. Antihistamines are given to try to combat this undesirable sensitivity. The periodic injections of house dust were administered again, but this time Periactine, an antihistamine drug was given also; either at 7 A.M. or at 7 P.M. Erythema and wheal responses were greatly depressed; the inhibitory effect lasted 6–8 hours after the 7 P.M. administration, but 15–17 hours after the 7 A.M. one. Drug companies could therefore save the public a considerable amount of money by informing them of the existence of this rhythm and recommending that they take only half an antihistamine tablet in the morning after waking for the same relief.

In a similar experimental paradigm, 1-gm doses of sodium salicylate (aspirin) were given to 6 adults at 7 A.M., 11 A.M., 7 P.M. and 11 P.M., and urine samples taken at 4-hour intervals thereafter to follow the rate at which the drug is passed from the body. The duration was found to be dependent on the time of administration. The response was rhythmic, the drug remained in the body about 22 hours when administered at 7 A.M., and only 17 hours when taken at 7 P.M. The difference is significant.

THE EFFECT OF A 10 Hz ELECTRIC FIELD ON HUMAN RHYTHMS

One of the periodic components in the geophysical environment is a 9–10 Hz electromagnetic field (see Diagram 4-1). Especially during bouts of fair weather, a rhythm in intensity is displayed with maximum amplitudes occurring during the afternoon and minimum ones at night.* The effect of this field on man has been tested in the following manner.

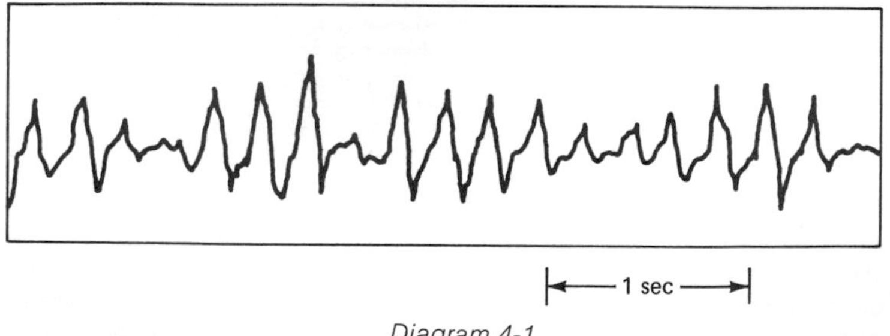

Diagram 4-1

* 10 Hz is a common frequency in nature; for example, the earth's crust vibrates at this frequency as does man's body surface, and the alpha rhythm of the human brain has this average frequency.

One of the two underground experimental bunkers described earlier in the chapter contained modifications not previously mentioned. Built into its walls were shielding materials designed to screen out most of the electrical and magnetic components of the physical environment. Also hidden within the walls were high-tension electrodes which, when energized, generated a 10-Hz electric field throughout the room, 1000 times greater than that of the earth's. The other bunker, which was unmodified, served as a control chamber. Neither the natural, nor the artificial 10-Hz electric fields are consciously perceived by man.

In the first series of experiments, 10 subjects were isolated in the shielded room, with the lights left on continuously, and their sleep-wakefulness and temperature rhythms studied. After the periods of their rhythms had been established, the artificial field was turned on for several days. The results of one subject, in and out of this field, are portrayed in Figure 4-24: the periods of both his rhythms over the first 10 days in the absence of the natural field were about 28.5 hours; with the artificial field turned on, the periods were shortened to about 25.8 hours. The average shortening for all 10 subjects was 1.7 hours in the field.

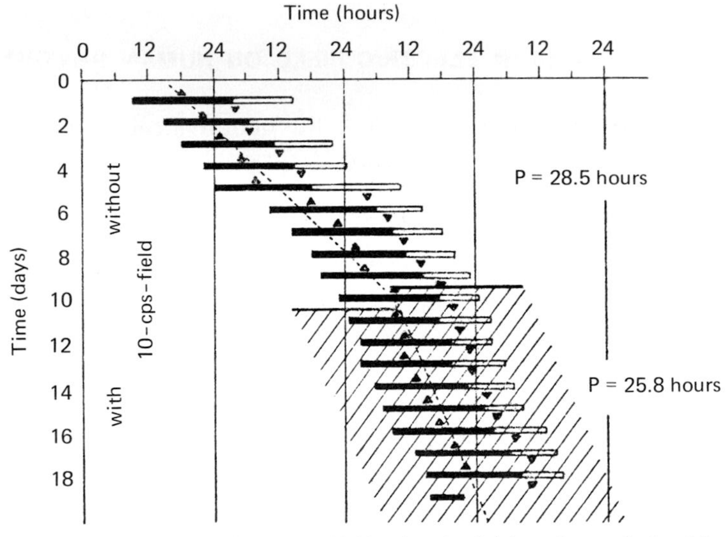

Figure 4-24 The effect of an imposed 10-Hz electric field on the periods of the sleep-wakefulness and temperature rhythms. The open bars indicate the times of wakefulness, the upright pyramids indicate daily temperature maxima, while the inverted ones signify temperature minima. Between days 1 and 10, both rhythms had an average period (P) of about 28.5 hours. During the remainder of the experiment, the 10-Hz field was imposed (shaded area) and the periods of both rhythms shortened to about 25.8 hours. From the data of R. Wever, *Pfluegers Arch.* **302,** 97–122 (1968).

Thus, an artificial 10-Hz field was shown to significantly influence biological rhythms. To discover if the natural field acted similarly, one had only to compare the results obtained in the unshielded bunker, in which the natural field was almost completely unaltered, with the results obtained in the shielded bunker. Other than the shielding, the two rooms were identical in every way, including lighting and temperature conditions. The activity rhythms of the 24 subjects in the shielded room had an average period of 25.00 ± 0.55 hours, while the 21 subjects in the unshielded room had an average period of 25.66 ± 1.02 hours. The difference was statistically significant. It should be pointed out, however, that the same people were not tested in both rooms; the difference between the two average periods is therefore that of different sample populations. Still, the data are consistent with the hypothesis that a 10-Hz field—both the natural and an artificial one—does significantly shorten the period of some circadian rhythms.

In the second set of experiments, the artificial 10-Hz field was turned on only between noon and midnight each day, while the subjects were isolated in constant light. All 10 subjects used responded in the same way: the circadian periodicity became temporarily entrained to the 24-hour period of field on and off, but later broke away (Figure

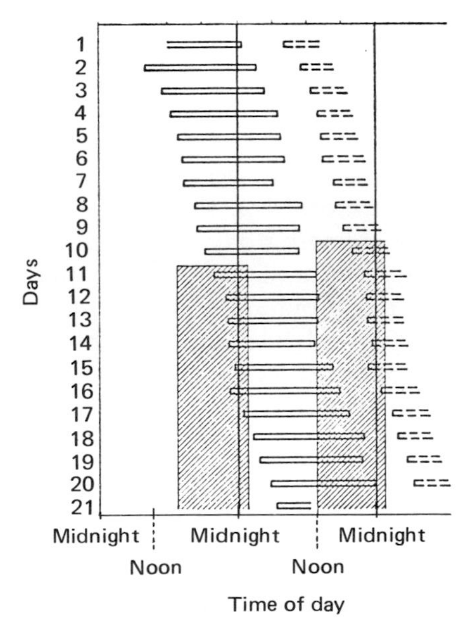

Figure 4-25 The short-term entrainment of the sleep-wakefulness rhythm by an on-off cycle of a 10-Hz electric field. The times of activity, in otherwise constant conditions, are indicated by the open bars. Shaded areas represent the times of "field on." Note that through days 12 to 16, the rhythm appears to be entrained by the 10-Hz field. Modified from R. Wever, *Z. Vergl. Physiol.* **56,** 111–128 (1967).

4-25). While the data suggest by analogy that the daily variation in the natural 10-Hz field could act as a weak entraining agent of biological rhythms, it should be stressed that the experimental conditions do not mimic the natural field, as the latter at no time completely disappears and is manyfold weaker.

Another interesting observation from this study is the fact that subjects whose various rhythms became desynchronized (Figures 4-20; 7-19), i.e., the period of various separate body rhythms differed and thus came out of phase with one another, had been maintained only in the shielded rooms. Furthermore, desynchronization did not take place in the shielded bunker when an artificial 10-Hz field was turned on.

LONGITUDINAL TRAVEL AND BIOLOGICAL RHYTHMS

In a rare incident of international cooperation, it was agreed to divide the face of the earth into 24 *time zones,* each equal in width to the distance the earth rotates in one hour (Figure 4-26). The rhythms of the people living in each time zone are synchronized to local time mainly by the ambient day-night cycle there. Therefore, when it is noon in New York City and the inhabitants are lunching, in Bangkok on the opposite side of the earth, it is midnight, and at least the early birds are all sound asleep. Travel between time zones requires resetting one's wristwatch and body clock.

Voyages before the air age caused no problems for man's rhythms because time zones were crossed so slowly that the rhythms easily adjusted to the gradually changing day-night schedules (and it is light-dark cycles that set the phase of human rhythms, just as all others). The jet age, however, has provided a new challenge for the biological clock, for this form of travel shaves hours off the length of a day as one travels eastward and piles on additional ones with westward travel. This puts considerable stress on the clock and the body processes it regulates. An example is given below.

Suppose an Italian businessman must attempt an important business deal with the managerial staff of a New York City firm. In making his travel plans, he finds it most convenient to take the 9 A.M. flight from Rome, which carries him across the 4200 miles and six time zones (Figure 4-26) to New York in 8 hours: he arrives at 5 P.M. his time, which is only 11 A.M. New York time. After completing the standard search for his luggage, he arrives at his meeting on time at 2 P.M. (EST). Thanks to the existence of human rhythms, he has a

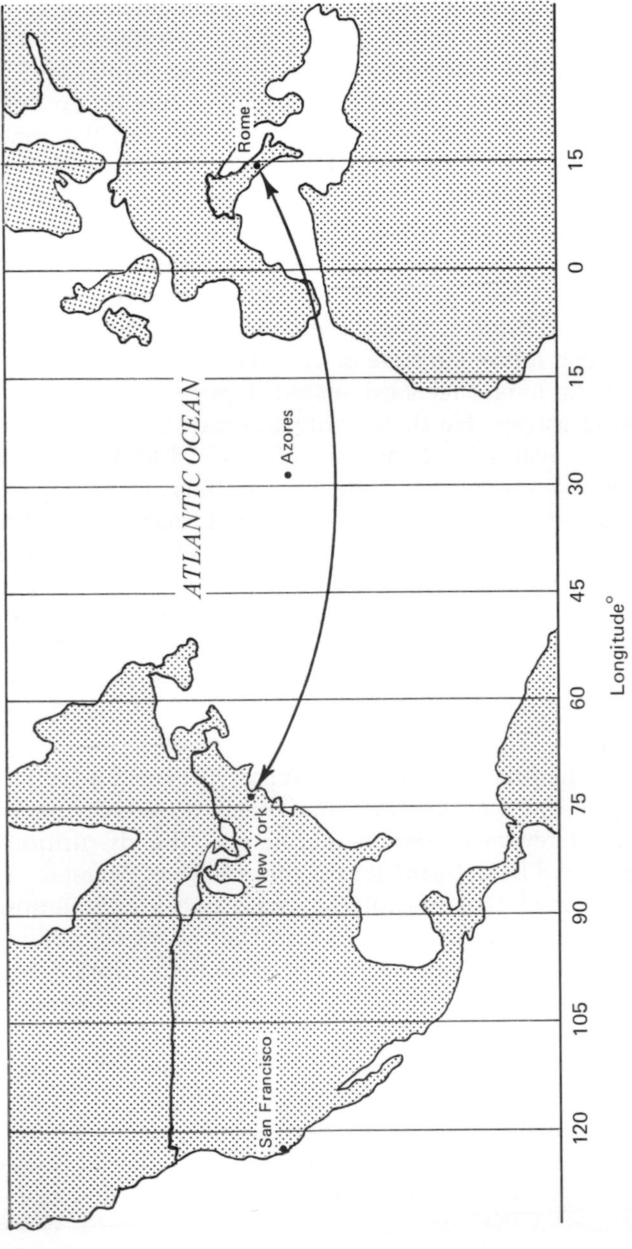

Figure 4-26 The time zone divisions of one face of the earth. The double-headed arrow depicts the route taken by the Roman businessman during his travels described in the text.

double advantage in this meeting: since it is 8 P.M. according to his biological clock, his temperature, performance, and mental acumen rhythms are at their peaks, while his New York business adversaries are experiencing a postprandial dip in efficiency (Figure 4-15). With this combination, the Americans are putty in his hands and he pulls off a great business coup. At 5:30 P.M. (EST), they all go out on the town to wind down. Both drinks and jokes flow, and the conquering businessman notices that the cocktails do not seem to dull the wits of his hosts (recall Figure 4-1, alcohol is cleared most quickly from the body at this time of day), while he, whose biological clock is registering 1 A.M., finds the experience most intoxicating and, on top of this, has grown very sleepy. The cocktail hours are followed by dinner and a visit to Times Square, where the revelry is not terminated until 1 A.M. (EST). Much to the surprise of our hero, when he finally returns to his hotel, he no longer feels exhausted (Figure 4-9) and, in fact, has difficulty falling asleep. He dozes only intermittently and tosses and turns through the rest of the New York night until 10 A.M. (4 P.M. Italian time), when he catches his flight back to Italy. In flight, at 5 P.M., the stewardess initiates the traditional cocktail hour and the businessman concedes to a sufficient number of rounds to uphold the reputation of his countrymen, even though his Roman watch is registering midnight. Exhausted, intoxicated, and with his biological clock in a state of utter confusion, he finally arrives home at 2 A.M. his wife's time, and awakens her to proclaim his triumphant business venture. She drowsily responds, "don't forget to put out the garbage." (See Scheme 4-1.)

In this story, the business deal was successful for the Italian, but if the distances or the air schedules had been different, the outcome could have been reversed, simply because of the disruption of the traveler's rhythms. I have heard it said that the then Secretary of State, John Dulles, claimed that one of the reasons his negotiations with Egypt over the construction of the Aswan Dam went so badly was because he had traveled back and forth across so many time zones in the weeks prior to the meetings that he was not functioning well at the conference.

The problem, therefore, is how to cope with the physiological and psychological difficulties that arise from having to live on a new schedule, while one's body rhythms are slowly adjusting from their home-phase relationship to the one of the new locale. Much of past travel by Americans was undertaken as touristic self-indulgence and the consequences of the surprise given to the biological clock when getting off a jet in Europe was mostly just a nuisance. To other segments of the population, though, such as diplomatic corps, interna-

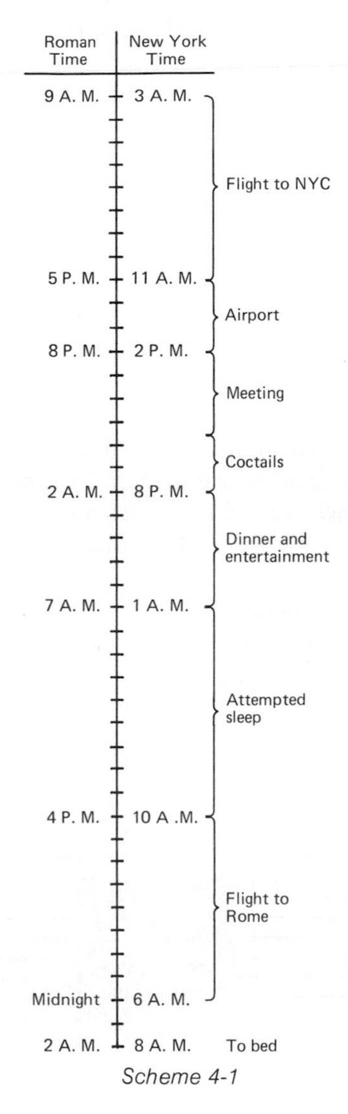

Scheme 4-1

tional businessmen, Olympic and professional athletes, racehorses, and the flight crews that transport them, it is a very serious matter; for the dollars, lives, and the happiness of millions of others may be profoundly influenced by the decisions and the performance of these people. For the latter reasons, a considerable amount of very expensive experimentation has been carried out in an attempt to discover just how quickly human rhythms can adjust to a new time zone.

After reading the first two chapters, it should be clear that all that really need be done in the way of experimental design is to find subjects with well-defined, large amplitude rhythms and, after determining the phase relationship of each under the local day-night cycle, subject them in the laboratory to a shifted lighting regimen. For instance, keeping with the previous Roman example, simply turn the light on and off 6 hours earlier. The subjects' rhythms should then be studied until they have reached an altered steady-state phase relation with the new light-dark regimen. This approach is especially desirable over actual translocation experiments because it is less expensive and eliminates a host of variables associated with the aura of visiting a new country. While it has been used quite successfully (Figure 4-27), the more difficult, more realistic approach of "go there and see what happens" is usually used.

Air caravans of equipment and test subjects are flown in radiating directions from the United States, and "field" labs erected in all sorts of exotic places on the face of the globe. The long-suffering students previously used in underground studies do not seem to be among those invited for these trips; instead, enthusiastic friends and relatives

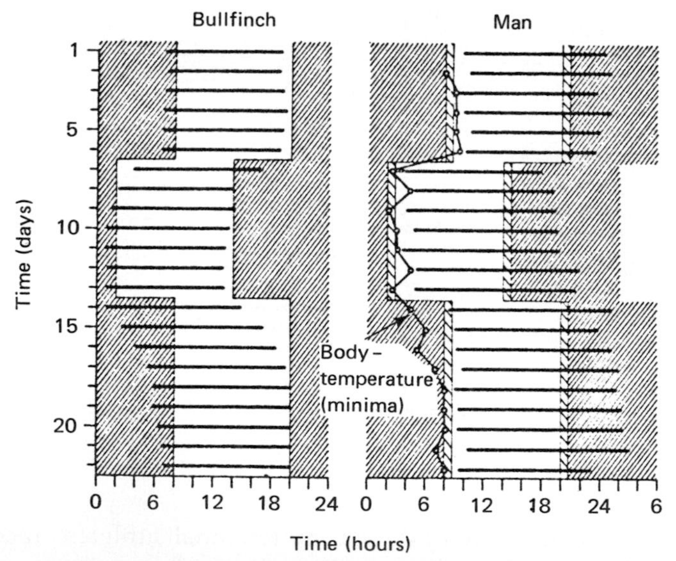

Figure 4-27 The phase shifts produced in a bullfinch and a man by a 6-hour advance and then delay of a light-dark cycle. The horizontal solid bars represent the time of wakefulness; the diagonal lines, the hours of darkness. On the seventh day, the light portion of the cycle was advanced by 6 hours (the same situation that would arise after a flight from New York City to Rome) and the rhythms of bird and man quickly adjusted to it. On day 14, the light was delayed by 6 hours (as would occur on the return flight from Rome). The man's sleep-wakefulness rhythm is seen to adjust more rapidly than his temperature rhythm. From J. Aschoff, *in* "Life Sciences and Space Research" (H. Brown and F. Favorite, eds.), pp. 159–173. North-Holland Publ., Amsterdam, 1967.

are chosen, who are mysteriously identified in subsequent publications by cryptic designations such as Subject EZ 2.5 (female, 20 years of age, fine physical condition, etc.). On reaching their destination the subjects are instructed to "regulate your living habits to those of the natives." Looking at the varability seen in the graphs of different studies, one gathers that this command is carried out with different degrees of zeal. In fact, so many variables are present in studies done like this, that the results are seldom clear-cut. The subjects are shipped not only to a new day-night schedule, they are also exposed to the excitement of a city unfamiliar to them, a new background of people (both the natives and the experimental-subject companions), new foods and drink, and the never-ending physiological and psychological measurements. This experimental design is not ideal. Therefore, rather than describing a large number of experiments and the ramifications and innuendos arising from the results, the author will discuss just one series made by a laboratory group in the Midwest so that the reader can get a feeling for the approach and the difficulties.

East to west, west to east, and north to south flights were undertaken. A few days before embarking on the trips, several physiological processes known to be rhythmic and several psychological performance tests were undertaken to establish a baseline for future changes to the new local time. The following physiological parameters were studied: (1) rectal temperature, determined at 30-minute intervals, (2) evaporative water loss from the left palm (a rhythmic body process), and (3) heart rate. Psychological tests included (1) reaction time—the time necessary to respond to visual or auditory stimuli, (2) decision time—the time needed to decide which of three possible responses to a stimulus was correct, and (3) subjective fatigue measured against a scale ranging from "extremely fatigued" to "extremely alert."

In westerly translocation experiments, six men were flown from Oklahoma City to Tokyo and four men from the same starting point to Manila. The flight to Tokyo (through 8 time zones) took 18 hours, and the one to Manila (through 9 zones) 23.5 hours. On arrival at the overseas destinations, the subjects were made to maintain their daily living habits in accordance with new local time and were subjected to periodic test measurements for 8–10 days. It was found that the temperature and heart-rate rhythms took 2–4 days to come into phase with local time, while the rhythm in evaporative water loss needed 8 days to come into synchrony. On the first day, reaction and decision times were significantly impeded and fatigue increased; but then all returned to the preflight Oklahoma City level. The older members of these groups (ages 40 to 48) experienced a higher level of fatigue on the first day and slower reaction times than the younger men.

After the return flight to Oklahoma City, it was surprisingly found that all phase shifts had been affected on the first day. Fatigue, reaction time, and decision-making times were increased on the first day of return, but not to the extent that they were on the first day at the westerly destination.

In an easterly displacement, four healthy males were flown by jet (15.5 hours) from Oklahoma City to Rome. Tests identical to those described above were undertaken before, during, and after the 12-day stay in Rome. The results were not particularly clear-cut; the temperature rhythm was in phase with local time in Rome on the first, third, and days thereafter, but out of phase on the second day; the evaporation rate became arrhythmic; and the graphs indicate that the heart rate was probably shifted to local time by the second day, but the authors interpret the shift as not being complete until day 8. Reaction times, decision-making times, and fatigue all increased on the first day in Rome, but only the latter to a statistically significant amount.

After the 18.5-hour return trip from Rome to Oklahoma City, the temperature rhythms of two of the subjects adjusted to local time after 2 days; the rhythms of the other two subjects were still not in phase with local time at the end of 5 days. The heart rhythm for all four subjects had not adjusted to the local time by the fifth day of the study. Again, the degree of fatigue (but not the decision-making time) was significantly increased on the first day after return to Oklahoma City.

A fourth flight, between Washington, D.C., and Santiago, Chile (an 18-hour flight), measured the effect of north-south translocation. There is a time difference of 1 hour between these two localities, but the experimental subjects were required to maintain their Washington, D.C., routine during the 12-day stay in Chile. As would be expected, there were no phase changes in body rhythms, though an increment of subjective fatigue was experienced after outgoing and return flights.

Without giving more detailed examples, the author will instead simply draw two generalities from a great deal of experimental work. First, the average person's clock can change its phase by only 2–3 hours per day, but the exact amount is a characteristic of each individual. For example, it takes my secretary 3 days at the end of daylight savings time to adjust to the loss of 1 hour and get to work on time; while my sleep-wakefulness cycle adjusts by the second night after a flight from New York to San Francisco across three time zones (Figure 4-26).

Second, with this in mind it is apparent that the time needed to adjust depends on how many time zones are crossed: the longer the journey the longer the rephase time. In general, 2–3 days are needed to adjust after a flight from the East Coast of the United States to

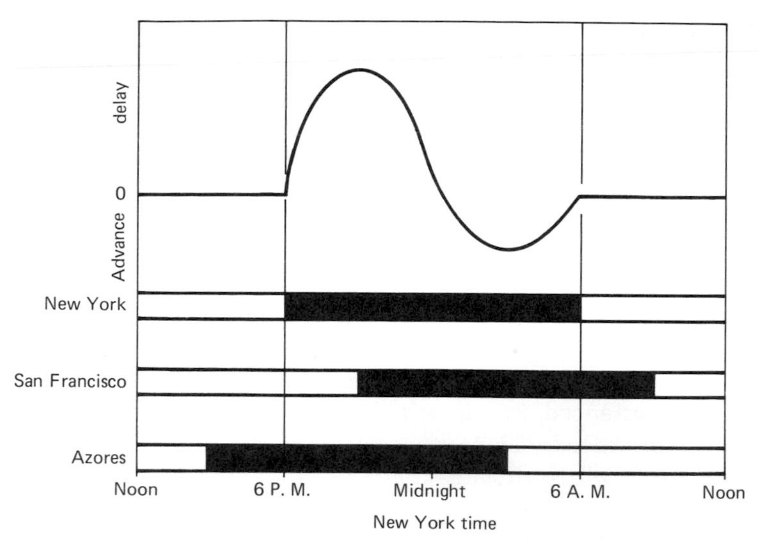

Figure 4-28 The illumination of the light phase-response rhythm of a New York City resident after geographic translocation westward to San Francisco, or eastward to the Azores. Because sunsets fall 3 hours later in San Francisco, the delay portion of the subject's response curve is illuminated. The opposite is true after a flight to the Azores. It can be seen from the shape of the phase-response curve that the 3-hour exposure to the delay portion produces the largest phase change. Therefore, a person's rhythms should adjust more quickly after westward travel.

Europe, and about 6 days after a flight half way around the world. Until entrainment is complete, most people, especially the older ones, feel tired and are significantly less efficient. The fact that all rhythms do not rephase at the same rate creates an internal desynchrony that is very upsetting to some people. (There is also a much smaller group of people—and one would hope that a large proportion of airline pilots belong to this one—not particularly upset by transmeridianal travels: they have, as a characteristic in common, the ability to be able to fall asleep at almost any time of day.)

In addition to these two points, it is likely that the time needed to rephase one's rhythms should depend on the direction of travel: adjustment after eastward translocation should be more rapid than westward. In the development of this point, recall the figure of the light phase-response curve, which has been reproduced and included as part of Figure 4-28. Note that it is asymmetrical* in form, which means that light falling on the segments of the peak above the zero value

* While a phase-response curve for light has never been developed for man, it must be asymmetrical with a large amplitude delay peak (see Figure 1-9, right-hand side), since human rhythms are almost always longer than 24 hours in constant conditions.

cause larger phase changes than those illuminating an identical segment length of the valley. Therefore, in westward flights, say from New York to San Francisco (Figure 4-26), you travel in the same direction as the sun is moving, which thereby increases the length of daylight (Figure 4-28). This extension of daylight into what used to be nighttime for you and your clock illuminates the delay portion of the response curve is illuminated. This produces a phase advance; but, in hand, if you had flown eastward to the Azores (Figure 4-26), you would find that the sunrise now occurs 3 hours earlier than your clock was used to (Figure 4-28). Here, the advance portion of the phase-response curve is illuminated. This produces a phase advance; but, in spite of the fact that the hours of illumination were identical in length to those experienced in the trip to San Francisco, a smaller phase change is produced. This is because of the shape of the phase-response curve which dictates that less of a change can take place. Because of the variable results thus far obtained in actual experiments, it has not been possible to verify or reject this speculation.

Four Remedies to Avoid or Minimize Jet Fatigue

For full alertness immediately after long distance transmeridian travel, the following suggestions should help:

1. If possible, schedule the trips as the Italian businessman did so that he was in peak shape for the negotiations (but then suffered the ensuing consequences).

2. Arrive in the new country a sufficient number of days in advance to allow your rhythms to adjust to the new local time.

3. If suggestion two is not possible (employers often call that kind of attempt at clock adjustment vacation time), try to follow Wiley Post's advice and preset your clock in advance. Set your wrist watch to the time of the country you intend to visit and then try to live that schedule while still at home. Then, spend any part of your new work-sleep schedule that corresponds to your firm's regular hours of business at the office and complete the rest of your work at home under your self-imposed artificial light-dark schedule. (Alternatively, a solution which is often equally impractical, attempt to maintain your home schedule after arriving in a new country.)

4. Not having done any of these—which is the usual case—pace your activity during the first days in a new time zone so as not to place additional stress on your body. In particular, avoid heavy eating and drinking because your "stomach clock" is not yet prepared for self-indulgences at the new times. And try to schedule important

meetings as close to the peak of your underlying efficiency rhythm as possible. For example, after a trip from New York to Paris, hold important conferences as late in the Parisian day as possible. Or, if you are already adjusted to Paris time and return to the United States, call luncheon meetings on the East Coast, or morning meetings if you have traveled to the West Coast. In other words, now that you know about your efficiency rhythm, try to take advantage of its existence.

SHIFT WORK

Industry, military, and paramilitary agencies, and the medical profession often work on a 24-hour basis. In an attempt at fair employment tactics, they routinely alternate their personnel between the daytime and nighttime workshifts. At the onset of one of these changes, the employees are abruptly subjected to at least an 8-hour phase change from their old working schedule. As far as their body rhythms are concerned, this treatment can be tantamount to transporting them to the day shift in New Zealand; but, after a sufficient number of days on the new schedule, their rhythms would be expected to shift appropriately. They usually do not.

This lack of adjustment apparently stems from two facts: (i) most employees take at least 2 days off work each week during which they return to their old sleeping patterns so as to rejoin the rest of the world's social activities, and (ii), as is especially true of unmarried nurses working the night shift, there are always opportunities for disrupting this altered pattern to join in the social life of the day-shift crowd at the expense of a little lost sleep. These violations of the routine are sufficient to prevent the clock from completing a complete phase shift. Since these people's rhythms do not adjust, the performance of most night-shift crews is usually worse than on the day shift. In jobs that require little talent, effort, or concentration, the reduced efficiency is tolerable. In other jobs where precision construction, a patient's life, or the national security are involved, it is definitely worthwhile taking an employee's rhythms into account.

MONTHLY RHYTHMS

The Menstrual Cycle and Related Parameters

The interval between successive new moons is called the synodic month and is 29.5 days in length. Many organisms, particularly marine invertebrates, commonly display monthly cycles (often in reproduction) that are synchronized to the phases of the moon.

Down through the ages, a certain segment of the human species has prophesied and proselytized that the moon has a direct influence on our daily lives. Thus, everyone's working vocabulary includes words like lunacy, moonstruck (and moonshine?). Additionally, because of the similarity between the cycle lengths of the egg-production rhythm in the human female and the period of the circling of our planet by the moon, the former is referred to as the menstrual (menstruus means month) cycle. The similarity between these two cycle lengths is even closer than believed, as is apparent from the work of H. Presser, who has reviewed all the past literature on determining the actual average length of the menstrual cycle. When all the data are presented side by side, it becomes quite obvious that the average menstrual cycle is much closer to 29.5 days than the 28-day figure given in most textbooks (and the length to which over 11 million American women regulate their cycle with the "pill"). (Pill-taking women especially, between the ages of 20 and 30, should note Figure 7-36.)

Combining our awareness of monthly reproductive rhythm in lower animals with our knowledge of the reason behind the menstrual cycle has lead some to speculate that the human female probably undergoes a monthly cycle in sexual desire, with the peak in urge coming around the middle of the menstrual cycle when the egg is in a fertilizable state. For years, the idea persisted as speculation; but, just recently, when participation in sex studies became as popular as being a contestant in a daytime quiz show, a great many studies on changing libido have now been carried out. Just two will be discussed here.

In a personal-interview study, 30 women who were normal physiologically, but had no sex life, were asked almost daily—and tactfully—by their psychotherapist about their desire for sex at that moment. The average of their responses is portrayed in Figure 4-29, where it is seen that libido begins to build through menses, remains high up to the approximate time of ovulation (around day 15), and then falls off, before ascending to a second major peak just before the onset of the next menses. (Other studies have shown that women's senses of touch, taste, smell, and hearing are also more acute at this time. They are also more willing to volunteer for all sorts of tasks during the ovulatory stage.)

In another study, carried out in a different manner so that the subjects could retain complete anonymity, 40 married women were employed. In return for a wage of $0.50 a day, they filled out cards indicating whether they had had intercourse in the previous 24 hours, and if so, whether or not they had achieved at least one orgasm. So that

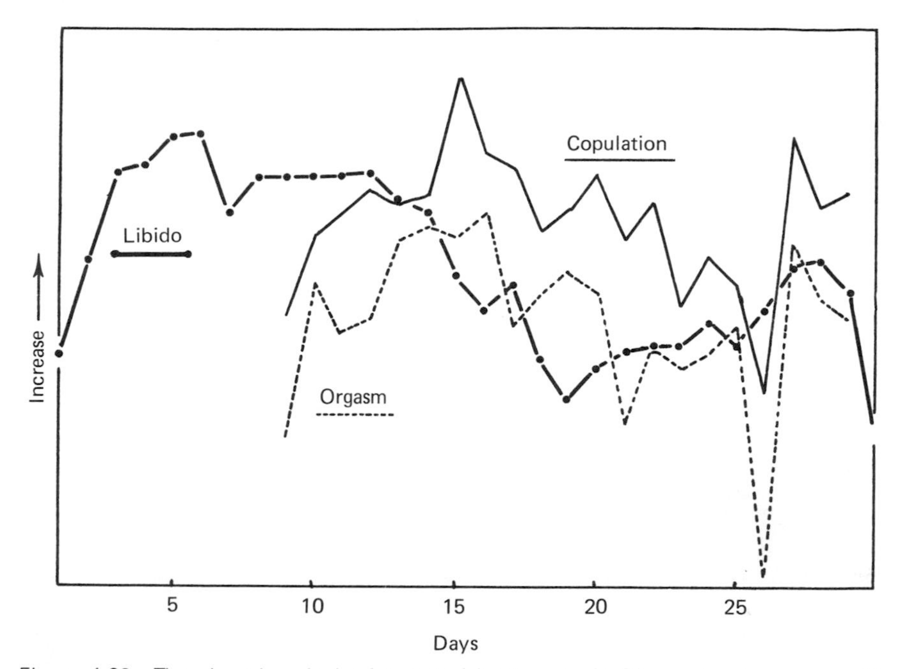

Figure 4-29 The changing desire for sexual intercourse in 30 women studied over 75 menstrual cycles (dashed line); and frequency of intercourse (bold solid line) and orgasm (tenuous solid line) in 40 different women over 73–115 menstrual cycles. Ovulations occurred around day 15. Drawn from the data of J. R. Cavanagh, *Med. Aspects Hum. Sex* **3,** 29–39 (1969), and pictured in J. D. Palmer, *Nat. Hist.* **74**(4), 53–59 (1970); J. R. Udry and N. M. Morris, 1968. *Nature (London)* **220,** 593–596 (1968).

there would be no confusion, the latter was defined as a "high peak of sexual excitement followed by sudden relaxation." No mention is made that the ladies suffered any confusion on this point. The results of this study have been added to Figure 4-29, where it is seen that the copulation and orgasm rate was highest about midcycle.* This, combined with the fact that none of the subjects was taking birth control pills, was sufficient to bring the study to a close at the end of 3 months.

Rhythmic Sensitivity to Pain

It will be remembered that at the beginning of the chapter a rhythm in the sensitivity of teeth was described. A similar rhythm has been

* Not all studies report data identical to the two sets reported here. Most, however, report a peak in increased libido just before the onset of, and during, the last days of menses.

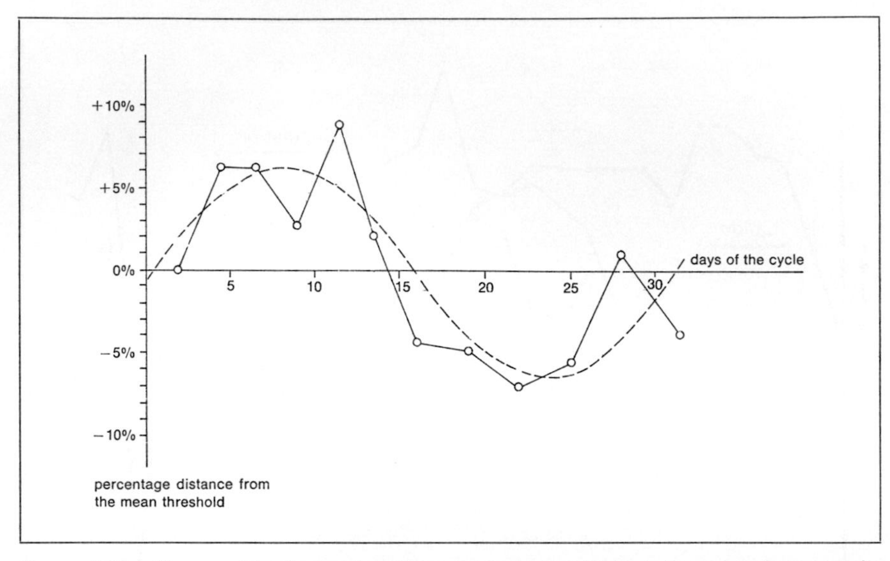

Figure 4-30 The monthly rhythm in cutaneous pain sensitivity in 8 women between the ages of 15 and 20 years. The dashed curve represents the percent deviation from a monthly average in threshold (solid line) to a set of pain-producing stimuli. The women were the most sensitive during the latter half of their menstrual cycles. From P. Procacci *et al.*, *Chronobiologia* **1,** 77–96 (1974).

discovered in the sensitivity of the skin to a hot spot of light produced by focusing light from a projection lamp through a lens. In addition to this daily rhythm, one with a period matching the menstrual cycle also exists. Eight young women submitted to the painful tests a sufficient number of times to produce the curve shown in Figure 4-30, which shows that the women were less sensitive (that is, had a higher threshold to pain) during the preovulatory stages of their menstrual cycle, than in the latter half.

Rhythmic Sensitivity to Color

In a study of the sensitivity of man's eye to different colors of the visible spectrum of light, serendipity added both a monthly and annual rhythm to the ever-increasing list of human cycles. As a spin-off from a rather dull, scholarly list of sensitivities to different colors was the surprising discovery that the relative sensitivity to orange and green was not constant—it varied with the day of the month and the season of the year. As can be seen from Figure 4-31, the relative visibility or luminosity of these colors is greatest during the days around

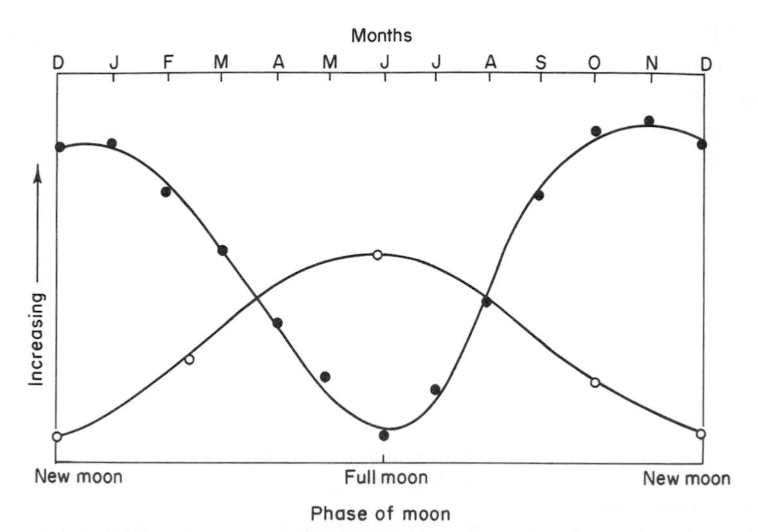

Figure 4-31 Monthly and annual rhythms in the changing relative luminosity of orange and red as interpreted by the human eye. The bell-shaped curve represents the average monthly change in two men's vision for a total of 24 months. Note that the eye was most sensitive on the days around full moon. The U-shaped curve signifies the change in their average monthly responses during the year and is greatest between September and February. Drawn from the data of A. Dresler, *Naturwissenschaften* **29,** 225–236 (1941).

full moon, and the average monthly responses are stronger during the fall and winter of each year. The changes are not large, but may be important for artists, interior decorators, and other discriminating people who react strongly to color and who are notoriously desultory in what pleases them from moment to moment.

SUMMARY AND CONCLUSIONS

1. The difficulties involved in using man as an experimental subject are discussed as are the paradigms used successfully.

2. Rhythms in the most basic human physiological functions are described and include sleep-wakefulness, body temperature, cell division, heart beat, and renal rhythms.

3. Rhythms in alcohol clearance from the blood, pain, color sensitivity, organ transplant, sexual desire, sensitivity to allergens, birth and death are also discussed.

4. The time interval required for these rhythms to develop after birth, and their persistence in caves, underground bunkers, and space are described.

5. Psychomotor rhythms in mental acuity, mood, performance rates, time perception, and dexterity are introduced.

6. "Jet lag" is diagnosed and prophylaxis prescribed.

7. A few monthly rhythms are described.

8. The conclusion drawn from these findings is that during the day, people are not the same from instant to instant; but, at the same time each day, people are much like they were the day before, and will be tomorrow.

Selected Readings

Aschoff, J. (1965). Circadian rhythms in man. *Science* **148,** 1427–1432.

Colquhoun, W. P., ed. (1971). "Biological Rhythms and Human Performance." Academic Press, New York.

Conroy, R. T., and Mills, J. N. (1970). "Human Circadian Rhythms." Churchill, London.

Ferin, M., Halberg, F., Richart, R., and Vande Wiele, R., eds. (1974). "Biorhythms and Human Reproduction." Wiley, New York.

Kleitman, N. (1963). "Sleep and Wakefulness." Univ. of Chicago Press, Chicago, Illinois.

Luce, G. G. (1972). "Body Time." Bantam Press, New York.

Mills, J. H. (1966). Human circadian rhythms. *Physiol. Rev.* **46,** 125–171.

Palmer, J. D. (1970). The living clocks of man. *Nat. Hist.* **79**(4), 53–59.

Reinberg, A. (1973). Chronopharmacology. *In* "Biological Aspects of Circadian Rhythms" (J. N. Mills, ed.), pp. 121–152. Plenum, New York.

Still, H. (1972). "Of Time, Tides and Inner Clocks." Stackpole Books, New York.

Strunghold, H. (1971). "Your Body Clock." Scribner, New York.

5

Clock Compensated Animal Orientation

SUN-COMPASS ORIENTATION

Bees

If one surrounds a beehive with eight potential feeding stands (Figure 5-1), one at each of the major compass points, and provides a source of sugar water at one of them, a truly remarkable sequence of events takes place. After a short interval, one or two bees locate the sham nectar by chance and carry samples of it back to the hive. Soon after, many bees return to the location to collect—so many, in fact, that something other than chance discovery must be operating. One wonders how the word spreads so quickly and effectively.

MESSAGE OF THE TAIL-WAGGLE DANCE

The answer is that the bee making the original discovery actually *tells* its hivemates of the existence, direction, distance, and richness of the new food source. It does this by performing the very incitive "tail-waggle dance," which can be observed if one cuts a hole in the hive and inserts a glass window. This dance is done on the vertical comb inside the dark hive, while the other bees crowd around and sense the

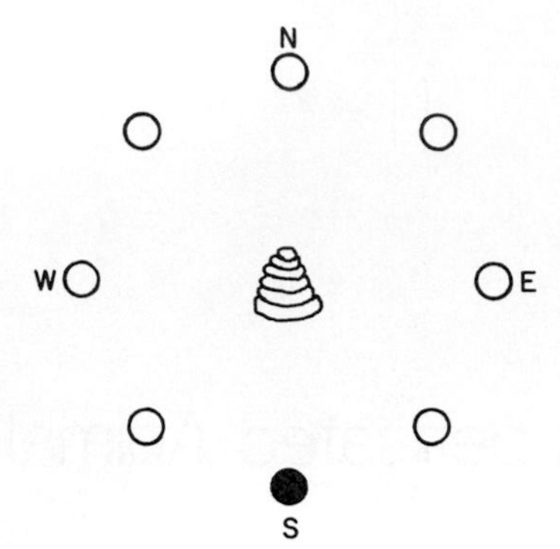

Figure 5-1. A beehive surrounded by eight feeding stations, each placed at a major compass point. The southern station contains a sugar-water solution as is indicated by the shading.

movements with their antennae (Figure 5-2). After "viewing" a performance, the previously naive bees are then able to leave the hive and fly directly to the food source.

The dancing bee's message is transferred as follows (Figure 5-3): In the dark hive, straight up on the comb always symbolizes the direc-

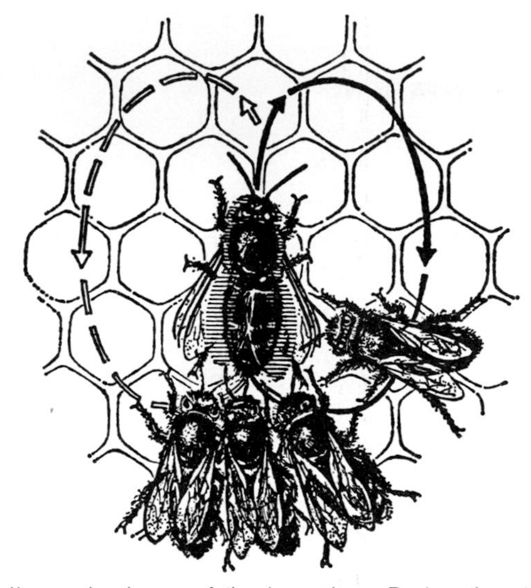

Figure 5-2 The tail-waggle dance of the honeybee. During the straight portion of its dance, the bee waggles its abdomen and makes a buzzing sound. To complete the dance, it then circles to the right, waggles throught the straight portion again, and then circles left. Other workers crowd around on the comb to sense the movements with their antennae. From K. von Frisch, "The Dance Language and Orientation of Bees." Belknap Press, Cambridge, Massachusetts, 1967.

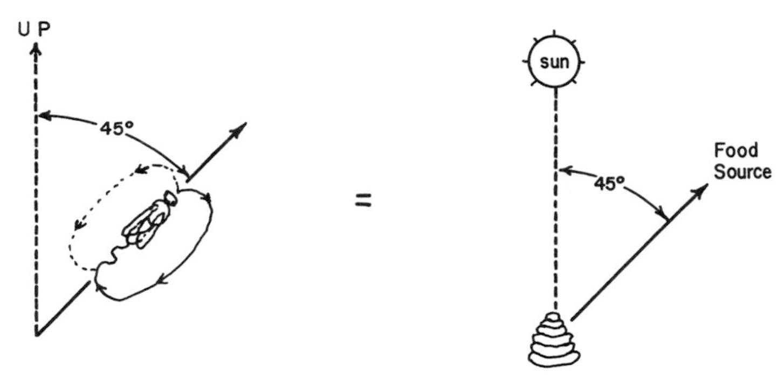

Figure 5-3 The information contained in the tail-waggle dance. The tail-waggle portion of the bee's dance is here shown assuming an angle of 45° to the right of vertical on the comb. This configuration signifies to onlooking bees that, on leaving the hive, they should locate the sun in the sky and set a course 45° to the right of it to locate the food source.

tion of the sun outside the hive. The angle assumed between straight up and and the tail-waggle portion of the bee dance is the angle that must be assumed with the sun to locate the food source. The message, once received by the other workers, permits them to go to the entrance of the hive, locate the sun 93 million miles away, assume the proper flight angle with it (shown as 45° to the right in the figure), and fly directly to the feeding station. This response pattern is called, sun-compass orientation.

THE ROLE OF THE CLOCK IN DIRECTION FINDING

The story is more complicated than this, however, because the sun is not a stationary reference point in the sky. It moves across the heaven from east to west at an average speed of 15° per hour, so that the bees must cope with a wandering reference point. Experiments such as the two to be described revealed how it is done.

At 9 A.M., the dance orientation depicted in Figure 5-3, directs a bee to the south feeding station (Figure 5-4), which contains a supply of sugar water. Suppose now that an experimenter plugs the hive entrance for the next 3 hours, so that the dancing bee cannot leave. During this interval, the sun will have moved across the sky to the point indicated by the tip of the solid arched arrow in Figure 5-4. If the trapped bee now wishes to relay the information necessary to find the source, it must change the angle of its dance on the vertical comb; for, if it used the old angle of 45° right—the angle measured when it last left the hive at 9 A.M.—its fellow workers would be directed to the southwest feeding station, which of course, lacks a reward. This

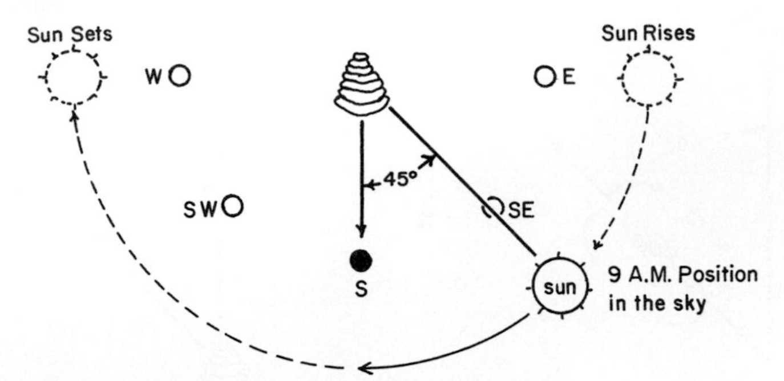

Figure 5-4 The orientation of the bee dance shown in Figure 5-3, if performed at 9 A.M., would direct other bees to the south feeding station. By noon, the earth would have rotated so that the sun would be positioned at the point of the solid arrow. Now to signal the south feeding station, the waggle portion of the dance would have to be oriented vertically on the hive.

does not happen, however; for, in spite of its 3-hour incarceration out of the sun, its dance on the comb at noon is straight up, signifying the south feeding station. The conclusion is inescapable; the bee's biological clock had measured the passage of time and this information was used to compute the new dance angle required to compensate for the movement of the sun.

The converse of this experiment has also been done. That is, wait until a bee has arrived at the feeding station and then hold it there for several hours (starting at 9 A.M.) by putting a tin can over it. Had the bee not been captured, it would have returned directly to the hive, for, being a mathematical microwizzard, after learning the angle required to find the food source, it had easily calculated the required angle home—in this case, the angle with the sun would have been about 135°. If, when the bee is released from its can 3 hours later, it should assume the old angle of about 135° with the sun, it would fly in a northeasterly direction and not get back to the hive (Figure 5-5, dashed line). Instead, it flies directly to the hive showing that its clock has compensated for the apparent movement of the sun across the sky during its stay under the dark can. So, as unexpected as it might seem at first, it is quite clear that bees use their clocks very skillfully in their daily feeding sorties. For unraveling this secret, and many more, from the bees, Karl von Frish got a Nobel Prize in 1974.

MARATHON DANCING

The action of the clock on the orientation of the bee dance may be seen in nonfeeding bees also. Bees do not venture forth on rainy days

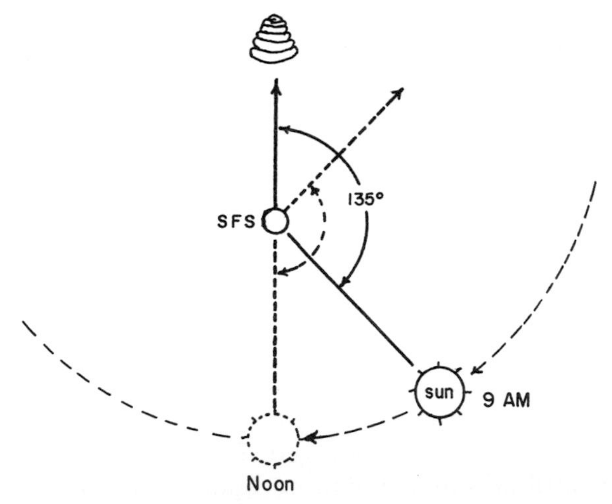

Figure 5-5 The role of the biological clock in maintaining the correct course back to the hive from the south feeding station (SFS). A bee returning to the hive at 9 A.M. must fly at an angle of 135° to the left of the sun (indicated by the solid lines). After it had made several trips, the bee was trapped at the feeding station and kept there in an opaque container until noon, when it was released. If the bee used the 9 A.M. orientation now, it would fly to the northwest (dashed line). Instead, it flies directly to the hive, which indicates that its clock had signaled the change in time and therefore the required angle.

or at night, but sometimes one will dance spontaneously in the hive. When this happens, the dance orientation indicates the direction of the last place the bee fed. In fact, some bees, dubbed "marathon dancers," seem to dance almost all the time and, because their clocks control the vector of tail-waggle segment, the bee's orientation on the face of the comb rotates counterclockwise at a rate of approximately 15° per hour.

Bees can "remember" a feeding direction for long periods of time. This tenacity was discovered accidently. Bees had been trained to visit a northwest feeding station, but before a planned experiment could be completed, the weather turned cold and the hive was plugged and brought indoors. Five weeks later, the colony was fed, and the presence of food stimulated some of the bees to begin dancing. The orientation of their dances signaled the northwest direction!

Birds

DIRECTION FINDING IN MIGRATORY BIRDS

Each fall and spring, more than 100 species of birds in the United States carry out their migratory flights to and from their breeding grounds. The timing of these travels is known to be controlled in many cases by their living clocks which measure the relative lengths of the changing day-night intervals and thus mark off the seasons. Seasonality, however, is the subject of Chapter 6.

In addition to being able to migrate hundreds or thousands of miles (the Arctic Tern carries out an annual circumpolar migration totaling 22,000 miles), bird travels terminate at very specific geographic locations. For example, careful studies have shown that each spring 75% of all robins return to within a 5-mile radius of their northern nest sites of the previous year (which is about the same precision required of an Intercontinental Ballistic Missile). This statistic also includes the fledglings of the previous year, who have never made the trip before. Even more amazing, it was found by banding and recapture that when recently hatched crows, Blue-winged Teal, and White Storks were held captive until all the experienced adults had emigrated south at the end of the breeding season, these naive birds also reached the wintering grounds when they were eventually released. Quite obviously, birds have a keen navigating capability. Some of the mechanisms of this uncanny skill have now been deciphered; the earth's magnetic field and even olfaction play roles. The best-known component, however, is birds' ability to use heavenly bodies to guide their travels.

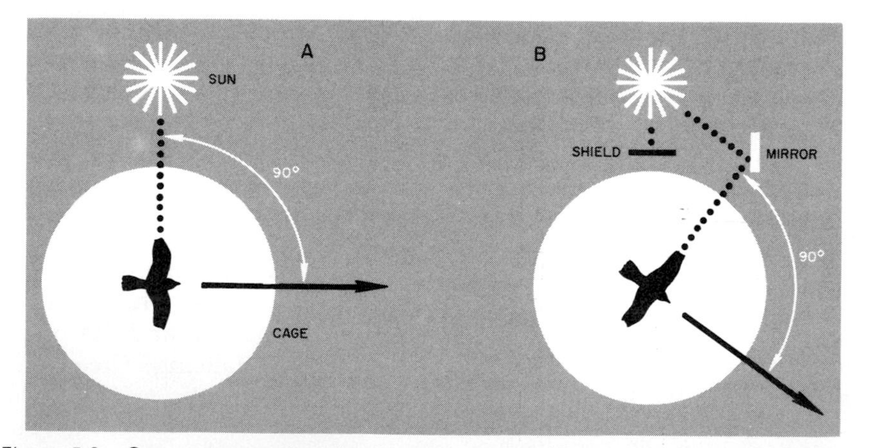

Figure 5-6. Sun-compass orientation in a caged bird. (A) Bird shown orienting in a desired direction which is 90° to the right of the sun. (B) To prove that the sun was being used for the orientation, it was screened from the bird's view and reflected into the cage from a different direction. The bird immediately assumed a 90° angle with the mirror image, which proves that the sun serves as a guide post. From J.D. Palmer, *Nat. Hist.* **75,** 48–53 (1966).

But before describing this, it should be pointed out that one of the greatest problems in studying migratory orientation is the fact that one's experimental subject always terminates each experimental observation by flying away forever. The late Gustav Kramer solved this problem after observing that recently trapped migrants, held in an outside aviary, seldom sit still. Instead, they stand on their perch, point in the direction of their normal migratory route (north in the spring and south in the fall), and flutter their wings—mimicking, presumably, what they would be doing if they could escape. This behavior is called *migratory restlessness*. On overcast days, the birds continue to flutter, but no longer orient their body axes in a particular direction. Kramer guessed correctly that they must be using the sun as a guidepost and devised a simple experiment to prove it. On a sunny day, he blocked the bird's view of the sun with an opaque screen and, using a mirror, projected the sun's image into the cage from another direction (Figure 5-6). The bird instantly realigned its body to the new position of the sun and fluttered with renewed vigor in that direction.

THE ROLE OF THE CLOCK IN DIRECTION FINDING

This discovery was made at the same time that the bee orientation studies were being carried out, but neither investigative group knew about the other's work. The ornithologists were also concerned with

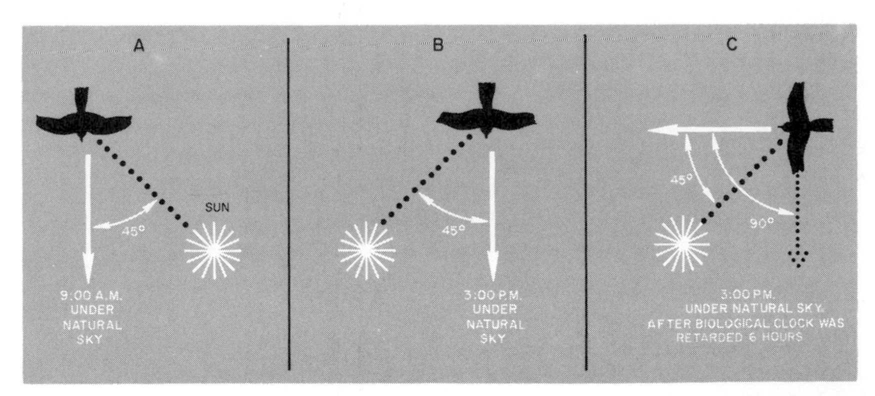

Figure 5-7. The role of the biological clock in bird orientation. A fall migrant flying in a southerly direction would be required to keep its path of flight at an angle of 45° to the right of the sun at 9 A.M. (A), and at 45° to the left at 3 P.M. (B). When the bird's clock was reset to 6 hours later in the day by artificial light-dark cycles, and its orientation retested at 3 P.M. (when the bird's clock was indicating 9 A.M.), the bird flew west (C). Therefore, it is the clock that is responsible for the bird's compensation for the movement of the sun across the sky. From J. D. Palmer 1966. *Nat. Hist.* **75,** 48–53 (1966).

how a bird compensated for the movement of the sun across the sky; but, by the time they got to this aspect of their studies, a great deal was already known about living clocks and how to reset them. Therefore, they needed only to perform the following experiment to demonstrate the clock's role in avian orientation.

A bird that was orienting to the sun was brought into the laboratory and subjected to artificial days in which "sunrise" and "sunset" were delayed by 6 hours. After the bird had entrained to this new cycle, it was taken out-of-doors and its orientation retested. As shown in Figure 5-7, the pretreatment direction had been south, which means that at 9 A.M. the bird had to keep its path 45° to the right of the sun, while at 3 P.M. the angle had to be 45° to the left. The bird was now tested at 3 P.M., and because its personal clock was signaling 9 A.M. at this time, it assumed an angle 45° to the right of the sun. Since the sun was now in the southwestern portion of the sky, the bird's heading became west. Therefore, just as with bee flight, the clock is a fundamental component of orientation.

THE MASTER CLOCK CONCEPT

There was a time when it was speculated that the clock involved with spatial orientation was not the same one used in the control of rhythms. To resolve this question, the locomotor activity and orientation responses of two starlings maintained in constant conditions were examined simultaneously.

Using an experimental design similar to the feeding stations around a beehive motif, the starlings were trained to search for food in a particular direction—one to the north and the other to the west. Between orientation tests, the birds were maintained in special cages in which their activity was continuously measured and recorded automatically. Activity and orientation were first measured in natural day-night cycles, then in continuous dim light and constant temperature for 3 weeks, and then back in day-night conditions again.

As expected, in constant conditions the period of the birds' activity rhythms deviated from 24 hours to become about 30 minutes less than that interval (Figure 5-8). After 11–12 days in constant conditions, the birds were briefly returned to the orientation apparatus under the natural sun and their directional preferences measured. The bird that had previously been trained to the north now oriented to the west, while the one trained to the west now searched to the southwest (Figure 5-8). These directions are close to the orientations that would be expected, as indicated by the change in phase of the activity rhythm.

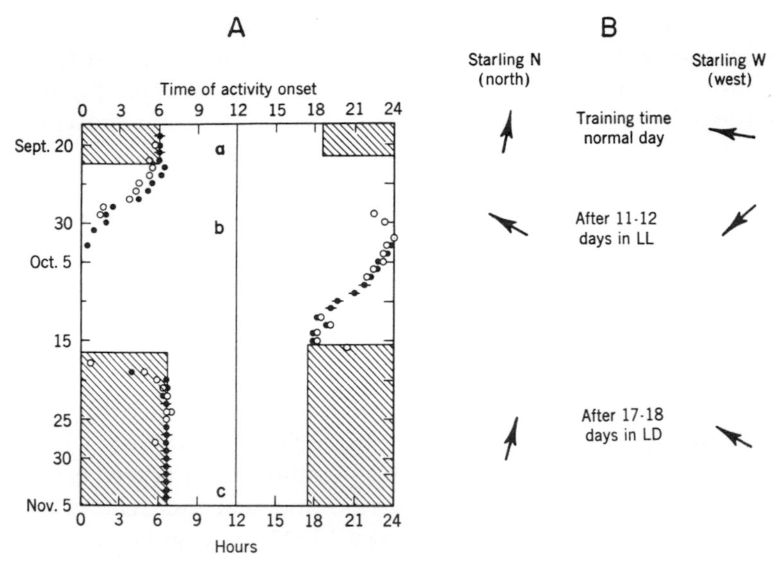

Figure 5-8 The corresponding changes in the periods of the activity rhythms (A) and orientational directions (B) of two starlings while held in constant conditions (Sept. 23 to Oct. 17). The hatched areas signify the hours of darkness. The times of onset of activity are indicated by the open circles (for the bird trained to orient to the north) and solid circles (for the bird trained to the west). *a*, *b*, and *c*, indicate the days of testing and the orientational direction chosen by the birds. See text for further description. From the data of K. Hoffmann, *Cold Spring Harbor Symp. Quant. Biol.* **25,** 379–387 (1960).

The birds were later returned to natural day-night conditions and 17 days later their orientation tested again. The preconstant conditions direction was again approximated. Thus, it appears that the same clock governing the birds' activity patterns is the chronometer for spatial orientation.

HOMING PIGEONS

While capitalizing on migratory restlessness solved the problem of losing each experimental bird as it flew off to complete its migration, one was still hampered by the restriction of being able to experiment only during the spring and fall migratory seasons. This problem was alleviated by studying homing pigeons who, of course, can find their way home at all times of the year. Studies consist of simply taking birds up to 600 miles from their loft and letting them go. This, of course, reintroduced the problem of following the bird. Therefore, in the studies to be described, about all that could be measured is the "vanishing direction," which is the bearing at which a bird is last seen through binoculars after it has been released in an unfamiliar locale. However, homing pigeons usually do appear again at the loft so that additional information about the directness of their flights could be gained by using elapsed times (speeds of 50 mph are common). Those birds that do not return are not wanted anyway.

In the most sophisticated and expensive studies, tiny radio beacons were glued on the backs of the pigeons and their travels followed by airplane. In this way, it has been found that pigeons usually fly in a very straight line back to the loft. Birds that deviate only a mile or two from a perfectly straight course of a hundred miles or more have often been observed.

The Sun Compass. Experiments identical to those described for migrating birds, such as resetting a bird's clock with altered light-dark cycles, have been done with homing pigeons and it is well established that the birds use the sun as a compass and that their clocks compensate for its movement across the sky. There is a great deal more to homing than just this, however; because of one of the redundancies built into the process, some doubt was temporarily shed on the existence of sun-compass orientation.

Use of the Geomagnetic Field. This question arose when it was learned that pigeons could still find their way back to the loft on days so overcast that the sun was never visible. It turned out that in the absence of the sun the birds simply switched to backup means of direction finding and used the earth's magnetic field to find their way home. To demonstrate this, small bar magnets were attached to the necks of some birds (which distorted the natural field around them) and sham "magnets" of the same size and weight, but made of brass, were attached to others. When all the birds were released on an overcast day, those with the brass bars disappeared from view in the direction of the loft, while those with magnets attached vanished in all directions (Figure 5-9). In other words, when the pigeons are unable to use sun-compass orientation, they simply use the earth's magnetic field.

PENGUINS

When studying pigeon homing, an observer has only minutes to measure and record its vanishing bearing and observe the bird's actions between release and its disappearance on the horizon. This observation sequence can be transformed into slow motion by using a flightless bird such as the Adélie penguin. As a result of its waddling

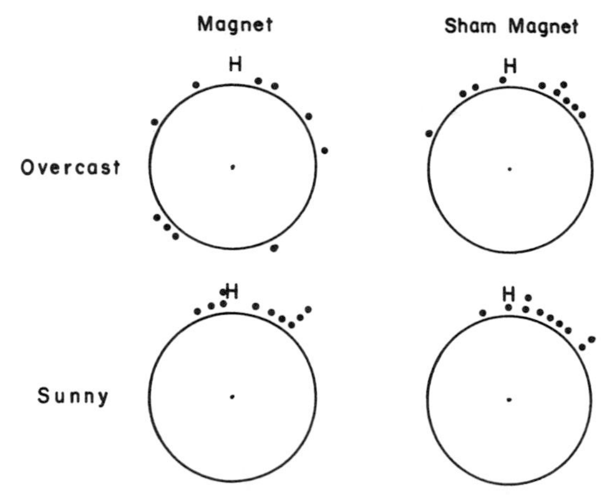

Figure 5-9 The role of the magnetic field in pigeon orientation. Birds were released away from the loft and their vanishing bearings recorded (and plotted as the points around the circles above). Some birds had small magnets attached to them while others had brass bars of the same weight. On sunny days (bottom) all the birds flew off in the general direction of their loft (H-direction of home loft), but on days when the sun could not be seen (top), only the birds bearing sham magnets oriented toward the loft. Those with magnets became disoriented because they could not use the geomagnetic field for direction finding. Modified from W. T. Keeton, *Proc. Natl. Acad. Sci. U.S.A.* **68,** 102–106 (1971).

pace, a single bird can be watched for half an hour or more before it finally drops over the featureless horizon of Antarctica. If orientational measurements are wanted over longer distances, one simply follows the tracks left in the snow as the penguin walks, or toboggans along on its belly.

If a bird is taken from a rookery and released at a strange, distant point, its immediate behavior is interesting. The bird either settles down and takes a nap (in which case the observers just have to sit in the bitter cold and wait), or it stands and looks around for a few minutes (as if getting its bearings) and then waddles off. If the birds from the Cape Crozier rookery (that part of the Antarctic shore facing the tip of South America) are translocated only short distances longitudinally, or all the way inland to the South Pole, and released on a clear day, they all choose north as their escape direction. If the day is so overcast that the sun cannot be seen, the birds are totally disoriented (Figure 5-10). Quite obviously, they are using the sun, and because they maintain a constant compass direction while the sun circles overhead, they must be using their clocks to compensate. Because these birds always go north when released, which is seldom the way back to

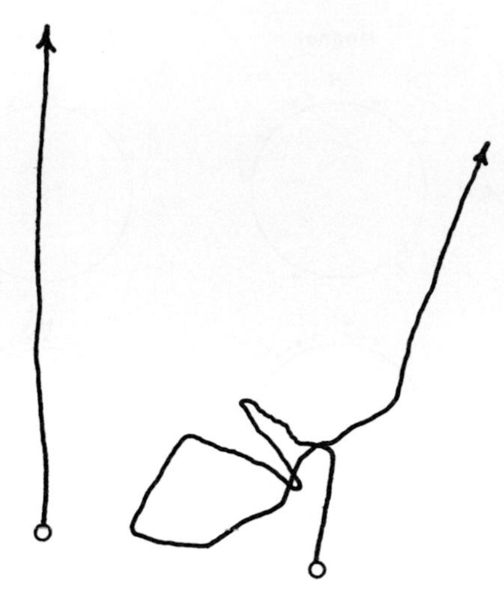

Figure 5-10 The use of sun-compass orientation in Adélie penguins. The path indicated to the left above, is that of a bird released (open circle) on a sunny day; using its sun-compass capability it moved off directly toward the sea. Another bird released on a cloudy day (right-hand path above) wandered about aimlessly until the sun came out from under a cloud (that instant being indicated by the straightening of the course). Modified from J. T. Emlen and R. L. Penney, *Ibis* **106,** 417–431 (1964).

the rookery, the response can hardly be called homing. But, north is an ideal direction to choose for a South-Pole penguin; because no matter where it is released, north is seaward, and the sea is the penguin's only source of food. Additionally, when they do reach the sea, they then begin to navigate, and penguins that have been displaced by as far as 1200 miles have returned to the very nest sites from which they had been taken—*10 months before*!

As one travels from the equator southward, the time-zone meridians come closer and closer together until they eventually unite at the South Pole. This means that in Antarctica, one can travel relatively short distances to the east or west and cover dozens of degrees of longitude and several time zones. For example, the Russian outpost at Mirny is only 2000 miles, but 6 time zones, away from the Cape Crozier rookery on the Ross Ice Shelf. When Russian penguins were transported to the rookery and released, they headed west (rather than north), showing that their clocks were 6 hours out of phase with the local residents'. If, however, these penguins were first held for about 3 weeks in an open pen before being released, they oriented to the north. Therefore, in spite of the fact that the sun never sets during the Antarctic summer, but instead, simply circles counterclockwise above the horizon, the birds' clocks appear to be entrained to it anyway.

STAR-COMPASS ORIENTATION

A great many migrating birds fly at night: the generalization being, the larger the bird, the more likely that the species migrates during daylight. Thus, most ducks, hawks, crows, and geese migrate during daylight, while smaller birds like the warblers and orioles fly at night using the hours of daylight to search for food. When night migrants are caged, their urge to move on is not confined and is manifest as *Zugunruhe* (Figure 2-21). When their cages are placed under the night sky, the birds stand on their perches and flutter in the direction of their desired migratory route.

That the birds' clock is not involved in this type of orientation has been demonstrated in the planetarium where within a few minutes time, star configurations 3, 6, or 12 hours out of phase with local time could be projected. The artificial changes made no difference to the birds—they did not alter their orientation.

This was really not an unexpected finding, because the rotation of the earth relative to the stars is not equal to the interval of a solar day, but to a siderial one. The latter is only 23 hours and 56 minutes, and this slight difference is very significant, for it means that the stars and constellations rise 4 minutes earlier each night, or 2 hours earlier in just one month. Obviously, a bird's 24-hour clock would fail to compensate under this situation and thus guarantees that the bird would become lost. (The first evidence has just been obtained that at least some organisms may have a siderial-day clock, see Figure 7-16).

The night-migrating birds it appears, can get all the directional information they need just from the patterns of the stars in the heavenly canopy. To be more specific, only those stars close to the North Star (which are those that do not pass under the horizon at any time during the night) seem to be important. In the planetarium, other portions of the sky can be switched off without disrupting a bird's orientation.

MOON-COMPASS ORIENTATION

In spite of its prominence in the night sky, the fact that the moon rises 50 minutes later each night precludes its use in orientation—unless an organism has a lunar-day clock to compensate for its movements. Additionally, the moon is visible on only half of the nights each month, while the stars are out every night. *A priori*, one would therefore expect that the moon is never used—and this is true for birds. There is, however, a known user, a group of crustacean amphipods that live by the seashore and possess lunar-day clocks.

Direction Finding in *Talitrus*

Especially common on the western shore of Italy, in a habitat somewhat similar to the one described for another crustacean, *Excirolana* (Figure 3-5), lives another "sand hopper," *Talitrus*. During midday, this animal lives buried in damp beach sands; but, in the late afternoon, it emerges and wanders about through most of the night. Although it is only a few millimeters in length, it routinely journeys as far as 100 meters inland. Being very susceptible to drying out, it must find its way back to the wave-soaked shoreline soon after sunrise. To carry out this migration, it uses its solar-day clock and the sun while still visible, and the moon and its lunar-day clock at night.

To demonstrate this orientation, the amphipods were captured and placed in the center of a large concave glass dish through which they could be observed from below. Low screens were set up all around the apparatus so that the surrounding landscape could not be seen by the animals, but full view of the heavens was permitted. On moonlit nights, the animals quickly gathered on the seaward side of the dish (Figure 5-11), but on moonless nights no particular direction was chosen. If orienting animals were prohibited from seeing the moon, they become disoriented, but when a flashlight was used to simulate

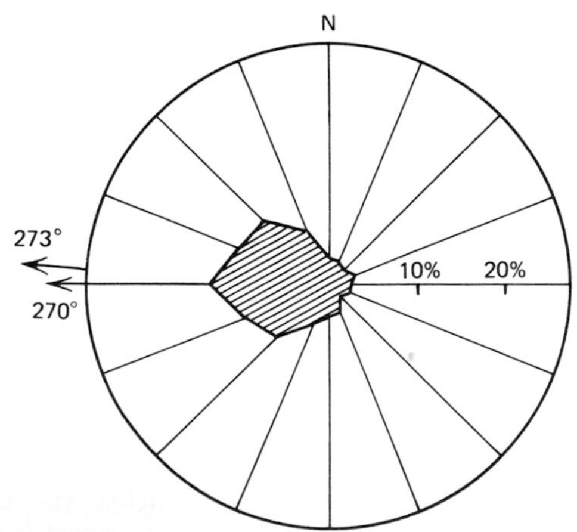

Figure 5-11 Moon-compass orientation by the amphipod, *Talitrus*. The proper escape orientation is 270° due west. The average path of 3108 animals in the orientation chamber was 273°. Modified from F. Papi and L. Pardi, *Z. Vergl. Physiol.* **41,** 583–596 (1959).

the face of the full moon, and presented from a different side of the apparatus, the amphipods did not catch on to the trick and assumed the proper orientation angle to it.

All-night observations of the animals in the orientation chamber revealed that they remained at the seaward side of the dish constantly, which means that their lunar-day clocks must have been compensating for the moon's travels across the sky. And, by testing the animals' orientation and finding it to be correct after they had been stored in constant conditions for up to 2 weeks, it was demonstrated that the lunar-orientation chronometer had continued to run without the animals being exposed to the moon.

During the early morning and late afternoon, *Talitrus* uses the sun to orient. Just as was found with lunar orientation, solar orientation is not affected by maintaining the organism in constant conditions before testing. In fact, in one experiment, in addition to being kept in constant conditions for 2 weeks, the sand hoppers were shipped from Italy to Argentina. Testing there showed that their clocks were still running on Italian time.

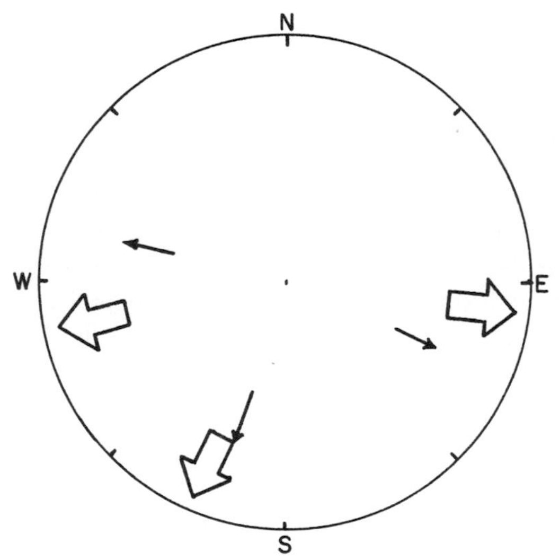

Figure 5-12 The average escape directions of three parent populations (white arrows) of *Talitrus* and their offspring (closest black arrows). The latter had been maintained in the laboratory for 1–3 months and had never seen the sun before the day of the orientation test. Obviously, the escape direction must be inherited. From the data of L. Pardi, *Cold Spring Harbor Symp. Quant. Biol.* **25,** 395–401 (1960).

ITS GENETIC BASIS

The escape direction (i.e., seaward) is genetically controlled. When animals living on the West Coast of Italy (where the escape route to the sea is, of course, west) were transported overland to the Adriatic shoreline and tested; the escape direction signaled was west, in spite of the fact that seawater and safety lay within a few feet to the east. In an equally dramatic experiment, animals from three separate populations, each with its own escape direction (because of the contours of the local shoreline) were brought into the laboratory to reproduce. The offspring were separated from their parents and not permitted to see the sun until the day of testing. Still, all chose the direction used by their parents (Figure 5-12).

SUMMARY AND CONCLUSIONS

1. The sun-compass orientation of birds, bees, and sand hoppers is described.
2. Bees use the tail-waggle dance to communicate food sources to their hivemates, and their clocks govern the orientation of the dance when the bees are prohibited from seeing the sun.
3. The orientation capabilities of migrating birds are studied in the laboratory by capitalizing on their migratory restlessness.
4. By subjecting birds to artificial light-dark cycles offered out of phase with the natural day, and then observing the alterations in their orientation to the sun, the role of the bird's clock in compensating for the movement of the sun across the sky was demonstrated.
5. Many of the details of the mechanisms of homing in pigeons have been worked out by following the birds in airplanes. Sun-compass orientation, the use of the geomagnetic field, and even olfaction have been found to be important means they use to find their way back to the loft.
6. Penguins are able to use the antarctic sun in their "homing" responses, in spite of the fact that it never sets during the summer, but just circles above the horizon.
7. Night-flying migrants are able to use star patterns to guide their journeys. A living clock is not used in this type of orientation.
8. Moon-compass orientation is possible only in those organisms which possess a lunar-day clock and, thus far has been demonstrated only in a shore-dwelling group of amphipods. These animals also have solar-day clocks and use the sun for direction finding during the day-

time. Both these clocks continue to function when the sand hoppers are maintained in constant conditions and during translocation experiments over thousands of miles of longitude.

9. There seems to be no question that the same clock that governs organismic rhythms is the one used for time compensation in spatial orientation.

Selected Readings

Emlen, S. T. (1967). Migratory orientation in the Indigo Bunting, *Passerina cyanea*. Part II. Mechanism of celestial orientation. *Auk* **84,** 463–489.

Emlen, S. T. (1974). Bird migration. *In* "Avian Biology" (D. S. Farner and J. R. King, eds.), Vol. 4, pp. 129–219. Academic Press, New York.

Emlen, J. T., and Penney, R. L. (1966). The navigation of penguins. *Sci Am.* **215,** 105–113.

Galler, S., Schmidt-Koenig, K., Jacobs, G., and Belleville, R., eds. (1972). "Animal Orientation and Navigation." Natl. Aeronaut. Space Admin., Washington, D.C.

Griffin, D. R. (1964). "Bird Migration." Anchor Books, New York.

Hartwick, R. (1976). Aspects of celestial orientation in Talitrid Amphipods. *In* "Biological Rhythms in the Marine Environment" (P. J. DeCoursey, ed.), pp. 189–198. Univ. of South Carolina Press, Columbia.

Hoffmann, K. (1972). Biological clocks in animal orientation and in other functions. *In* "Circadian Rhythmicity," pp. 175–205. Centre for Agricultural Publishing and Documentation, Wageningen, Netherlands.

Keeton, W. T. (1974a). The mystery of pigeon homing. *Sci. Am.* **231,** 96–107.

Keeton, W. (1974b). The orientation and navigational basis of homing in birds. *In* "Advances in the Study of Behavior" D. S. Lehrman, R. A. Hinde, and E. Shaw, eds. Vol. 5, pp. 47–132. Academic Press, New York.

Matthews, G. V. T. (1973). Biological clocks and bird migration. *In* "Biological Aspects of Circadian Rhythms" (J. T. Mills, ed.), pp. 281–311. Plenum, New York.

Palmer, J. D. (1966). How a bird tells the time of day. *Nat. Hist.* **75,** 48–53.

Palmer, J. D. (1967). Geomagnetism and animal orientation. *Nat. Hist.* **76,** 54–57.

Renner, M. (1959). The clock of bees. *Nat. Hist.* **68,** 434–440.

von Frisch, K. (1967). "The Dance Language and Orientation of Bees." Belknap Press, Cambridge, Massachusetts.

von Frisch, K. (1974). Decoding the language of the bee. *Science* **185,** 663–668.

Walcott, C. (1974). The homing of pigeons. *Am. Sci.* **62,** 542–552.

6

The Clock Control of Plant and Animal Photoperiodism

PLANTS

The Cause of Seasonality: Daylength

Most plants do not produce flowers all year long, nor do all plants flower at once. Instead, there is an annual progression of floral change with the seasons. Bloodroot and buttercups appear in the spring, iris and columbine flower in the summer, and goldenrod and ragweed introduce the fall hayfever season. The events behind this orderly scheduling are now well known.

Early speculation as to cause centered on temperature change and maturation rate. The former was the most obvious suspect; it was thought that plants must grow vegetatively until a proper seasonal temperature finally prevailed and this then acted as the stimulus for flowering. The speculation was easily disproved by growing irises in the hothouse all winter long at summer temperatures. The plants waited to flower until May and June, which is their normal reproductive season, inspite of receiving several seasons of optimal flowering temperatures before that time.

Maturation rate, the other suspect, is obviously involved in flowering; for, as is true with children, until a certain amount of fun-

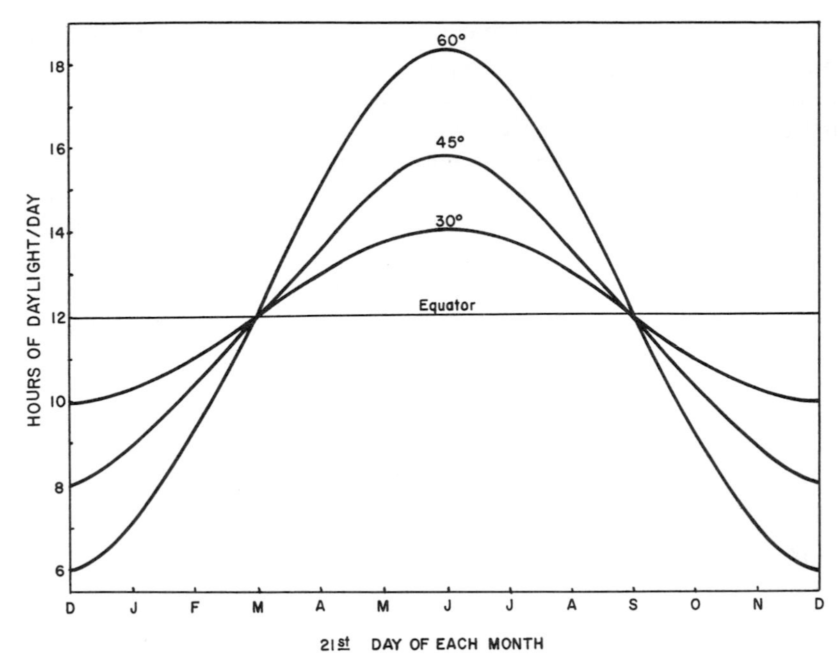

Figure 6-1 The changing day lengths over the year as a function of latitude.

damental development has taken place, reproduction cannot commence. It seemed logical to assume that the germination time of the seeds combined with the maturation rate thereafter, must be the determining factors behind the time that a plant blooms. While important, of course, this is not the cause of seasonality, as was demonstrated by planting Biloxi soybean seeds in the early spring, again in June, and then again in July. By fall, the spring planting was 5-feet tall and the later plantings proportionally shorter; yet, all groups bloomed simultaneously in September, the usual reproductive season.

Several other possibilities were explored and disproved, until the correct environmental parameter was finally found. It was day length. And, in a way, this is not surprising, for the seasonal change in photoperiod (that is, the length of the illuminated portion of each day) is the most regular changing feature (Figure 6-1) in the environment in the temperate latitudes. Plants capable of measuring the relative lengths of day and night are therefore always advised of the present season.

The original discovery in 1919 that plants have this capability is credited to two United States Department of Agriculture plant physiologists, Wightman Garner and Harry Allard.* This was before the time of unheeded warnings on cigarette packages and these men were—in the public's interest—trying to develop ways of improving tobacco production. In particular, they were attempting to get established a variety of tobacco which appeared spontaneously in a field as a mutant. It was called Maryland Mammoth and grew to a height of 10–15 feet and produced as many as 100 leaves/plant—but, it would not set seeds in the field. The only way to keep the new strain going was to bring it into the greenhouse before the first killing frost. Thus protected, it would flower in the late fall. The two botanists reasoned correctly that maybe the greenhouse temperature kept the plants alive until the days finally shortened sufficiently to provide the stimulus for flowering. To test this, in July, one group of potted Maryland Mammoths were placed in a dark house each night at 6 P.M. and not returned to daylight until 8 A.M. the next morning. These plants, which received only 10 hours of light per day soon flowered, while those living under the long days of July remained vegetative. Because the plants could measure the day length, the botanists matter of factly named the activity, a "length-of-day response"; but this lackluster appellation was soon changed to "photoperiodism."

Garner and Allard returned to the Biloxi soybean and found they could cause it to flower anytime by exposing it to 12-hour photoperiods (the September day length), and could cause poinsettias to turn anytime by subjecting them to 10-hour Christmas season photoperiods. They also did the converse experiment, and kept a plant that required short day lengths to flower, on days artificially lengthened by electric lighting. The subject, a cosmos, grew to a height of 15 feet, but never produced a single flower bud.

The plants just described bloom during the short days of fall and were therefore named "short-day plants." Others, such as hollyhocks, black-eyed Susans, and irises only bloom during the long days of summer, so they were named "long-day plants." The group including tomatoes and roses, which flower over large ranges of photoperiods

* It is unquestionably true that the first paper on the subject by Garner and Allard—which was a masterpiece of foresight and breadth of comprehension (and was almost not published because the reviewers felt it was "insufficiently novel")—heralded the beginning of all work on the importance of day length on seasonality. However, speculation on the importance of relative day lengths began in 1852, and Tournois, in 1912, induced flowering in the eternally popular plant *Cannabis,* by covering the plants with boxes so that they received only 6 hours of light per day. While championing another cause, he was killed early in World War I, and his work went mostly unnoticed.

whenever environmental conditions are propitious, were named "day neutrals."

These early studies stimulated a great flurry of work in other laboratories, all of which confirmed Garner and Allard's pioneering findings and thus firmly established the short-day, long-day, and photoperiodism terminology. Then an experiment was performed which showed that all of this was somewhat incorrect.

The Importance of Darkness

Short-day and long-day plants were placed under short days. This, of course, permitted flowering in the former and prohibited it in the latter. The dark portion of each cycle was then interrupted by a short interval of light (see Diagram 6-1).

Diagram 6-1

This simple treatment inhibited flowering in the short-day plant and permitted it in the long-day plant. This means that it is the length of the dark period that the plants are actually measuring, and so, disrupting it with a light break spoils the measurement. Therefore, the terminology that has become so thoroughly ingrained is incorrect; short-day plants are really long-night ones, and long-day plants really short-night plants.

The Role of the Clock in Measuring Night Lengths: Light Sensitivity Rhythms

Just how do plants measure the length of the interval of night? The answer to this was finally provided by experiments using light interruptions at all hours of the night. It was found that these interruptions do not always mean the same thing to a plant and that the final response produced is dependent on the time during the night that the light is offered. The experimental design used was as follows:

One set of long-night plants was given cycles of 6 hours of light alternating with 18 hours of darkness, a treatment that has the potential of causing them to bloom. Other sets of plants received the same light-dark cycles, but a specific segment of the dark portion (different

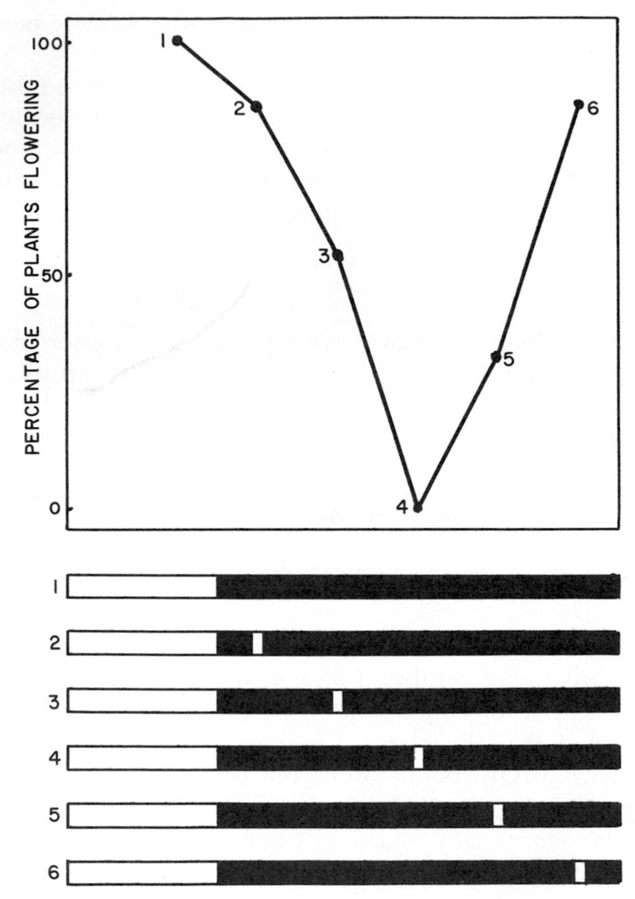

Figure 6-2 Inhibition of flowering in a long-night plant—the changing sensitivity to brief intervals of illumination given at different times throughout the night. The number next to each point on the flowering curve identifies the interrupted light-dark cycle, below, that produced it.

for each set of plants) was interrupted each night by a 15-minute interval of light. As can be seen in Figure 6-2, flowering was completely prevented by brief illumination during the center of the dark period, while at other times it was just impeded. This kind of experiment outlined the changing sensitivity to light inhibition during the night.

The discovery was examined in greater detail by using a long-night plant (*Kalanchoe,* a common house plant) that would flower when exposed to light-dark cycles of 10 hours of light alternating with 62 hours of darkness, cycles totaling 3 days in length. In experiment after experi-

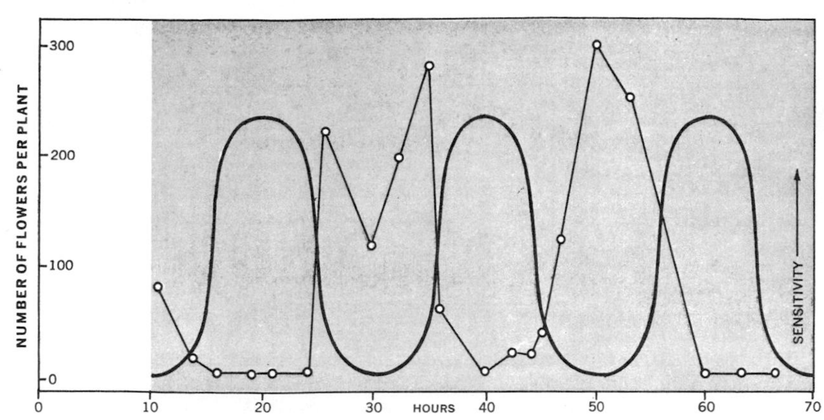

Figure 6-3 The daily rhythm in sensitivity to light-flash inhibition of flowering in the house plant, Kalanchoe, a long-night plant. Twenty-one groups of plants were subjected to light-dark cycles consisting of 10 hours of light alternating with 62 hours of darkness. At the times indicated by the open circles on the curve, the dark portion of one group of plants was interrupted by a 2-minute flash of light. The time that the flash was given determined how many flowers/plant formed. Maximum inhibition was found at 24-hour intervals, which suggested the presence of a daily rhythm in sensitivity to light (indicated by the superimposed solid curve). From J. D. Palmer, *Nat. Hist.* **80,** 64–73 (1971).

ment, a different segment of the dark interval was systematically interrupted by 2-minute light flashes and flower inhibition was recorded. As seen in Figure 6-3, the effectiveness of the light breaks was a function of the time in the dark period at which they were given. More interesting, peaks of inhibition came at 24-hour intervals, which signified that the plants must undergo a daily rhythm in sensitivity to inhibition by light. This sensitivity rhythm has been diagrammatically added to Figure 6-3.

Similar results have been found with short-night plants. In a typical experiment, 13 groups of henbane plants were subjected to an unnatural light-dark schedule of 9 hours of light alternating with 39 hours of darkness, a condition that inhibits this species from flowering. Following the usual design, definite segments of the night of each group were illuminated with single short flashes which, in some cases, permitted flowering to take place. The response pattern again clearly indicated the involvement of a rhythm in sensitivity to light (Figure 6-4).

These results, and a great many more just like them, have lead to the conclusion that it is the underlying light-sensitivity rhythm that is responsible for measuring the length of the night and is therefore the governing mechanism of the flowering response in long- and short-night plants. This rhythm is *diagrammed* in Figure 6-5, where it is

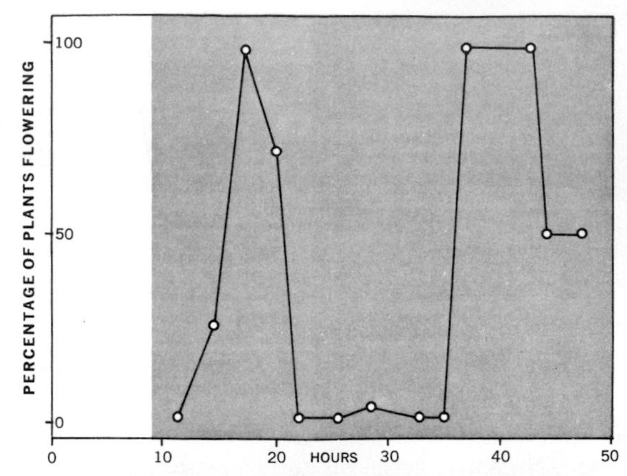

Figure 6-4 The changing sensitivity to short light interruptions of the dark period in a short-night plant, the henbane. The open points indicate the time of the light-flash interruptions and the percentage of flowers produced. Note that the peaks of flower promotion appear at 24-hour intervals, which suggests an underlying rhythm with this frequency. From J. D. Palmer, *Nat. Hist.* **80,** 64–73 (1971).

seen that the sensitive phase is not exposed to daylight in the spring and fall, but is during the long days of summer. Because exposure of this part of the rhythm stimulates flowering in short-night plants and inhibits it in long-night plants, seasonality in the flowering process is generated.

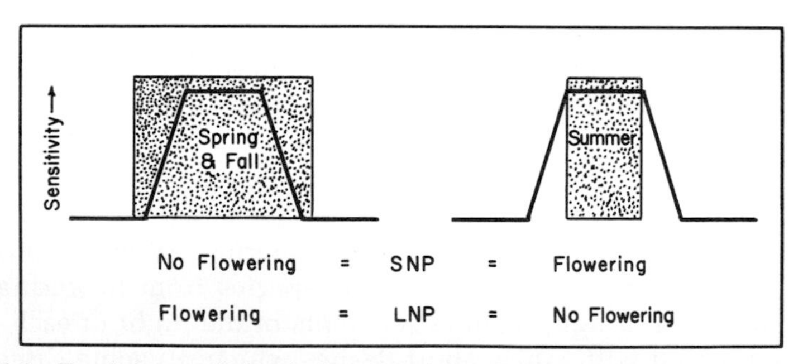

Figure 6-5 The light-sensitivity rhythm underlying the flowering process in short- and long-night plants. In short-night plants (SNP) during spring and fall, the nights are of such length that the sensitive phase of the flowering-response rhythm is never illuminated. Since exposure is a requirement for flowering, the plants remain vegetative. During the short nights of summer, at least the "shoulders" of the sensitive peak are illuminated and flowers are thus produced. Long-night plants (LNP), on the other hand, are prohibited from flowering if the sensitive peak is illuminated; therefore, they remain vegetative in the summer. It should be emphasized that this figure is just a hypothetical representation and, as such, an oversimplification of reality.

The Master Clock of Plants

After getting this far in deciphering the control of seasonality, the next question is whether the same clock governs both the flowering process and daily rhythms, or whether each have their own timepiece. The following observations have been used to suggest that only one master clock is involved.

Erwin Bünning, the man who first speculated that plants use the same clock for everything, published figures such as 6-6 in support of his conclusion. In Figure 6-6, it is seen that the peaks of the sleep-movement rhythm (Chapter 2; Figure 2-1) in the soybean are in virtual synchrony with the peaks of the light-sensitivity rhythm. (The plants were maintained in cycles of 8 hours of light alternating with 64 hours of darkness and the hours of darkness probed with light flashes to produce the flower-production curve. The sleep-movement rhythm was measured during the 64-hour interval of darkness.) The sameness of phase was interpreted as showing that the same clock must be ruling both processes. This condition is, however, not a necessity to show a relationship; there are, for example, 4 rhythms in *Gonyaulax* (Figures 2-3; 2-10; 2-11) all out of phase with one another; yet, all evidence indicates that they are all governed by the same clock. The fol-

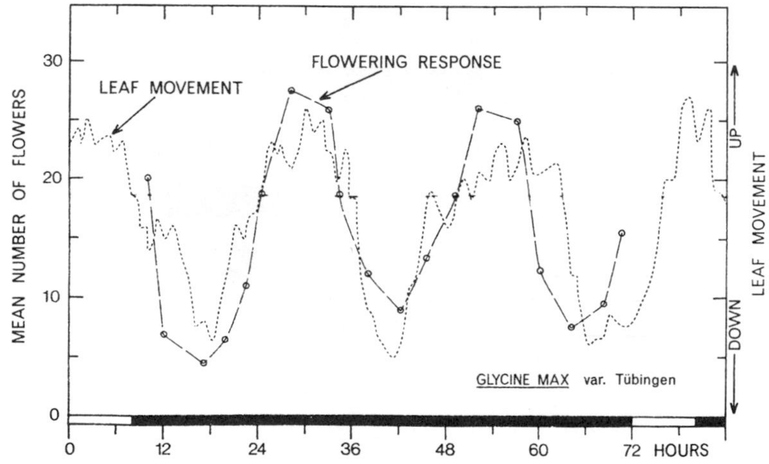

Figure 6-6 The synchronous nature of the sleep-movement rhythm and flower-inhibition rhythm in the soybean, a long-night plant. The flower-inhibition rhythm was elucidated by systematically interrupting the 64-hour dark interval with 30-minute light breaks. The sleep-movement rhythm was recorded during the dark portion of the 3-day light-dark cycle. From the data of E. Bünning, *Photochem. Photobiol.* **9,** 219–228 (1969).

lowing experimental evidence shows, in a much more convincing way, the common control.

The sleep-movement rhythm of the house plant *Coleus* was studied under 7 different 24-hour light-dark cycles, ranging from 2 hours of light alternating with 22 hours of darkness, to 14 hours of light and 10 hours of darkness (Figure 6-7). The phase of the rhythm was altered by some of the cycles: the times at which the leaves dropped maximally came 5 hours after the onset of each dark period in cycles containing between 16 and 22 hours of darkness, but only 3 hours after lights off in cycles with dark periods of 12 or 10 hours (Figure 6-7).

The variety of *Coleus* used is a member of the long-night group and would flower if the nights were not shorter than 12 hours. The plants were therefore divided into 3 groups and placed in 24-hour cycles in which the light period was 4, 8, or 12 hours. The darkness was then probed with short light pulses in order to determine the point of maximum inhibition of flowering. In the first two groups, the maximum came 10 hours after the onset of darkness; but, in the last group, it came 8 hours after this point (Figure 6-7). The phase change produced in the light-sensitivity rhythm was therefore identical to the one that had been produced in the sleep-movement rhythm. And, since the

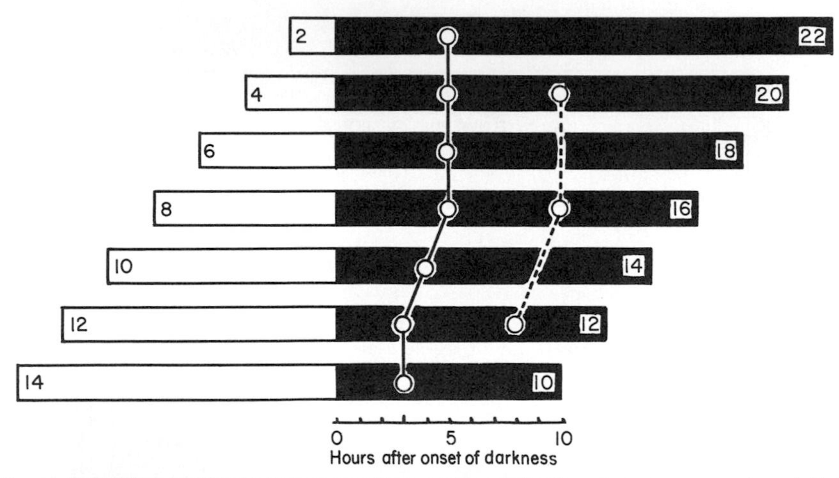

Figure 6-7 The similarity between phase changes in the sleep-movement and light-sensitivity rhythms in the house plant, *Coleus*, produced by altering the relative lengths of the light and dark portions of the ambient illumination cycles. The open circles connected by a solid line are the minima of the sleep-movement rhythm, while the dashed line connects the points of maximum flower inhibition by short bursts of light. The open part of each bar signifies the hours of light on, and the shaded part, darkness. See text for details of experimental precedings and interpretations.

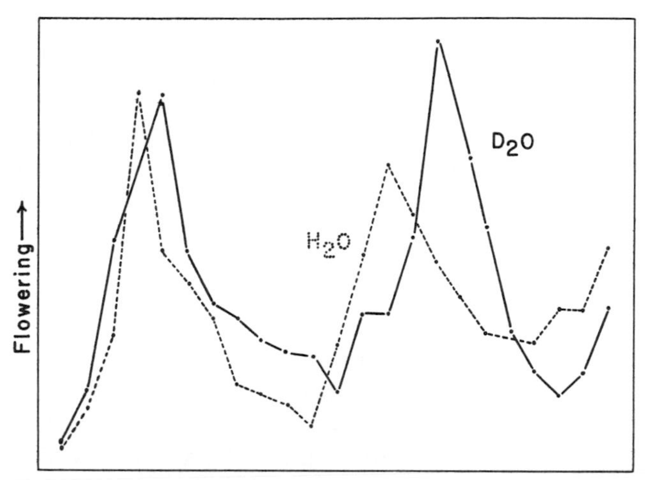

Figure 6-8 The lengthening of the period of the flower-inhibition rhythm by deuterium oxide, in *Chenopodium*. Dashed line represents the rhythm when plants received plain water. The solid curve shows the lengthening of the period after treatment with 35% D_2O. The abscissa is 72 hours in width. Drawn from the data of W. Brenner and W. Engelmann, *Z. Naturforsch., Teil C* **28,** 356 (1973).

same treatment brought about the same responses in two different temporal displays, it suggests that the same clock must underly both of them.

One last example, the results of which are indicative of a common clock mechanism, will be given. A long-night plant, *Chenopodium* (a member of the goosefoot family), was given 35% deuterium oxide, a substance known to lengthen the periods of biological rhythms (Figures 2-20; 2-22). A flowering inhibition curve was derived for the plant while it was under the influence of heavy water and was compared to the normal response. Figure 6-8 shows that the sensitivity rhythm was significantly lengthened, just what would be expected from a standard clock-driven rhythm.

ANIMALS

Seasonality in animals is governed by annual clocks and/or photoperiodism. Most of the information on the clocks underlying the photoperiodic response comes from studies on the control of diapause and polymorphism in insects and of testes recrudescence and migratory preparations in birds.

Insects

THE CLOCK OF THE PARASITIC WASP *Nasonia*

Some insects, in their development from the egg to adult, go through a stage termed diapause, which is an interval of complete dormancy. The parasitic wasp *Nasonia* is one of these insects, and the timing of its entrance into diapause is under photoperiodic control. When females are exposed to short nights, they produce larvae that complete their development to adults without a delay; but if the females are kept under long nights, they soon develop larvae that include a diapause interruption in their development.

To determine whether the bioclock is involved in the response, the now classic technique of giving ultralong light-dark cycles and scanning the long nights with short light interruptions was used. Two such experiments were run, one in which the cycles consisted of 12 hours of light alternating with 36 hours of darkness and the other of 12 hours of light and 60 hours of darkness. In separate experiments, the dark phase was systematically interrupted by 2-hour light pulses. Figure 6-9 shows that the peaks of greatest dispause inhibition in both light-dark cycles came at 24-hour intervals. The light-sensitivity rhythm described previously for plants is apparent here also and functions in the same way it does in long-night plants.

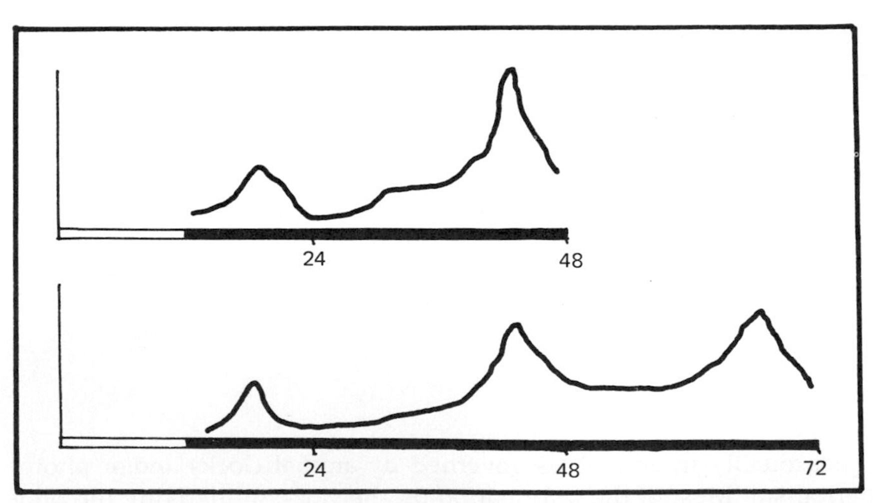

Figure 6-9 The diapause-inhibition rhythm in the parasitic wasp, *Nasonia*—a long-night insect. In separate experiments, the ultralong dark periods were interrupted by 2-hour light breaks which caused the relative inhibition of diapause in the larva. From the data of D. S. Saunders, *Science* **168,** 601–603 (1970).

THE CLOCK OF THE APHID *Megoura*

The previous example is one of the best in the literature, but is not representative of all insects. In other groups, the evidence points to a clock that measures intervals in a way analogous to an hourglass. The vetch aphid *Megoura* falls into this category. This polymorphic insect produces either virginoparae (parthogenic) or oviparae (sexual) offspring, with the final selection determined by the ambient photoperiod. When kept under very long nights, such as 8 hours of light alternating with 64 hours of darkness, no virginoparae are formed. This trend could be reversed in the usual way by probing for the sensitive spots during the hours of darkness with flashes of light. However, only one responsive site could be located, and it occurred at a time 8 hours after the onset of darkness (Figure 6-10). If the typical light-sensitivity rhythm had been present, similar peaks at hours 32 and 56 (after dark onset) should have also been located. Therefore, the clock envisioned at work here is something akin to an hourglass that starts running by the transition of light to darkness and runs its course

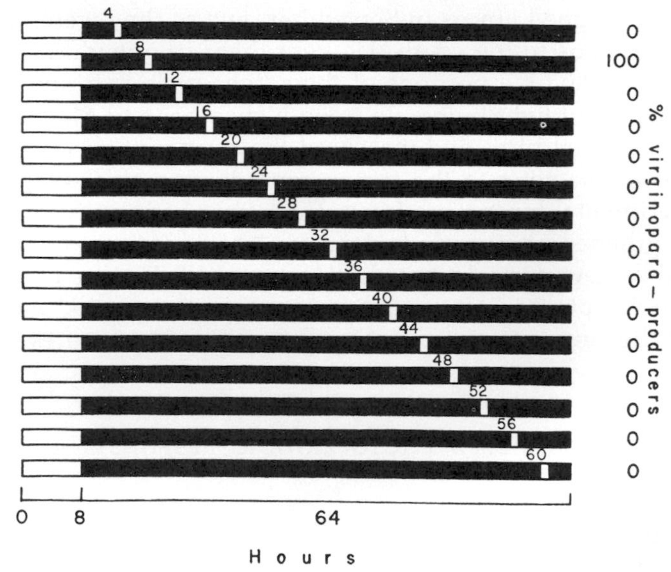

Figure 6-10 Night-interruption experiment with the vetch aphid *Megoura*. The times that 1-hour light interruptions were given are signified by the diagonal chain of numbers and open spaces across the figure. At only one interruption (8 hours after light-off) were virginoparae produced. See text for interpretation. From A. D. Lees, *in* "Circadian Rhythmicity," pp. 87–110. Centre for Agricultural Publication and Documentation, Wageningen, Netherlands, 1972.

by hour 8, at which point a light-sensitivity moment is produced. The hourglass is not inverted again until the next light-to-dark transition.

Birds

Each year, a great many birds fly north in the spring to a favored breeding ground where they set up territories, mate, and raise a brood or two of young. This completed, their gonads regress to an immature state; they often molt their plumage; accumulate a thick layer of fat under their skins; and fly back south to their wintering grounds. Their fat deposits which are depleted by the long flight southward are eventually replaced and, in spring, they repeat their trip back north. Thanks to the pioneering work of William Rowan in 1925 (and of those investigators following him), it is now known that changing photoperiods can act as the stimulus for the initiation of many of these events.

Rowan was familiar with Garner and Allard's work on plants, and he designed his first experiments after theirs—in spite of the potentially dire consequences for many of his research subjects. Working in Canada, he trapped juncos in September, just as they were beginning their emigrations to the south. He kept them in outside aviaries and extended the daytime portions of their light cycle to a spring length with two 50-watt light bulbs. Between November and March, he periodically sacrificed birds and measured the sizes of their testes: all had increased in diameter to the spring reproductive size. When control and experimental birds were freed, those that had received the artificially lengthened photoperiods departed (thinking it was spring, they presumably flew north to breed). The controls all stayed in the area and were recaptured. All of this springlike activity took place in the midst of the severest winter in 12 years, with temperatures dropping to −35°F.

THE CLOCK-CONTROLLED PHOTOPERIODIC RESPONSE OF THE HOUSE FINCH

For a great many birds the stimulus to migrate and for gonads to mature is the change in relative day length. And, in some birds, it is known that their bioclock is what is used to measure the relative lengths of day and night. A clear-cut example comes from studies of the very common, house finch *Carpodacus,* a bird in which short nights are known to cause the testes to increase in size and mature

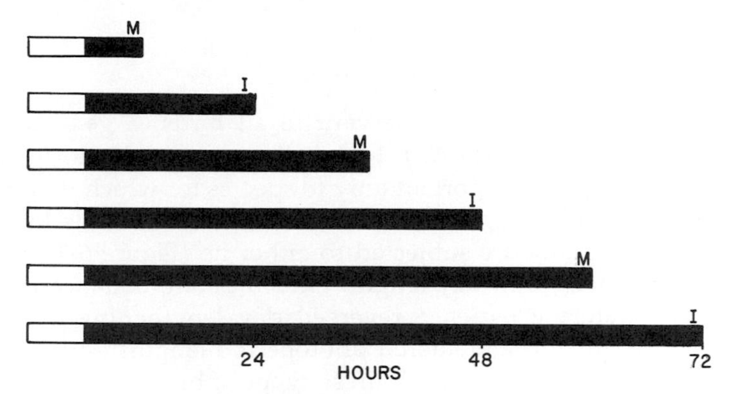

Figure 6-11 The influence of extended light-dark cycles on the reproductive maturation of the house finch, *Carpodacus*. M stands for maturation of the testes; I, no development of the testes. From the data of W. Hamner, *Science* **142,** 1294–1295 (1963).

into sperm-producing organs. To reveal the action of the clock, caged finches were placed in "photocyclers," which are glorified steel filing cabinets possessing motors that automatically run the drawers in and out. The drawers were opened—thus exposing the birds to light—for 6 hours, and then, depending on the experiment, closed for 12, 24, 36, 48, 60, or 72 hours—which provided 6 different light-dark cycles. After about a month of these cycles, the left testis of each bird was removed for measurement and observation and, in most cases, the birds were then released.

The results of the experiment are portrayed in Figure 6-11. In light-dark cycles totaling 24, 48 or 72 hours, the testes did not mature. In cycles totaling 12, 36, or 60 hours, they did. The conclusion is twofold. First, the length of the photoperiod is not important, since all the birds received the same 6-hour periods and some did, while others did not, mature. Second, the data indicated the presence of an underlying rhythm in sensitivity to light. The interpretation is as follows: the 6-hour light intervals that began again after each 12, 36, or 60 hour dark periods must have illuminated an inductive portion of a light-sensitivity rhythm. Those that fell after 24, 48, 72 hour stints did not. To ensure that this interpretation was correct, cycles of 6 hours of light alternating with long intervals of darkness (some up to 66 hours), with short light interruptions offered just after hours 12, 24, 36, and 48, caused testis recrudescence only when given just after hours 12 and 36. The sensitivity rhythm is surely there.

THE ANNUAL CLOCK OF EQUATORIAL AND CIRCUMPOLAR DWELLERS

Photoperiodism cannot be a servant to all birds, a generalization stated succinctly by the late A. J. Marshall—a rare and colorful scientist—as, "day length is important only to species for which it is important that it should be important." For example, those birds living on or close to the equator are subjected to either no (Figure 6-1), or very small, annual changes in day length. And the long-distance and circumpolar migrators experience reversed day lengthening each time they cross the equator, and altered photoperiod lengths with each degree of latitude crossed. Yet, both these types of birds often reproduce on a seasonal basis. Where adequate studies have been made, this periodicity is often found to be controlled by an annual clock (Figure 6-12). The Slender-billed Shearwater (*Puffinus*) provides the most remarkable example. After breeding near Australia, it flies up the Asiatic coast, crosses over the Aleutian chain of isles to North America, travels down the California coast, and crosses the Pacific back to Australia. Eighty-five percent of the millions of returnees lay their single egg during a 3-day interval centered around November 25. Individual birds, identified by banding, have been found to lay their eggs on November 24 for 4 successive years. As if this display is not sufficiently astounding, birds held in captivity for a year, in an artificial environ-

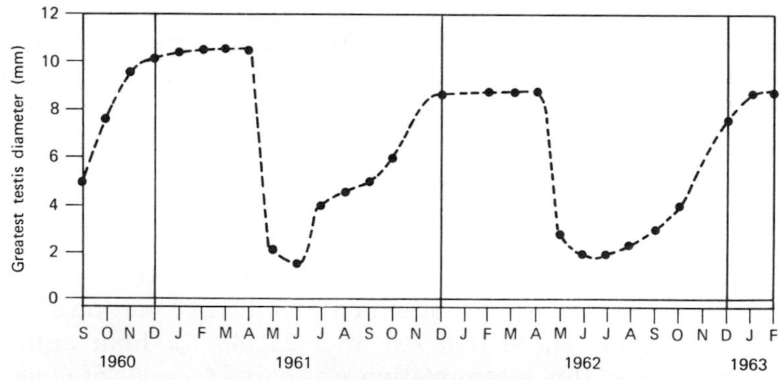

Figure 6-12 The annual rhythm in the average testis size of 10 equatorial finches (*Quelea*) during a 2.5-year study in the laboratory. During the entire time, the birds were maintained under cycles of 12 hours of light alternating with 12 hours of darkness, and a constant temperature. At monthly intervals, the birds were surgically opened, the left testis measured, and the incision closed again. A clear cut annual rhythm is apparent. From B. Lofts, *Nature* (*London*) **201,** 523–524 (1964).

ment with a variety of lighting schedules, all became sexually mature in synchrony with the wild population.

Mammals

Reproduction in mammals can also be adjusted to season by photoperiodism. The goat and ram are known to be "short-day" animals that breed in the fall, while the snowshoe rabbit, vole, and ferret breed during the lengthening days of late spring. The golden hamster is another well-studied member of the long-day variety and will provide us with a final example with which to close the chapter.

When exposed to normal day-night cycles, the testes of this common children's pet will enlarge and begin producing sperm if the animal is exposed to photoperiods longer than 12.5 hours. However, 6-hour photoperiods can also be effective, if they are coupled with unnaturally long dark periods. For example, hamsters that have been maintained in light-dark cycles of 14 hours of light and 10 hours of darkness (and therefore possess enlarged sperm-producing testes) were subjected, in the same, now tedious, way to experimental light-dark cycles of 6 hours of light alternating with dark periods of 18 hours, 30 hours, etc. In the cycles that totaled 36 or 60 hours in length, the testes were maintained in the mature state. In those that totaled 24 or 48 hours, the testes shriveled to immaturity. The interpretation of these results is identical to that for the house finch. The exotic light-dark cycle manipulations revealed the presence of a 24-hour underlying rhythm in sensitivity to light. The living clock again makes its presence known.

SUMMARY AND CONCLUSIONS

1. Many seasonal events in plants and animals are timed by changing day lengths.
2. The bioclock drives a rhythm in sensitivity to light which is characterized by peaks of light sensitivity alternating with insensitive phases. In plants, if any part of a sensitive peak of a long-night plant is illuminated (as happens during the short nights of summer) flowering is inhibited. If the same peak of a short-night plant is illuminated (as it is in the spring or fall), it brings about flowering.
3. The light-sensitivity rhythm can be rephased by light-dark cycles, and its period is lengthened by deuterium oxide. Evidence like

this is used to support the hypothesis that the same clock that governs all other physiological rhythms is used in photoperiodism also.

4. Not all photoperiodic responses in insects are governed by an escapement-type clock. A clock similar in action to an hourglass is used by some.

5. Seasonality in birds can be produced by a bioclock measurement of night lengths and by annual bioclocks. Other mechanisms are also used by birds, but these are outside the subject matter of this text.

Selected Readings

Beck, S. D. (1963). "Animal Photoperiodism." Holt, New York.

Bünning, E. (1936). Die endogene Tagesrhythmik als Grundlage der photoperiodischen Reaktion. *Ber. Dtsch. Bot. Ges.* **54,** 590–607.

Bünning, E. (1973). "The Physiological Clock," 3rd ed. Springer-Verlag, Berlin and New York.

Cumming, B. G. (1971). The role of circadian rhythmicity in photoperiodic induction in plants. *In* "Circadian Rhythmicity," pp. 33–85. Centre for Agricultural Publishing and Documentation, Wageningen, Netherlands.

Garner, W. W., and Allard, H. A. (1920). Effect of the relative length of day and night and other factors of the environment on growth and reproduction in plants. *J. Agric. Res.* **18,** 553–607.

Lees, A. D. (1971). The role of circadian rhythmicity in photoperiodic induction in animals. *In* "Circadian Rhythmicity," pp. 87–110. Centre for Agricultural Publishing and Documentation, Wageningen, Netherlands.

Menaker, M., ed. (1971). "Biochronometry." Nat. Acad. Sci., Washington, D.C.

Palmer, J. D. (1971). The rhythm of the flowers. *Nat. Hist.* **80,** 64–73.

Pengelley, E. T., ed. (1975). "Circannual Clocks." Academic Press, New York.

Salisbury, F. B. (1963). "The Flowering Process." Macmillan, New York.

Saunders, D. S. (1973). Circadian rhythms and photoperiodism in insects. *In* "The Physiology of Insecta" (M. Rockstein, ed.), 2nd ed., Vol. 1, pp. 461–533. Academic Press, New York.

Sweeney, B. M. (1969). "Rhythmic Phenomena in Plants." Academic Press, New York.

Withrow, R. B., ed. (1959). "Photoperiodism and Related Phenomena in Plants and Animals," Publ. No. 55. Am. Assoc. Adv. Sci., Washington, D.C.

7

Evidence for External Timing of Biological Clocks

FRANK A. BROWN, JR.

COMPARISON OF THE INTERNAL AND EXTERNAL HYPOTHESES

Are biological clocks timed by a fully autonomous, internal, physicochemical oscillator system? Or are the clocks dependent upon physical fluctuations in the organism's environment to which the organisms are responsive? These are two opposing hypotheses, and it is not yet known which is correct. Supporters of the internal timing hypothesis have sought long and hard, but have come up with little more than guesses concerning mechanisms that might possess the well-known clock properties. Supporters of the external timing hypotheses, while finding organisms fantastically sensitive to a variety of parameters of the terrestrial electromagnetic fields, have been unable to demonstrate that periodisms in any parameters of these fields are essential to the running of the clocks. In short, the clocks run steadily, but it is not yet known upon what their accuracy, temperature compensation, or independence and drug compensation or independence depends.

The internal-timer advocate identifies the observed periods of the numerous biological rhythms with the clock's periods and finds that the periods in "constant conditions" very commonly are slightly different than their natural geophysical rhythmic correlates. This together with the fact that different individuals each may simultaneously have a different period led to the hypothesis that the timer was independent of all environmental rhythms. Indeed, it seemed individual and private. And furthermore, these deviating periods could be altered very slightly, usually by differing constant temperature levels, different constant illumination levels, and a few substances such as heavy water. The argument seemed invincible for a good number of years. No one thought to question the basic assumption that the observed rhythms had to reflect by their periods those of the clocks.

The supporters of the external timing hypothesis make initially clear their own view that the period of the observed rhythms and the period of the clocks that time them may differ from one another. They differentiate more sharply between clocks and clock-timed rhythms. They point out that if the organisms can sense fluctuations in the natural geophysical environment by any means whatsoever, then the organism has available a clock system with these natural periods, completely temperature and drug independent. The rhythms that so commonly deviate from the clock periods in constancy of those factors that normally determine the phases of the observed rhythms they believe to result from a steady phase resetting of the clock-timed rhythms, either to earlier or later times. They postulate that the observed very small temperature and drug effects and the effect of differing levels of illumination result from influences of these factors on the amounts of periodic phase shifts and, hence, on the observed periods of the rhythms.

As commonly occurs with all controversies over questions for which either there are no answers possible, or no method currently available to resolve the questions, the controversy over the nature of the clock has been one laden with emotion. Investigators cling to their preferred hypothesis and interpret their data in terms of it, while usually ignoring or belittling work purporting to support the potentially equally likely alternative. So far, all the data that have been presented on the "free-running" clock-timed rhythms are capable of being accounted for by either hypothesis alone, or by some combination of the two.

In this chapter, a case will be made for the external timing of the biological clocks by presenting the history and major findings that have been offered in support of the hypothesis. An attempt will be made to disclose many things that appear to demand the operation of an external influence operating in "constant conditions" upon the rhythmic systems of organisms. Attempts will be described to discover the nature of effective subtle environmental factors, and parameters of them, that can and do influence organisms. Roles they can play will be reported as well as something of their periodic characteristics. Finally, it will be shown how, theoretically, in terms of properties of the rhythms agreed to by about everyone, all the periods deviating slightly from the natural physical ones, the circadian, circatidal, circamonthly, and circannual periodisms that animals, plants, and microorganisms show, can be timed by clocks that are accurately geophysically timed. This would solve, at once, some of the toughest problems concerning the clock's properties. It will be shown that the very small observed effects of genotypes of some chemicals and of light and temperature levels in constant conditions can be easily explained. Indeed, by hindsight, they should have been predicted.

In short, the author will try to present a case for an environmental dependence of the biological clocks so tight and so rational that it should be extremely persuasive. It will be pointed out how each and every known property of the clocks would be nearer their ultimate solution with the adoption of such environmental timing as the working hypothesis than would be the case for the opposing or autonomous internal oscillator concept. An attempt will be made to resolve the nub of the controversy, of whether the living clocks do or do not depend steadily upon the environmental rhythms in favor of the view that they do.

Others have, for longer or shorter times, espoused the hypothesis that the clocks are externally timed. However, the modern view of this hypothesis was born in, and has been supported by experimental evidence and systematically developed, mainly in our own laboratories. It has now become sufficiently well established that it can be supported as, at least, an equally likely candidate for the actual mechanism ultimately to be disclosed.

What initially brought me into the field was the then ongoing studies during the period 1930–1934 of crustaceans by J. H. Welsh at Harvard and of insects by O. Park at Northwestern. These investigators were clearly giving reason to postulate that the mechanism that was used in the timing process was essentially, if not fully, independent of temperature. As a graduate student, the author had been thoroughly impressed by W. J. Crozier with the relatively large rate changes produced by the kinetic effects of temperature upon rates of all biological reactions. Yet, the hypothesis of virtually all workers on circadian rhythms, following its adoption by W. Pfeffer in 1915, was that the rhythms were timed by independent internal timers. The evidence, therefore, then coming in made it appear that with this timer was the first case of a Q_{10} near 1.00 over a substantial temperature range—itself an incredible phenomenon. This seemed to be a really challenging problem, an intrinsic time-measuring device that was independent of temperature. And yet, to be a suitable timer for life, this property was essential in the natural temperature-varying environment of the organisms.

RHYTHMS IN THE FIDDLER CRAB

It was nearly 15 years later, in 1948, that the author first encountered the problem of influences of the "clocks" on the regulation of the endocrine system in crustaceans and turned over his full time to studying the mystery. Though the nervous and endocrine systems were regulating ones for the individual, something superimposed was in turn regulating these. Our first studies were related to our endocrine ones that investigated the rhythms of endocrine-controlled color change in populations of fiddler crabs, *Uca* (Figs. 2-30; 7-1). Not only

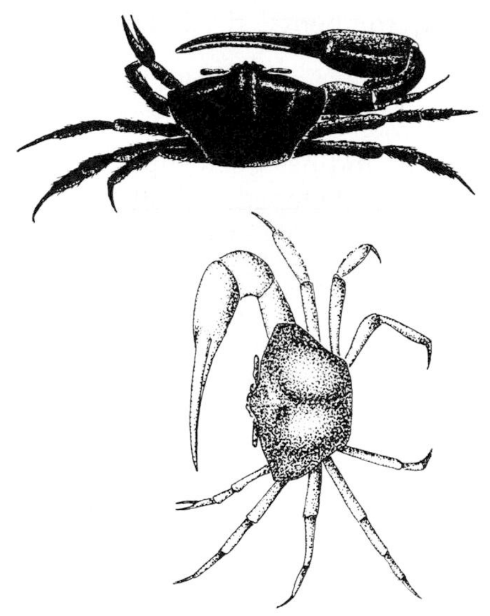

Figure 7-1 A drawing from a photograph of fiddler crabs, *Uca pugnax*, showing one crab in its dark, daytime phase and the other in the blanched, nighttime phase. By subjecting one of the two crabs to a reversed light cycle, the phase of the color-change rhythm was reversed, enabling crabs in both phases to be photographed together. From F. A. Brown, Jr., J. W. Hastings, and J. D. Palmer, "The Biological Clock: Two Views." Academic Press, New York, 1970.

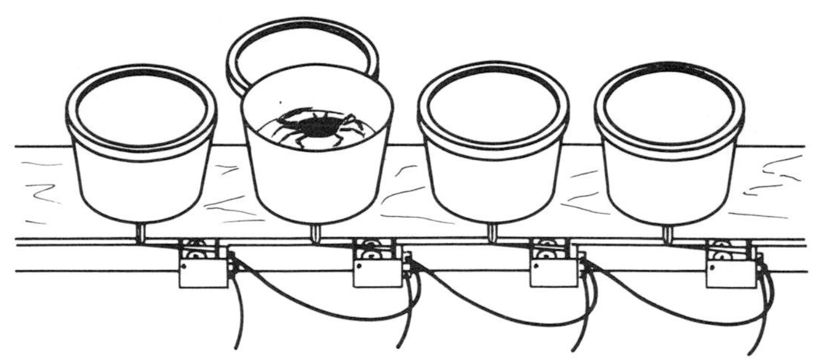

Figure 7-2 A simple actograph for fiddler crabs. Each light plastic container is supported by a single point-fulcrum centered on the bottom. As a crab circles within the dish, the latter rocks around, closing an attached microswitch which completes a circuit causing a pen deflection on an event recorder. In this manner, one obtains a complete record of the variations in the crab's running activity. From F. A. Brown, Jr., J. W. Hastings, and J. D. Palmer, "The Biological Clock: Two Views." Academic Press, New York, 1970.

were the 24-hour rhythms found to persist in constant conditions in the laboratory, but their mean period in darkness and whether at 6°, 16°, or 26°C was over periods of even weeks of darkness indistinguishable from 24 hours. And yet, the color change cycles could readily be reset by altering the times of light and darkness within the 24-hour cycles and the new settings would persist in darkness. H. M. Webb, then a graduate student in the author's laboratory, in 1950 published the first paper indicating a 24-hour variation in the ability of light to reset the phases of the 24-hour rhythms. Four years later, a 24-hour cycle of light response was discovered during a portion of which the color-change cycle could be delayed and another portion during which light advanced the cycle. Beautifully performed later studies by P. DeCoursey in 1960 at the University of Wisconsin were to demonstrate clearly the existence of the phase-response system, which is shown to be extremely important in the external timing hypothesis.

After viewing an apparent modulation of the 24-hour rhythms, it was quickly determined that a lunar tidal rhythm was simultaneously present in the constant darkness and was interacting with the 24-hour one of the crabs. A result of the interactions was a persistent semimonthly rhythm exhibiting an astonishing accuracy, the equivalent in the great precision of the two shorter periods upon which it appeared to depend (Figures 7-2, 7-3, and 7-4).

While the preceding studies were in progress, the working hypothesis was that the clocks were internally timed. The period preci-

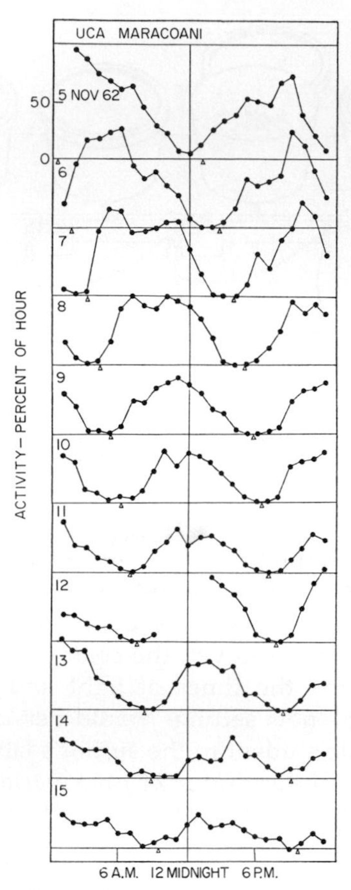

Figure 7-3 Average daily patterns of spontaneous motor activity in *Uca maracoani* from November 5 through November 15, 1962. Values are expressed as the percentage of the hour that the crabs were active. Triangles indicate predicted times of high tide on the beach where the crabs were collected. From F. H. Barnwell, *Biol Bull.* (*Woods Hole, Mass.*) **125,** 399–415 (1963).

sion and Q_{10} of 1.00 over a large temperature range that these studies disclosed was amazing. It was extremely difficult to discard an old hypothesis that had not been questioned for about 35 years. But by carefully viewing the steadily incoming data from the crabs and by working with other animals and plants, many newer findings which increasingly strained the plausibility of the concept of independent internal timing of the cycles had to be considered and, eventually, compelled us to abandon this earlier working hypothesis. There was soon to be no alternative to a conclusion that organisms in constant condi-

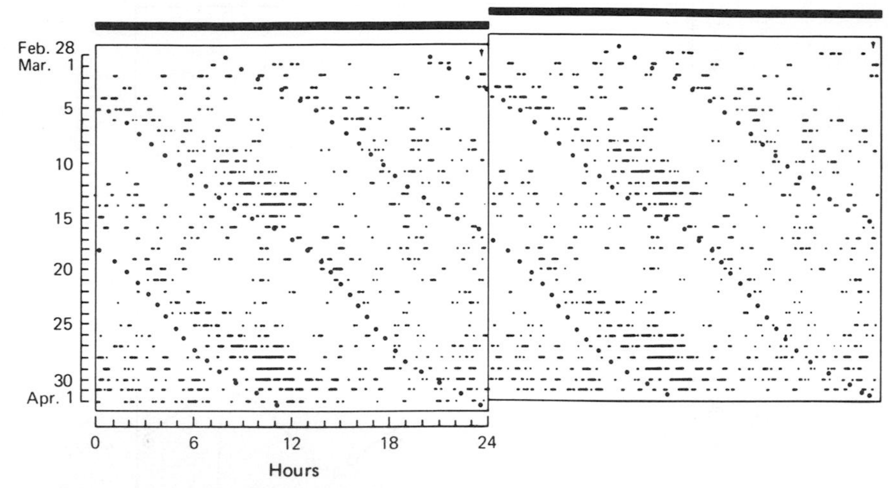

Figure 7-4 The spontaneous locomotor activity (black marks) of a Costa Rican fiddler crab (*Uca princeps*) during a month's sojourn in continuous darkness. The small circles mark the times of high tide on the crab's home beach. Note the two bursts of greater activity (hours) which occur at the same time of day with a semimonthly interval. The record is repeated and displaced upward one day to facilitate the viewing of the passage of the patterns of activity from one day to the next. Unpublished record by F. H. Barnwell, University of Minnesota, Minneapolis.

tions were actually steadily still responding to subtle, pervasive atmospheric parameters.

TRANSLOCATION EXPERIMENT WITH OYSTERS

Our first findings in this expansive new domain of environmental input to organisms came through our studies with oysters in continuous dim illumination. Fifteen oysters shipped in a light-proof container by air from the Milford, Connecticut, Fisheries Laboratory were placed in seawater in several large, glass vessels. Threads were fastened to stainless steel saddles attached to their valves. The threads activated pen writers on a moving strip of paper. After analyzing all the data for the first two weeks, two peaks in activity were noted in the lunar day, and these indicated maximum shell opening in the population of oysters at the times of the Connecticut high tides (Figure 7-5). When the data for the second two weeks were analyzed in exactly the same manner, the double tidal cycles were seen to have moved to a time about 3 hours later in the lunar day, a relationship which appeared to persist for the third and final fortnight of observation.

It was learned shortly that the peaks for the second two fortnights

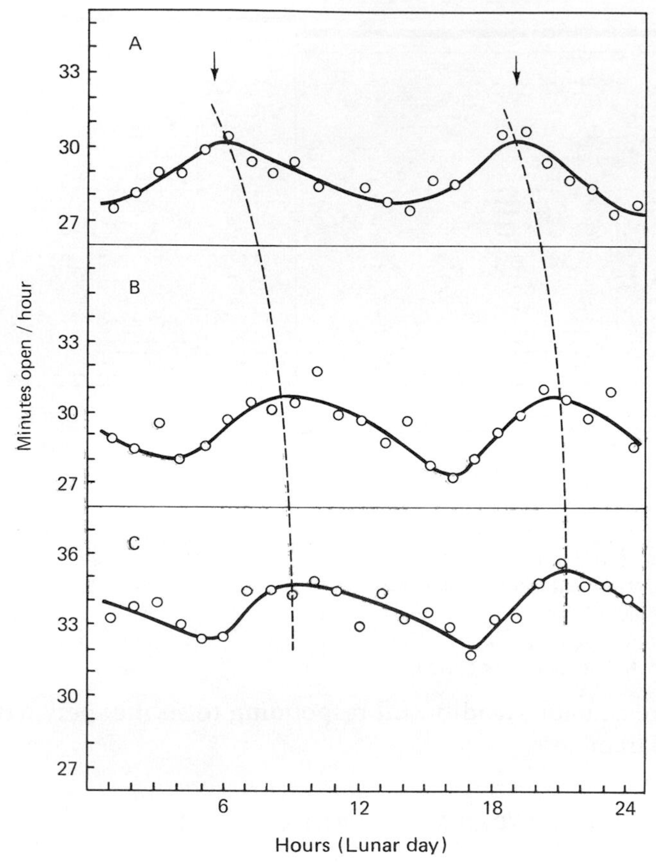

Figure 7-5 The average number of minutes of open shells of 15 oysters transported from Milford, Connecticut, to Evanston, Illinois, and maintained in constant conditions. The average number of minutes open for hours of the lunar day, the horizontal axis, is plotted. (A) For the first 2 weeks in Evanston (arrows indicate time of high tide in oysters' home water); (B) for the next 2 weeks; (C) for the last 2 weeks of the study. The time of maximum shell opening is seen gradually to shift to, and then stabilize at, the times of upper and lower transits of the moon. From F. A. Brown, Jr., *Am. J. Physiol.* **178,** 510–514 (1954).

had drifted to, and had become fixed at, the times of upper and lower transits of the moon. Could this be fortuitous? Or were perhaps the oysters substituting the moon-caused high tides of the atmosphere when away from the oceans for the stronger ocean tidal ones of their home seashore? This phenomenon was confirmed by M. F. Bennett in 1963 with crabs collected from two beaches with a 4-hour difference in tidal times. In the laboratory, in separate containers, the crabs synchronized their cycles to times of upper and lower lunar transits.

STATISTICAL ANALYSIS OF TIME SERIES

Some account should be given at this point to one of our methods for analyzing data. This is especially needed in view of the criticisms given our earlier results by L. Cole of Cornell University, J. Enright of Scripps Institute of Oceanography, and others. Our general method harked back to one first used by the great French mathematician and astronomer, P. LaPlace, who postulated early in the last century the existence of lunar tides of the atmosphere—tides now known to be present all over the earth. These lunar tides are of the order of 1/15 to 1/20 of the amplitudes of the solar-day tides whose range is augmented by a natural atmospheric vibrational period of 12 hours, together with the influences of the heating effects of the sun.

A technique the author employed over a number of years involved the filtering out of periods from others by superposition of periods of the desired mean frequencies. This requires large numbers of hours of continuously recorded data, long enough intervals of recording that the cycles of other major frequencies may be randomized out. On the assumption that the major biological periods conforming to atmospheric ones would involve the 24-hour solar day with its 12-hour tidal variation and the 24.8-hour lunar-day with its 12-hour and 25-minute mean tidal period, the shortest period of continuous recording to differentiate the approximate lunar tidal cycle, essentially bereft of the solar-day one, is a natural semimonth. Since only whole days are counted, the time arbitrarily chosen is 15 or 30 days.

Conversely, to extract a lunar-day cycle from a complex variation containing a solar-day, or a solar day from a complex containing a lunar-day pattern, a single 30-day period or some simple integral number of 30-day periods must be employed. In many of our studies, monitoring of organisms has been carried out even for a simple integral number of years. By using a complete year of hourly data, any annual modulation of the solar-day and lunar-day cycles is eliminated. Such an annual modulation can be disclosed by superimposing the solar-daily or lunar-daily cycles separately, month by month, over a simple integral number of years.

Such superposition of periods for filtering out others of known frequencies has been powerfully applied to geophysical data on geomagnetic fluctuations by S. Chapman and J. Bartel and has contributed much of the major current knowledge of these patterns of change. The problems of separation of lunar and solar effects in geophysical data have been recently refined further by P. A. Bernhardt in a 1974 issue of *Journal of Geophysical Research*.

At ground level in the turbulent troposphere of the temperate latitudes, methods such as the foregoing have extracted the information from an extremely variable mass of data, including a number of periods. It is now known that far greater regularity of these cycles exists for the ionosphere. The relatively regular lunar-tidal cycle (12.4-hour) of rise and fall of the ionosphere involves about a mile or so. More than 7 miles occur for the solar tidal cycles.

One should emphasize that the geophysical cycles are not of constant length, but are systematically variable. This is due to the noncircular orbits of the earth, its moon, the systematic rotation of the lunar orbital plane, and many other factors. The method employed avoids the pitfalls succumbed to by Enright and Cole. Enright's method was based on the false assumption that the form and period of a cycle—if rhythmic—must be present in even only 5 or 6 days of data and also employed a method of clearly inadequate sensitivity to permit his conclusions or to exclude simultaneous lower-amplitude cycles of geophysical frequencies also being present. Also, Cole's employment of a stochastic method not employed for biological data in which he imposed a statistically significant periodicity on random numbers is clearly inadequate as well. It is very surprising that these reports are still occasionally referred to despite common knowledge of these faults and others. If cyclic variations are actually environmentally imposed, then, as the systematic changes in day-to-day differences in ocean tidal frequencies indicate simply and clearly (see Figures 7-3 and 7-4), the periods are quite variable. Only hypothetical independent oscillators have regular periods of the mean lunar-day or solar-day.

In our studies, therefore, similarities and differences in the forms of the mean fluctuations between organisms and concurrent variations in other organisms and in physical environmental factors are the ones of importance. These, together with cross correlations, day by day, between a specific organismic and concurrent physical or other biological variations, have constituted the major methods. In concurrent day-to-day correlations, great care must always be exercised to avoid the well-known pitfall of the almost invariable correlation present between any two variations sharing a common period. This was accomplished through correlating only a single parameter for each daily cycle, such as cycle range, or deviation from daily mean at one point in the day. Significant positive or negative correlations with factors from which the organisms are shielded point to common superimposed periods of other frequencies or to parallel fluctuations due to responses to, or correlations with, some common effective factor for them both.

ON THE ABSENCE OF CONSTANT CONDITIONS

To return from our digression on statistical methods, the second discovery, at about the same time as the finding with the oysters and crabs, was that between the same month of 1954 and 1955 a large mean 24-hour cycle of oxygen consumption in crabs in recording respirometers had inverted, with a modest alteration between the cycles in their forms. This was hardly to be expected for crabs from the same beach and treated in exactly the same way in the laboratory. This puzzle was solved within the following 2 to 3 years when it was discovered that eight species of organisms, including the crabs, over the 20 months they were investigated, showed 24-hour and lunar-day (24.8-hour) variations which, in every case, either paralleled or mirror-imaged for parts or wholes of the cycles the contemporary fluctuations in primary cosmic radiation being determined by J. Simpson and his associates at the University of Chicago (Figure 7-6). The coincidences that had occurred by this time were becoming so numerous and striking that chance seemed to be ruled out.

The longest uninterrupted study was an hourly one over 11 years, for the potato sealed in automatically recording respirometers (Figure 2-7). Each year, a characteristic mean cycle was evident with sharp maxima at 7 A.M., 12 noon, and 6 P.M. (see Figure 2-8). This cycle form was confirmed by E. F. Lutsch working for 3 years in my laboratory with O_2 consumption of seedlings of beans taken from a single 100-lb bag (Figure 7-7). With both the potatoes and the beans, there was also a large annual variation in O_2 consumption (Figure 7-38). In both, there was also an annual modulation in the forms of the daily patterns. The same characteristic form of daily and annual variations were soon described for other species as well.

Any residual doubts left concerning the continuous inflow of pervasive information were abolished by long series of studies of a number of species, their spontaneous activity, or their O_2 consumption while sealed in constant conditions including all *pressure* and *temperature* changes. The mean rates of O_2 consumption over the 5–7 A.M. and 5–7 P.M. hours of the day for potatoes and crabs correlated with very high statistical significances, separately, with the barometric pressure changes occurring between 2 and 6 A.M. and P.M., respectively, on the same days (Figure 7-8). The day-by-day ranges in variation in the O_2 consumption—from midnight to noon—also correlated highly significantly with the day-to-day mean daily outdoor air temperatures. The organisms were responsive day after day, even to details about ambient meteorological changes that were reflected by the rising and fall-

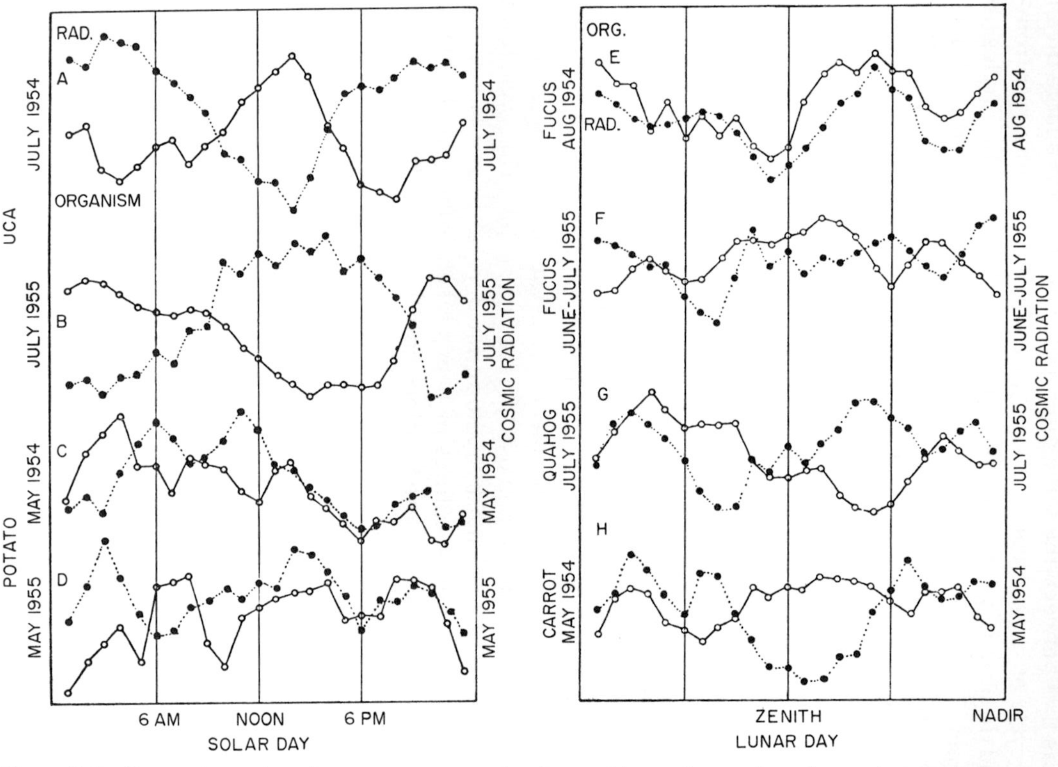

Figure 7-6 Comparison of contemporary mean solar-day and lunar-day cycles of organisms' activity and primary cosmic radiation. Crabs, seaweed, and quahogs, studied 1000 miles east of point of radiation and other metabolic studies, were adjusted for suntime difference to render them synchronous. From F. A. Brown, Jr., *Cold Spring Harbor Symp. Quant. Biol.* **25,** 57–71 (1960).

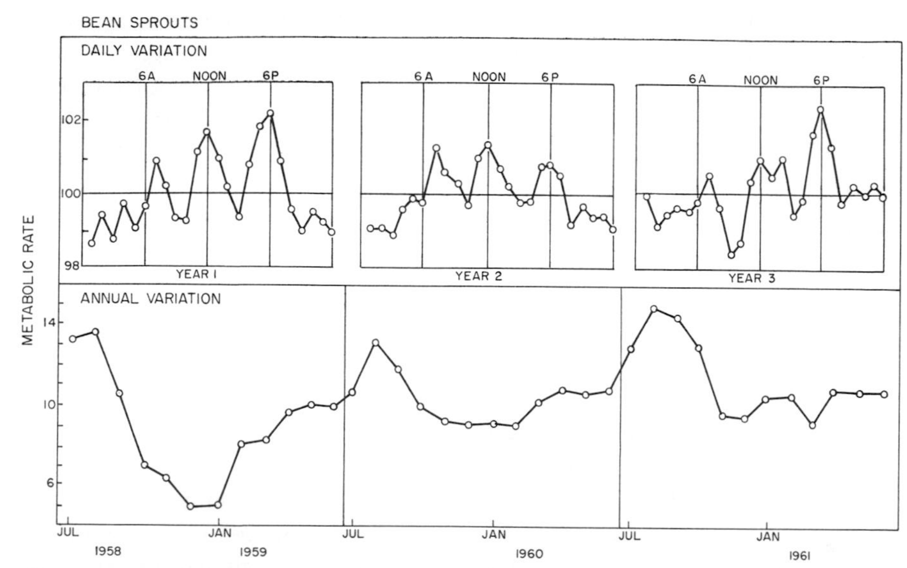

Figure 7-7 Mean daily and annual fluctuations in O_2 consumption in sprouting beans over a 3-year period of study. The daily variations are expressed as percentage variations from a straight-line daily upward trend. The annual variations are in arbitrary respirometer units. Drawn from data of E. F. Lutsch, Doctoral Dissertation, Northwestern University, 1962.

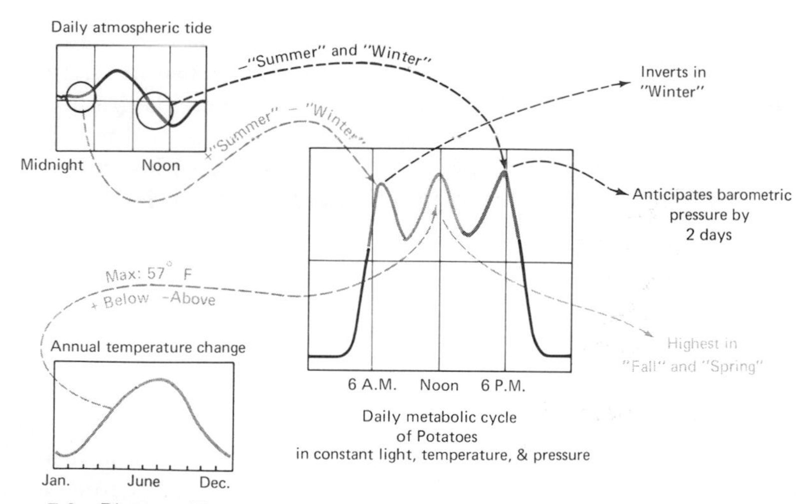

Figure 7-8 Diagram illustrating the specific times of day for the potato (*Solanum*) when specific metabolic deviations correlated with concurrent daily parameters of barometric pressure changes and of temperature levels. These continued over a 3-year study even though the potatoes were shielded from all direct influences by these two factors. From F. A. Brown, Jr., *Ann. N.Y. Acad. Sci.* **98,** 775–787 (1962).

ing of barometric pressure and temperature, from which they were fully shielded. Since all these organisms could not be expected to have inherited a chart of the weather fluctuations to occur during their lifetimes, they were clearly deriving a precise mean characteristic oscillatory pattern, day by day, from their external environment.

RESPONSIVENESS TO VERY WEAK MAGNETIC FIELDS

With the assurance that information was steadily reaching the organisms from the environment, the author commenced in 1959 an investigation of what the effective environmental parameters could be. Geomagnetism was suspected to be an important factor for two major reasons. First, the respirometers had involved deep copper drums within which the organisms were placed, drums which could be expected to attenuate to some degree all electric fields and electromagnetic ones up to the frequency of x-rays. Magnetic fields, on the other hand, would freely pervade them. Second, the most plausible explanation of the striking correlations with the fluctuations in primary cosmic rays, many miles up in the atmosphere, was that the organisms were responding to a major factor controlling the intensity of the rays entering the earth's outer atmosphere. This last could be clearly geomagnetism.

The first studies, with mudsnails, dealt with their responses to altered directions of weak experimental horizontal magnetic fields. Not only were these animals responsive to fields very close in strength to the earth's field, but their responsiveness to these fields gradually altered with time of day, and with phase of moon, or to the relationships between times of day and times of tides. These last were quite what one might expect if the magnetic fields were, indeed, effecting daily and tidal rhythms of the organisms.

Further studies, with the mudsnails and particularly with planarian worms, disclosed that a monthly rhythm could be abruptly abolished by suitable application of an experimental magnetic field, that organisms were maximally able to differentiate direction of the horizontal magnetic vector at the strength of earth's field to which they appeared to be adjusted, and that they could readjust in about 30 minutes when strengths deviated by no more than twentyfold or one-fifth the strengths of the earth's. The worms and snails appeared to be extremely well adjusted to their ambient magnetic fields. For the planarians, the sign of a response to magnetic fields reversed between 5 and 10 gauss. Responsiveness to magnetic fields appeared to be, in short, natural rather than adventitious.

More recent studies by others, including M. Lindauer and H. Martin at Munich, W. T. Keeton of Cornell University, W. Wiltschko and R. Wiltschko at Frankfurt, and A. Stutz at San Diego, have indicated that organisms as diverse as honeybees, flies, European robins, homing pigeons, and gerbils are influenced by magnetic fields no stronger than geomagnetism. The honeybees and gerbils, evidence suggests, are able to perceive field changes of the order of 10^{-4} to 10^{-5} times the strength of the earth's field. The European robin appears clearly able to identify its migratory direction by means of magnetic fields in conjunction with the gravitational vector. Homing of pigeons can be very highly significantly impeded by the attachment to them of small bar magnets, an alteration not effected on control birds carrying brass bars of the same weight and shape. Pigeons have been shown by C. Walcott and R. P. Green of the State University of New York to be reversed in their flight direction following abruptly reversed direction of their ambient magnetic fields produced by minute Helmholtz coils about their heads. No longer can any doubt remain that very weak magnetic fields not only can be perceived by organisms, but can also influence their behavior.

Recent studies with bean seeds and planarians have suggested that these widely different kinds of organisms can both distinguish between a weak magnetic field slowly rotating clockwise and one rotating counterclockwise. For the beans, a clockwise magnetic field seems to interfere with their responses to unidentified subtle atmospheric parameters that influence the day-to-day fluctuations in their rates of absorption of water in constant conditions of all obvious factors. Counterclockwise magnetic fields do not. For the worms, these appear to affect differently the strength of their negative response to light. Could this be for organisms an adaptation to the spinning earth in which the changes in numerous factors of the environment, both subtle and obvious, are unidirectional?

A series of experiments performed a few years ago by Y. H. Park and the author on planarians led to results that gave strong inferential support to geomagnetism being involved in the clock mechanism. This series was planned and executed with the full appreciation that there was an ambiguity in the timing phenomenon that could not be resolved at the present time. One could not distinguish between the two alternatives: (1) the rhythms were timed by an autonomous endogenous oscillator possessing the clock periods, and (2) the organisms were able to read time through sensitivity to changes within the cycles in the external environment. The first could be shown true only by exclusion of the second, and vice versa. And yet, it had not been possible to date to exclude either one.

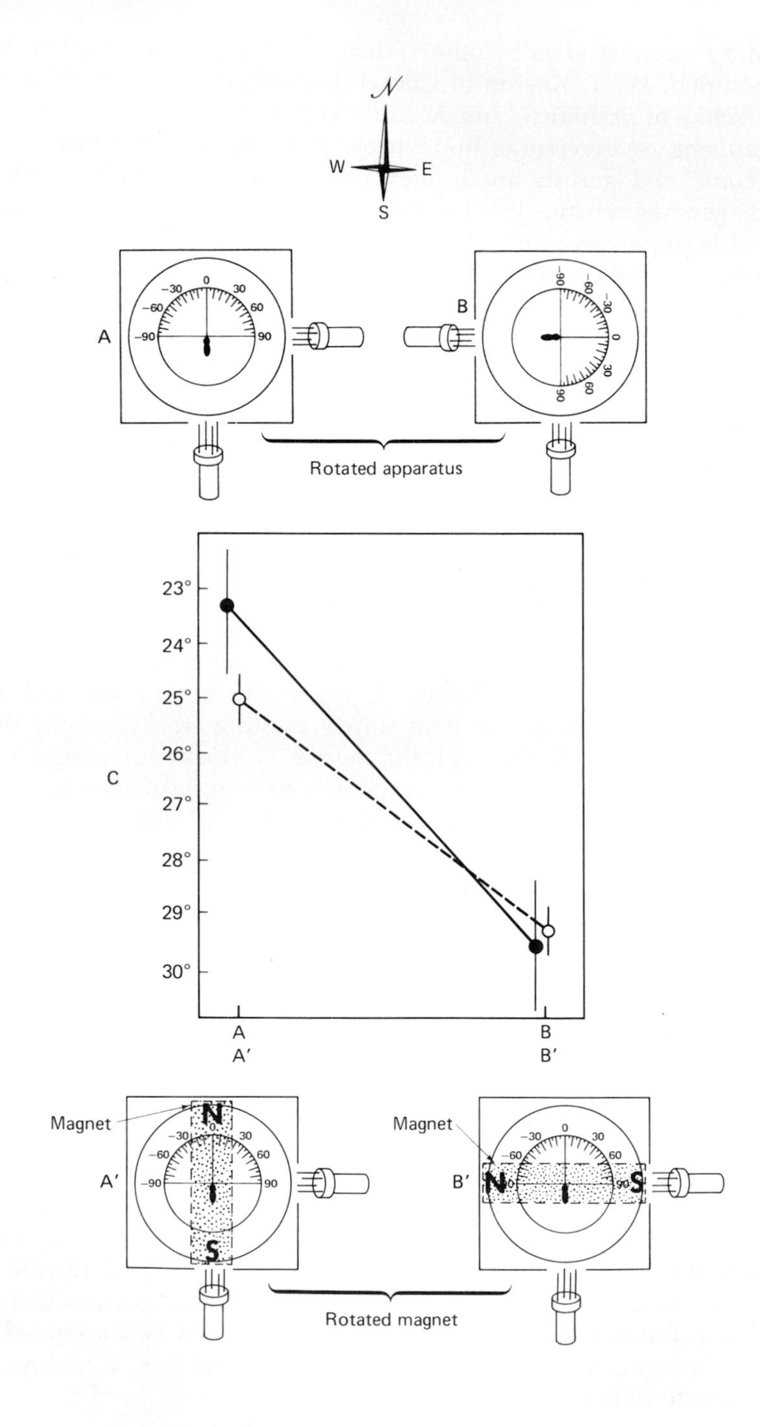
N
W
E
S
A
B
-90
-60
-30
0
30
60
90
Rotated apparatus
23°
24°
25°
26°
27°
28°
29°
30°
C
A
A′
B
B′
Magnet
A′
N
S
Magnet
B′
N
S
Rotated magnet

The experiments by Park and the author were designed to transpose the "clock phenomenon" from its insoluble temporal coordinate to a compass one. Then "clock-type questions" could be asked, but on a spatial basis, and definitive answers could be anticipated. The worms were assayed for the strength of their left turning away from a light source initially 90° to their right and a second light directly behind them. The effects on this response of alterations in orientation to the geomagnetic field as well as to experimentally altered magnetic fields produced by a straight bar magnet beneath the place of assay were determined. The strength of the light response was determined as the amount of left-turning at the end of a 1-inch course in the lights.

When the entire apparatus was directed north, a mean turning of about 25° was obtained. On the other hand, when the apparatus was directed toward the east, the turning was about 29°. The standard errors, illustrated in Figure 7-9, point to the clear statistical significance of the differences, which can only be interpreted as follows: the animal was able to distinguish between the north and east headings. Since the total ambient environment of obvious factors for the organism was identical for the north and east orientations of the apparatus, the animals must be depending upon subtle information.

Next question: How do the worms distinguish the different directions? In an attempt to answer this question, the following steps were taken. The apparatus was oriented toward the north, but this time the field of a bar magnet was imposed on the experimental conditions. This magnetic field was applied with only a horizontal vector which was oriented in the same direction as the earth's field, but augmented to 20-fold. The mean path of the worms running in this condition was found to be 23.5°. The larger standard error indicated that the behavior of the worms showed greater variability in this unnaturally strong field, but the mean path taken by the worms was not significantly different from the mean path obtained for north-directed worms in only

---

Figure 7-9 Experiments demonstrating that flatworms can use magnetic direction to orient themselves geographically. A and B show the apparatus oriented north and east, respectively, in the earth's magnetic field. The averages of the worm paths obtained in these two orientations are indicated in C by the terminal points of the dashed diagonal line. A′ and B′ show the apparatus remaining oriented to the north in both cases, but with the earth's magnetic field augmented about twentyfold by placing a bar magnet beneath the apparatus. The magnet was oriented to parallel the earth's field in A′, but rotated 90° counterclockwise in B′. The average paths taken by the worms in these two conditions are indicated in C by the unbroken diagonal line. The size of the standard errors are portrayed by the vertical lines bisecting the points. Note that the worms act as if the whole apparatus had been rotated to the east in condition B′, indicating that they can use magnetic direction to orient in space. From F. A. Brown, Jr., J. W. Hastings, and J. D. Palmer, "The Biological Clock: Two Views." Academic Press, New York, 1970.

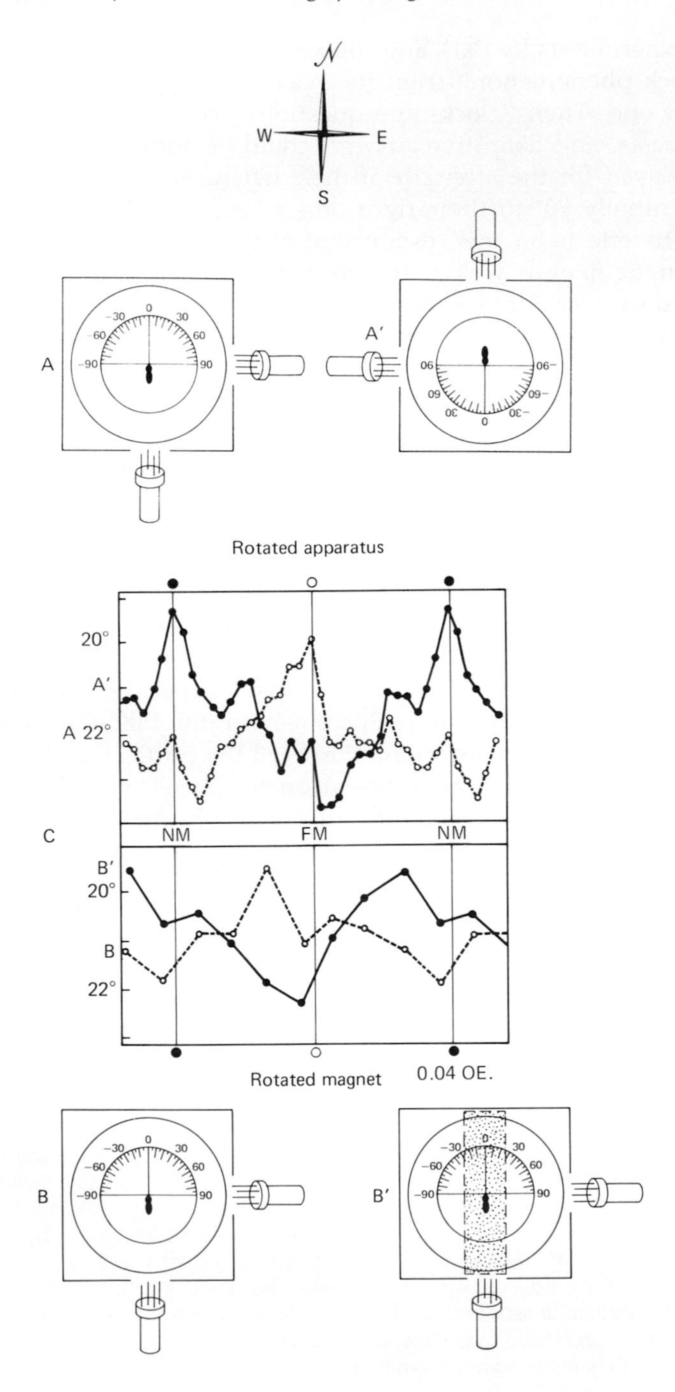

N
W
E
S
A
A′
-90
-60
-30
0
30
60
90
Rotated apparatus
20°
A′
A 22°
C
NM
FM
NM
B′
20°
B
22°
Rotated magnet
0.04 OE.
B
B′

the earth's field. When worm turning was measured with the apparatus still directed north, but with the same magnet rotated a quarter-turn counterclockwise to give the organism the magnetic illusion that its field was now directed to east, the mean path was altered to 29.5°. Rotating the magnetic field counterclockwise by 90° relative to the apparatus appeared to be essentially the equivalent of rotating the apparatus 90° clockwise in the earth's field. The organisms were able to distinguish north from east by using the horizontal vector of a magnetic field. In other words, a living organism possesses the capacity to employ a subtle geophysical field component to distinguish geographical directions; the organism thus possesses a good magnetic compass.

Another kind of experiment demonstrated that this compass capacity is related to the biological clock mechanism. Each morning, the animals were headed north (Figure 7-10) to establish the form and phase of the worms' monthly turning rhythm; as was expected the maximum turning to the left occurred at the time of new moon. Immediately, on completing the foregoing observations each day, the apparatus was rotated 180° and again average turning was measured (Figure 7-10). This treatment immediately reset the organism's monthly rhythm by one-half cycle, i.e., 180°. These results suggest that the monthly turning rhythm is somehow dependent on a subtle geophysical factor that is related to geographical direction. To determine whether this phase change could be brought about by the same subtle factor, geomagnetism, by which the organism distinguished direction, the following experiment was done. The apparatus was directed to the north to establish the form and phase of the monthly turning rhythm and then, at once, the observation was repeated after reversal of the magnetic field by an appropriately positioned bar magnet. In making these measurements, some of the observations were made in the early afternoon, though most were made in the morning. In this series, the times of maximum and minimum turning appeared slightly displaced relative to the purely morning study when the apparatus had been similarly

---

Figure 7-10 Altering by magnetic fields the phase of the monthly variation in turning of the flatworm, Dugesia. A and A′ are the geographical orientations of the experimental apparatus used to determine the monthly variations in turning rate shown in the top two curves of C. The monthly patterns are 180° out of phase with one another. B and B′ describe another experiment in which the geographical orientation of the apparatus remained the same, but with a magnet placed under B′ to reverse the direction of the magnetic field and to give the worms the magnetic illusion that the apparatus had been turned to the south. As seen in the lower two curves of C, the monthly pattern is again shifted in phase by 180° by the reversed magnetic field. In C, the solid circles and NM indicate, the day of new moon, and open circles and FM the day of full moon. From F. A. Brown, Jr., J. W. Hastings, and J. D. Palmer, "The Biological Clock: Two Views." Academic Press, New York, 1970.

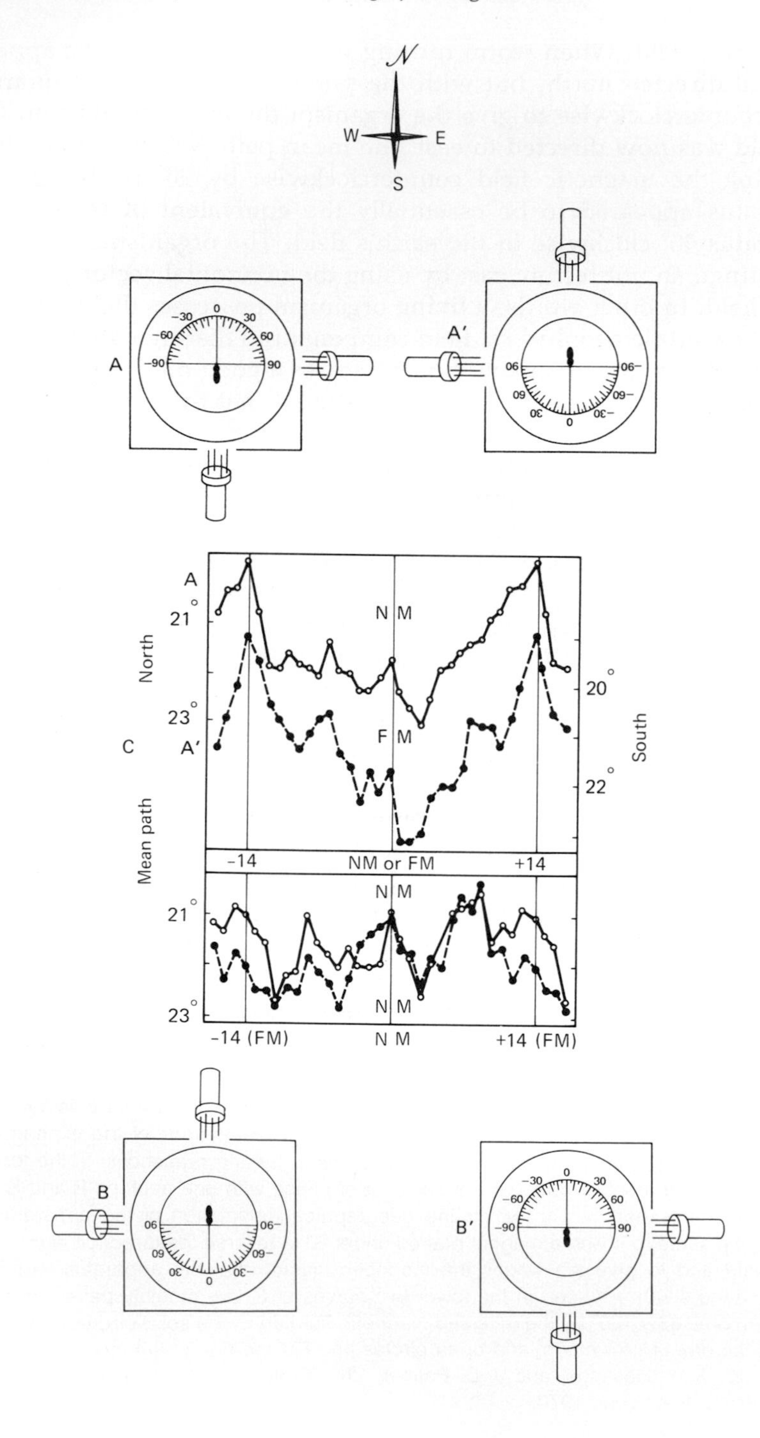
N
W
E
S
A
A′
21°
23°
North
C
Mean path
N M
F M
20°
22°
South
−14
NM or FM
+14
−14 (FM)
N M
+14 (FM)
B
B′

directed north. Of most importance here is that when a bar magnet reversed the magnetic field (with the strength of the ambient reversed field only a quarter the normal one) there was produced at once a 180° change in phase of the monthly rhythm. The 180° phase change in the monthly variation was effected by a 180° rotation of our lighting pattern relative to a magnetic vector, whether natural or artificial. A spatial vector component in the earth's subtle electromagnetic complex (magnetic) was clearly related to one biological clock-timed rhythmic behavior.

Finally, the experiments disclosed another capacity of a living system, an extraordinary one whose implications even transcend the phenomenon of "living clocks." The order in which the experiments were performed made a difference in the results obtained. This is depicted in Figure 7-11. One of the monthly curves in the upper pair of curves has been displaced by a semimonth to show that the two, though 180° shifted relative to one another, possess essentially the same form. New moon becomes the equivalent of the former full moon, and first quarter the equivalent of third quarter. Over the same period and on the same mornings, exactly the same experiments with other worms were done except that the order was reversed: first south and then north. Now the forms of the monthly cycles are quite different from those obtained under exactly the same conditions, other than for the reverse sequential order. This is evident from Figure 7-11. It is clear that the worms' monthly pattern of response to the asymmetrical illumination for each of the geographical orientations strongly tends to be retained even after that orientation of the worm has been altered. The "north" pattern persists even after the worms are redirected to south, and the "south" pattern of response persists even after the orientation has been redirected to north. The organism "remembers" the previous geographical direction of the light relations and continues to respond to them as if they still possessed the same geographical relations even after they no longer do so. The organism

---

Figure 7-11 A persisting effect of previous treatment on the form of the monthly variation in orientation of flatworms. A and A′ are the same as A and A′ in Figure 2-27 and the upper two monthly patterns in C are the same as in Figure 2-27 C, except that one has been displaced by a semimonth. Note that the forms of the two patterns are closely similar to one another. B and B′ indicate that the same orientations as A and A′ were used again, but with the order reversed: the worms were first directed south, followed at once by redirection to north. As seen in the lower two curves of C, the monthly pattern is very different from those of A and A′, but similar to one another. The worms appear to "remember" the initial geographical relations of the lights and to respond to the lights in their new relations just as if the apparatus had not been reoriented. From F. A. Brown, Jr., J. W. Hastings, and J. D. Palmer, "The Biological Clock: Two Views." Academic Press, New York, 1970.

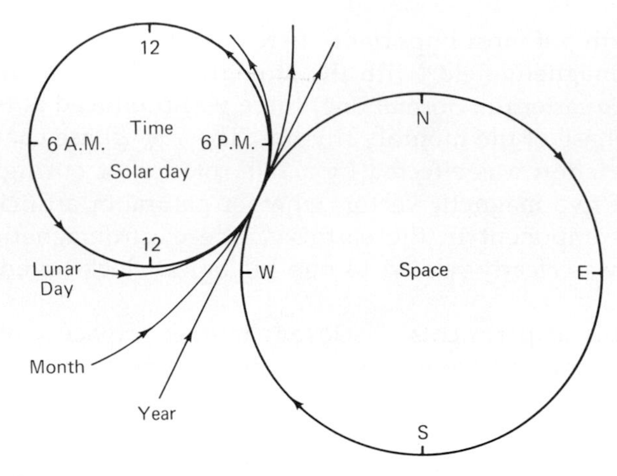

Figure 7-12 Subtle geophysical parameters vary continuously in space and time, forming a space-time continuum. At any instant in time there is a variation in space, here represented by the 360° geographical cycle. Each point in space is varying with time including solar-day, lunar-day, monthly, and annual periodic components. The demonstrations (1) that an organism can distinguish geographical directions by subtle geophysical parameters, (2) that involved parameters are related to the organisms' timing of their monthly rhythms, and (3) that organisms can "remember" directions of light stimuli in their relation to subtle parameters providing directional information suggest that an organism possesses just the appropriate capacities to "remember" light events in relation to the temporal cycles occurring in those same subtle parameters. From F. A. Brown, Jr., J. W. Hastings, and J. D. Palmer, "The Biological Clock: Two Views." Academic Press, New York, 1970.

possesses the capacity to associate, and retain an association, between such an overt stimulus as light and subtle pervasive factors providing geographical directional information.

The subtle geophysical parameters vary continuously in both space and time. The same three-dimensional geophysical parameters that vary with time of day, phase of moon, and time of year also vary continuously with orientation in space at any given instant. Theoretically, the same parameters that can provide geographical compass information can also provide temporal information within the coordinates of the natural geophysical cycles. In Figure 7-12, space is represented as a 360° compass circle, every point along which is fluctuating with time—illustrated by other circles such as the solar and lunar days. Diagrammatically, the space-time continuum can be observed. In the preceding three experiments, in effect, the ambiguous *temporal* problem is transposed to *space* and the ambiguity is removed.

Now it is known, first, that animals are able to differentiate points along the 360° compass cycle by using subtle geophysical parameters alone, e.g., the horizontal vector of terrestrial magnetism. Second, when the magnetic field is rotated 180°, the flatworm's monthly turning rhythm is abruptly reset by half a cycle. In other words, a 180° rotation in a subtle spatial factor resets by 180° a "biological clock." Subtle geophysical factors are thereby demonstrated to be intimately related to biological timing. Third, an organism stimulated by light from a particular direction "remembers" (at least for many minutes) that a light had come from that particular geographical direction; this occurs by associating the geographical pattern of illumination with subtle geophysical information that was providing geographical direction. In other words, the organism possesses the capacity to encode within itself information via obvious stimulus factors upon a 360° spatial cycle of variation.

The subtle geophysical characteristics that are unique to a particular compass direction, and with which the worm associated the light, are varying continuously in time with contained solar-day, lunar-day, monthly, and annual recurring patterns. Therefore, the organism can be presumed to be able to associate light information with subtle factors that are varying with time. For example, when viewing our space-time diagram, assume the organism has associated a light flash with the compass point (here shown as northwest by west where the time and space cycles become one). One can establish that the organism can distinguish this point on the spatial cycle because there is no time ambiguity and, in doing so, simultaneously establish that an organism can distinguish this point in time as well. In brief, since organisms have been shown to respond to specific geophysical directional information and the latter information is varying cyclically with time, it seems quite possible that these same subtle fields are also providing the organisms with their timing information. If the primary timing for organisms depends upon information steadily provided from the environment, then such remarkable properties of clock-timed rhythms as virtual independence of temperature and insensitivity to chemical disruptions become readily understandable.

In summary, the very fact that an organism possesses the three fantastic capacities described in these three experiments—the very capacities that one might wish to develop if one were posed the problem of designing a clock system such as organisms are known to display—argues eloquently for this being the actual basis of the whole mysterious clock system, the solar-day, the lunar-day, the monthly, and the annual clocks. All could even be explained without invoking at all an independent internal timer for the natural geophysical periods, or, as will be seen later, for the "circa" periods as well.

E. L. McBride and A. E. Comer at Mt. Holyoke College have very recently made an exciting contribution to the biological clock phenomenon. Working with the much investigated circadian rhythm of bean sleep movements, they reported that they had entrained free-running rhythms to 24 hours either in normal or 180°-shifted phase relations by alternation of a higher and lower field strength at a level very slightly above the earth's natural one. The higher strength in a 12:12 hour cycle was given at 8 A.M. or 8 P.M..

RESPONSIVENESS TO VERY WEAK ELECTRIC FIELDS

As a result of the demonstrations in 1958 by Lissman and Machin at Cambridge University of a remarkable sensitivity of some fish to extremely weak electric fields, the author decided to examine mud-snails and planarian worms for a comparable capacity. The worms and snails were both found to respond readily to static electric fields of the

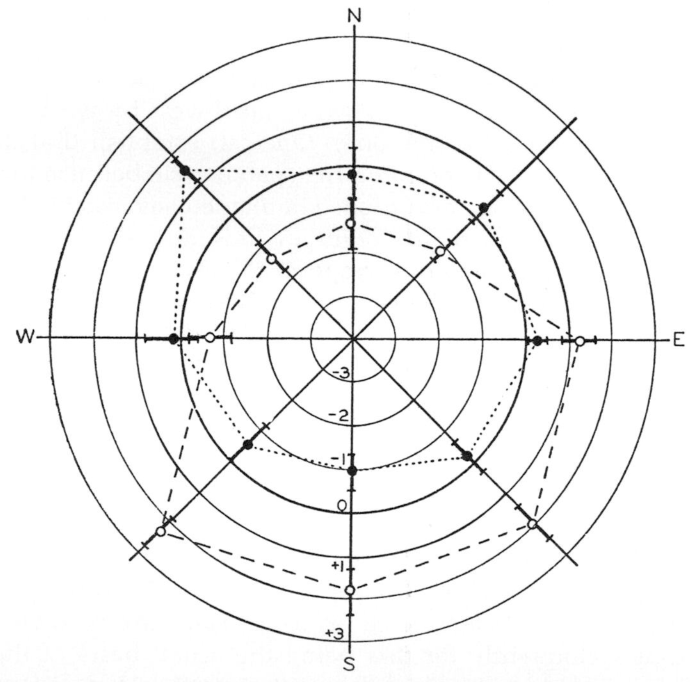

Figure 7-13 Comparison of the compass-direction effect upon response to electrostatic gradient for morning (dashed line) and afternoon (dotted line) hours. Degree of left turning is indicated by concentric circles inside heavily inked one, and right turning by concentric circles outside of it. From F. A. Brown, Jr., *Biol. Bull.* (*Woods Hole, Mass.*) **123,** 282–284 (1962).

order of 0.1 μV/cm. An extensive investigation of this capacity for the planarians showed the responsiveness to vary systematically with orientation both in space and time. The effect on the strength of a negative phototaxis in the worms varied with the geographic direction in which the fields were applied at right angles to their moving bodies. In the morning, the effect was greatest for worms moving northward and least for worms moving southward (Figure 7-13). This compass directional variation essentially reversed its form over the noontime period; afternoons the effects were most for southbound and least for northbound worms. In brief, not only were organisms able to perceive electrostatic fields at the levels of strengths of their natural ambient geophysical environment, but clearly the responsiveness of the animals was related in some manner to their clocks and compasses.

More recent studies in the laboratories of A. Kalmijn have unequivocally established that sharks can respond to fields no stronger than 0.01 μV/cm. Indeed, in those fish that generate special electrical pulses, such a fantastic electroperception is employed for sensing their environment in terms of its differential conductivity and for communication. In fish such as sharks, which apparently lack an ability to generate special fields, their capacity to perceive the weak fields generated and projected into the environment by muscular and neuronal activities of others is employed for finding their otherwise hidden prey. The American eel has been shown by S. Rommel and J. McCleave of the University of Maine to possess such electrical sensitivity that it probably could respond to the fields generated as the ocean currents flow through the earth's magnetic field.

RESPONSIVENESS TO BACKGROUND RADIATION

Finally, the author sought to determine whether an organism could sense small differences in strengths and direction of such high energy radiation as, for example, the hard monochromatic gamma radiated from ^{137}Cs. The most extensive studies were performed with the planarian. This worm showed responsiveness to horizontal gamma fields ranging in strength from 2 to 25 times the natural background radiation. The nature and strength of the response varied with strength of the gamma field (Figure 7-14). To fields stronger than about four times background, the worms oriented increasingly negatively to the direction of the source. To fields weaker, the worms oriented toward the source. The worms appeared to show a maximum positive directional response to a gamma source when the field was close in strength to the ambient background. The response, as with magnetic and electric

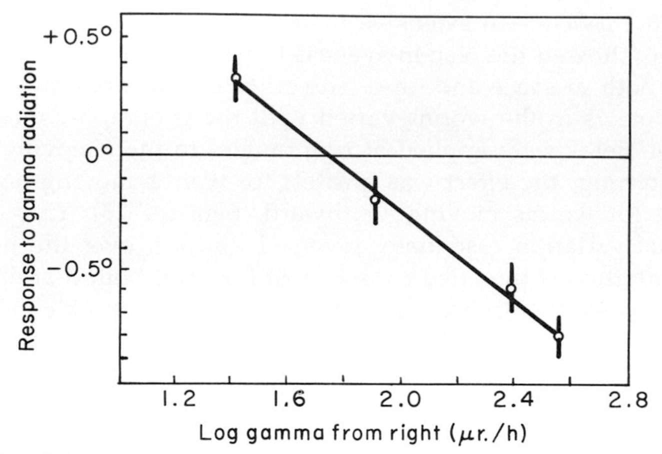

Figure 7-14 Relationship between sign and strength of orientational response of *Dugesia* to a cesium source and the strength of the radiation field. Each point is the mean of the differences between worm paths in the gamma field and paths in the dummy field. Standard errors of the means are shown. From F. A. Brown, Jr. and Y. H. Park, *Nature* (*London*) **202,** 466–471 (1964).

fields, varied with the geographic direction of the gamma beam, and with phase angles within the natural geophysical cycles of the month and year. Also, the response of mice to a fivefold increase in ^{137}Cs gamma traced out the same daily pattern as that of the concurrent 24-hour pattern of O_2 consumption in potatoes (Figure 7-15).

A. Rothen of The Rockefeller University, New York City, has very recently reported a 24-hour variation in a highly penetrating electromagnetic radiation. This radiation, apparently from the sun, produces a 24-hour variation in a slide nickel coated in a strong right-angle magnetic field—a slide made for use in immunological reactions. The rhythm persists in glass or plastic vessels and, slightly depressed, in 0.1-cm steel or aluminum containers. The rhythm is abolished and held in the initial "active" stage when the slide is protected by 3.5 cm of lead. Under 1 cm of lead, it can pass into the "inactive" daytime phase, but can not regenerate the "active" phase during the night and hence the rhythm is abolished and held in the daytime phase. It can be held in the nighttime "active" phase during the daytime under the action of weak gamma radiation from radium. Since a magnetic field parallel to the slide inactivates it or puts it into the daytime phase of the cycle, the natural fluctuations in high energy radiation are believed to produce the rhythm through influences upon magnetic characteristics of the slide. Also, the rhythm is described to exhibit an annual modulation following the times of sunrise and sunset, as well as the changing angle of the sun's declination. Some fully conceivable

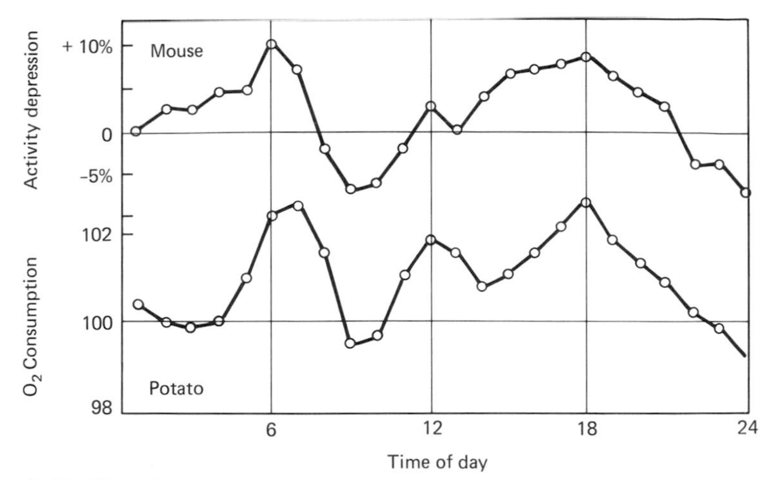

Figure 7-15 The mean daily pattern of variation in effect of a fivefold gamma-field increase on spontaneous activity of mice in natural illumination in the laboratory as compared to a simultaneously recorded mean daily respiration curve for potatoes sealed in respirometers. Whether the increased radiation increases the amount of activity or decreases it is seen to depend on hour of the day. More than 4000 mouse-hours of data and 100,000 potato-hours of data contributed to these curves. From F. A. Brown, Jr., Y. H. Park, and J. R. Zeno, *Nature* (*London*) **211,** 830–833 (1966).

biological equivalent of this phenomenon could, conceivably, contribute toward a timer for the circadian, and even the circa-annual rhythmic system of organisms.

In short, investigations to determine what subtle atmospheric parameters could be conveying information to the organisms sealed in their constant conditions of all ordinary factors indicated that there was not one factor and parameter, but many. Organisms were sensitive to essentially both ends of the electromagnetic spectrum, the static magnetic and electric fields, and the opposite end, the hard gamma. Sensitivities have been described to other frequencies. These fields affected the organisms as functions of both geographic directions of the organisms and the applied fields and of time within the natural geophysical cycles. For all these fields the organisms appeared to be poised for response at the natural environmental strengths, either to single fields or combinations of them including interactions among them. By no means of least importance for our clock problem, all of these fields are known to reflect in their fluctuations the natural cycles on planet earth including those related to relative motions of earth, sun, and moon, and even the stars. The stars are not added without evidence. The 11-year continuous study of the potatoes had disclosed during its first 2 years about a 1% unimodal sidereal variation of 23 hours and 56 minutes. As the earth rotated on its axis relative to the

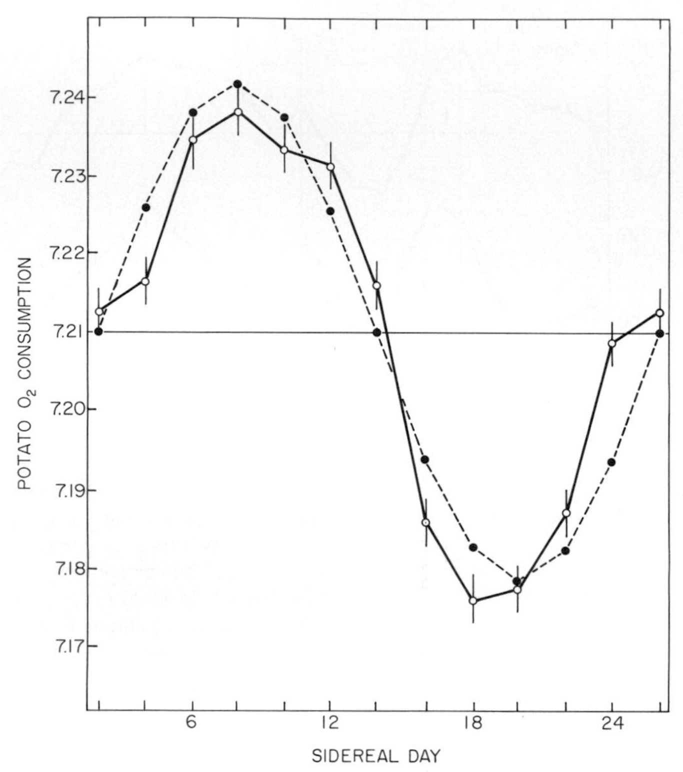

Figure 7-16 Sidereal variation in O_2 consumption of potatoes (solid line) in constant conditions of all obvious factors for an 11-year period (February 1, 1956, through January 31, 1967). Each point is the average of more than 110,000 potato hours. This exhibits a lag correlation ($r = 0.94$) with the annually varying axis of the earth relative to the sun (broken line, plotted with a 2-hour delay).

stars, a maximum of O_2 consumption occurred when the constellation of Gemini was at upper transit, and minimum when Sagittarius was. The average cycle for the next 9 years reproduced the form, phase, and amplitude of this cycle with dramatic similarity, as shown in Figure 7-16.

GEOELECTROMAGNETIC FIELDS AND "CLOCKS"

Where does all this now leave us relative to the clock problem? First, the organisms are not, and have never been, in any studies of their rhythms in what has been constant conditions for them. Second, it is obvious that the total composite of information reaching them by way of the pervasive fields possesses within it, for example, information relative to sunrise and sunset, moon rise and moon set, and the celestial longitude of the earth at any and every day during its annual journey around the sun. Organisms would certainly appear to be most intimately and continuously associated with their environment, including all its periodisms.

The organism as a dynamic electromagnetic entity has fields which in a real sense are continuous with the fields of fully similar character of its environment. There is a continuum of fields from that genotypically and phenotypically structured island which comprise the individual into the geophysical environment and vice versa. The fields of organisms, evidence is beginning to indicate, spread among other organisms close by and are responsible for subtle interactions between them, as with the shark and its prey or—as Carol Chow and the author discovered—with beans absorbing water in neighboring dishes. Organisms have no dearth of information through subtle means of events occurring in the world beyond them; indeed, they have the contrary problem of sorting, filtering, and interpreting a plethora of information. Upon such a foundation of detail, the relatively long periods of the clocks can comprise long wavelength, environmental "ground swells" to which the inherited biochemical, physiological, and behavioral patterns have developed and remain attuned. Indeed, these external cycles are postulated here to provide normally the integrative timing of life by action of the fields which, collectively, can never be fully screened out.

In the host of kinds of animals, plants, and microorganisms, there are innumerable processes and phenomena that exhibit rhythms which are normally correlated with the periods of the day, tides, months, and years. These are either behavioral ones which adapt the organism directly to the natural cycles of obvious and relevant factors of their environment or are supporting physiological or biochemical processes for the adaptive ones. These cycles are the ones which, collectively, provide the organism with their "circa" periodic fluctuations. When one isolates and investigates separately portions of these normally clock-timed rhythms, none of the clock characteristics are evident unless the isolated portions are still living organs, tissues, or cells, nucleated or nonnucleated. If nonliving, the Q_{10} values are two to three or even higher for the abstracted phenomena. Various drugs, or chemical stimulants or depressants, can alter very greatly their rates. There has been no clear suggestion presented to date that any of these components of the cycles constitute the remarkable clock or any part of it. This is true, despite the fact that not only single cells, but even enucleated living fragments of cells, can retain for many days continuing clock-timed rhythms of phenomena such as photosynthetic potential. The timing system suggestively resides in the plasma membranes of cells, or in cell particulates. The electromagnetic fields of these could be interacting with those of the earth's atmosphere.

The clocks are evidently a property of the whole organism or isolated living parts of it. They cease to exist only after death. This is perhaps not surprising since among the various species the vast number of clock-timed rhythmic events that they have evolved and use, usually adaptively, could not perhaps be expected each and every one to have included intrinsic clock properties. It is easy to contemplate a possibility that none of them constitutes a clock, that the "clock" was already there in the physical environment, and the organisms developed over it a temporally structured organization to use it. In other words, the temporal organization of life cannot itself serve as a clock, but can do so only when it is in continuing interaction with the environmental periodisms.

Many processes comprise the endodiurnal organization of life; these are normally period synchronized to the solar day. Others, in littoral oceanic species, may possess, in addition, an endotidal set of phenomena that normally are geared to the lunar tidal cycles with their mean periods of 24 hours and 50 minutes. Others, such as the many lunar correlated breeding cycles in the oceans and monthly or circamonthly primate menstrual rhythms, have cycles which naturally correspond closely to the synodic monthly ones whose mean 29.5-day geophysical fluctuations probably normally time them. For many, possibly most organisms, there are circannual variations, with biological cycles whose longer running times are adjusted to the year.

The recurring patterns of behavioral, physiological, and biochemical changes, in large measure of genotypic origin, even if accurately timed by atmospheric periodisms, could not be rigidly maintained as such in the organism. Natural selection would have been expected to have rendered the cycle forms more or less plastic and adjustable to the environments. The forms would be alterable by specific cyclic events within the organisms' immediate microhabitat. Indeed, what little evidence is available suggests that the detailed patterns of the rhythms display many recurring minor irregularities which can exist in the cycles even after the generating environmental factor is known no longer to be operative. As examples of this, think of the effects of a single light flash on an ongoing rhythmic variation in the bean seedling reported by E. Bünning of Tubingen University (Figure 7-17), or the effects of associating a particular time of day with the availability of a sugar-water solution, on the daily activity pattern of honey

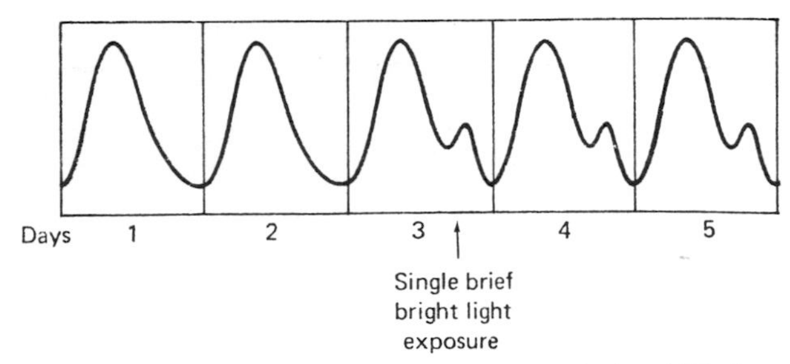

Figure 7-17 Tracing of a bean-seedling sleep-movement rhythm. With the exception of a brief exposure to a bright light on day 3, the plant was maintained in constant dim illumination and constant temperature. Note the leaf response to the bright light and the fact that this "response" recurred each day thereafter without any further stimulus. Redrawn from E. Bünning, *Naturwiss. Rundsch.* **9,** 351–357 (1956).

bees by M. Renner of the Zoological Institute in Munich. (Figure 2-28). These simple illustrations serve only to emphasize that the multiplicity of events to which the organism is exposed everyday cannot only be responded to directly, but at least many relevant ones for the organisms' survival may become simultaneously encoded into their ongoing daily rhythmic system. Clearly, the daily recurring patterns include not only a major inherited cycle usually with periods of rest and activity, but include larger or smaller effects of the specifically changing events in the organisms' private lives.

All the foregoing are adaptive, enabling the organism to anticipate and prepare for specific events which, in all likelihood, will occur at the same points in the succeeding daily cycles. An event recurring at the same time on successive days would be expected to effect a progressively stronger influence on the recurring cycles. An event occurring in one cycle and gone in succeeding ones should rapidly fade away. Possibly, our usual habit of having three meals a day at essentially the same hours could be shown to have imposed its influence upon aspects of those systems of our body concerned with dealing with the ingested foods.

LABILITY OF THE RHYTHMIC CYCLES

Slightly pattern-modified daily cycles, reflecting an environmentally imposed pattern on a 24-hour larger amplitude genetic cycle, help to deal with many of the problems faced by living things. But what about the effects of the changing times of sunrise and sunset through the year? What about active or passive movements of organisms over the planetary surface? Even poorly motile oceanic forms could be swept

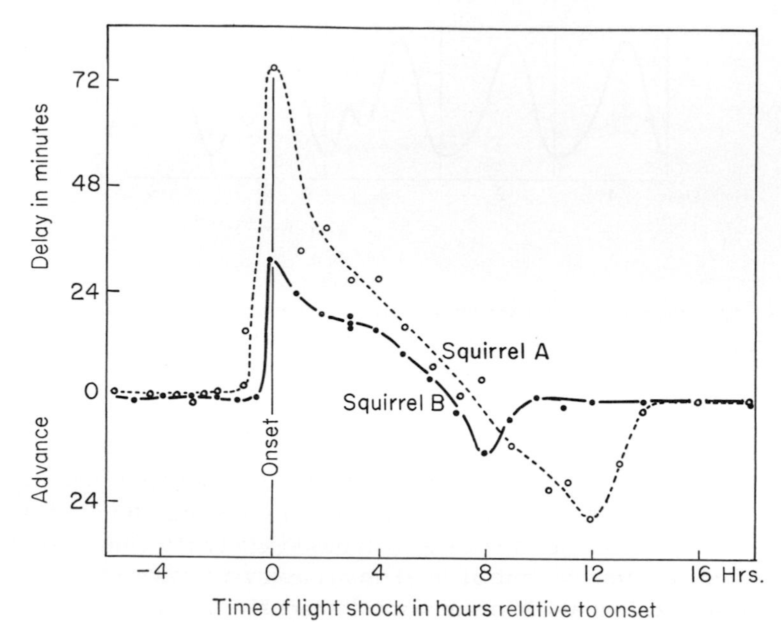

Figure 7-18 Light-sensitivity curves for two flying squirrels, A and B. Activity onset is indicated. From P. J. DeCoursey, *Science* **131,** 33–35 (1960).

thousands of miles in the major ocean currents. Changes could also include the spreading of a species over a large geographic area, or the migrations of individuals even from the northern to the southern hemispheres. Natural selection also took care of these problems for living things.

All organisms, except perhaps the prokaryotes, are able to phase shift whole daily patterns to either earlier or later times of day. To accomplish such a feat, a phase-response system made its appearance. Figure 7-18 shows the first ones described. This system is ideally adapted to resetting the whole of inherited cyclic patterns to carry whatever imposed or learned information they might have to earlier or later times. Eastward travel, of course, demands shifts to earlier times, and westward to later times. The phase-response system can shift, gradually and systematically, an ongoing cycle, no major portion of which could be omitted, or be accelerated or depressed in rate beyond viable limits.

The phase-response system is inherited and is associated temporally with the major genetic activity-rest pattern in just such a manner that the activity and rest are permitted to lie within the day-night cycles in the optimal manner for which the individual is otherwise adapted.

Change the times of day of light onset and termination and the 24-hour rhythm is systematically phase shifted over a few cycles until the relationship of the rhythm to the new times of the light changes are the full equivalent to those of the earlier. The cycles are thus labile, they can be phase reset, and, meanwhile, the organism remains alive and functional though, evidence suggests, apparently usually somewhat stressed by the shifting process. During the shift, the several consecutive cycles are, transiently, either slightly longer or shorter than 24 hours, depending upon the direction of the shift.

PHASE MAPS AND PHASE DISSOCIATION

During the phase shifting of the daily cycles, the normally 24-hour rhythms of all the bodily processes become fully phase-associated with the new light cycles after differing numbers of days. In other words, the rate of phase shifting varies among the many differing processes or phenomena within the individual. This leads to a temporary dissociation of some processes from others. Up to a week or two is usually required for all to have completed their shift with all now having returned to the same relative phase-map relationship to the light and to one another that they had before the change in the lighting schedule.

A comparable phase dissociation has been reported for the "free-running" cycles similarly apparent in constancy of light and temperature. In this latter case, however, the free-running of the different processes or phenomena may persist at their different rates of daily change. There is no terminal point for readjustment (Figure 7-19). This similarity between the free-running, transient cycles during the period immediately following light cycle shifts with its common dissociation of aspects of the individual endodiurnal organization and the free-running cycles in constancy of all *Zeitgeber* fields, also often with dissociations, clearly suggests a similarity in mechanism of production of the two.

All the circaperiodisms are known to be phase-labile cycles. Special, obvious parameters of the environment (*Zeitgeber*, phase setters, synchronizers) which reflect clearly that phases of the environmental cycles of relevance for the organism are normally involved in determining the phase settings of the biological cycles into their characteristic adaptive relationship to the environmental ones. For the circadian cycles, there is a hierarchical order with light changes the dominant *Zeitgeber*. Lesser in effect are such *Zeitgeber* as temperature, sound, and social cycles.

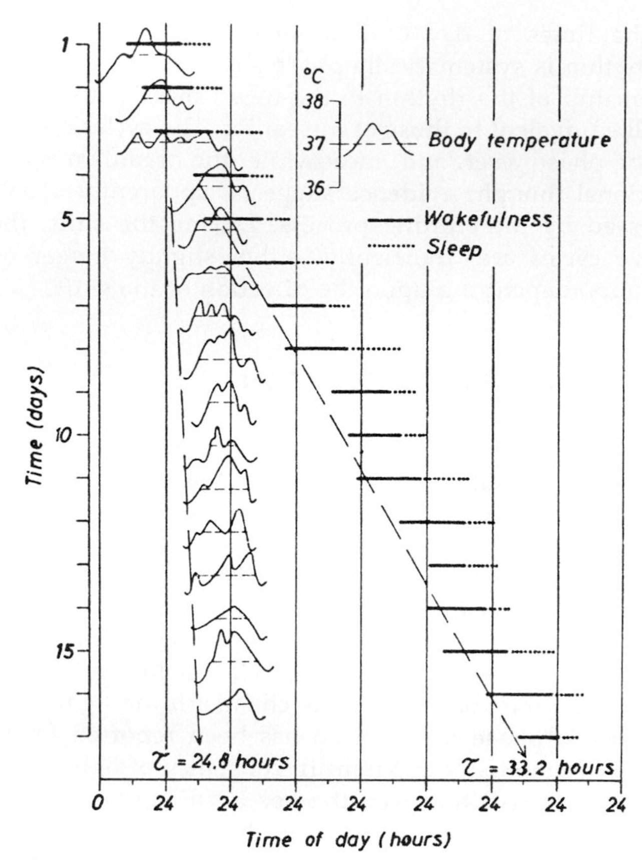

Figure 7-19 Circadian rhythms of body temperature and activity of a human subject enclosed in isolation without time cues. From J. Aschoff, *in* "Life Sciences and Space Research" (H. Brown and F. Favorite, eds.), North-Holland Publ., Amsterdam, pp. 159–173. 1967.

Pattern and phase lability of the clock-timed, largely genetic, tidal rhythms is perhaps even more important to the organism. The times of high and low tides may differ even between two beaches in the same geographic area. Near Woods Hole, Massachusetts, within a 10-mile range, the tidal times may differ by more than 4 hours. In long bays, the tidal times typically become later as one moves up the bays. The ocean tidal waves move inexorably, but at a relatively low speed. The tidal times depend not only on the major wave, but also on the tidal current flows they engender. Organisms are known to be adapted to the tidal times on their own beaches, but much less is known about the means than for the times of days. The longer cycles of the semimonth and month can also be phase labile within limits. In the field, these are usually synchronized to the particular phases of the natural cycles of their environment, but again there are only suggestions about how this is accomplished.

The phase setters reported as sometimes effective for the tidal cycles have included wave action, sand swirling, temperature, and pressure (Chapter 3).

For the monthly rhythms, variations in nighttime illumination by the moon has been suggested to be a synchronizer for some of them. The more common semimonthly variations would seem often, but not always, to be set by the times of tides relative to times of day on the home beaches. Beats between the beach-set, bimodal lunar-tidal cycles, and the daylight phased 24-hour ones could readily determine the detailed relationship of the semimonthly cycles relative to moon phase.

The circannual cycles are phase synchronized to adaptive times of year usually by the gradually changing relative lengths of day and night or photoperiodism. They can be phase-shifted somewhat relative to time of year by experimentally altered photoperiods. It has been postulated that the 24-hour cycles can comprise the yardstick upon which the relative lengths of light and darkness are measured (see Chapter 6).

As noted earlier, for the phase determiners or *Zeitgeber* to serve their roles in the adaptive setting of the cycles, a phase-response system is required. This, though an obvious necessity in some form for all the cycle lengths, has been most thoroughly investigated for the circadian system and will serve our purposes for illustration of the working of the phenomenon. There must be a variation in organismic phase-setting responsiveness that remains intimately associated temporally with the biological cycle to be phase determined. Such a phase-response cycle appears to be a genetically contributed component of the inherited circadian cycle. The effect of the *Zeitgeber* is to advance the cycle slightly during part of the cycle and to delay it a little during another portion. The relationship of the phase-response curves to the times of activity within the cycle will differ greatly, for example, between a nocturnal and a diurnal creature. The phase response of the nocturnal organism traps the activity period into the dark portion of the day and for a diurnal one, into the illuminated portion. Indeed, the times of day and durations of activity can be genetically determined by way of the activity-phase-response complex. This complex differs to some extent from individual to individual (Figure 7-18), which results in small differences in phase relations of the activity to the *Zeitgeber* cycles. Although this is importantly genetic, not much is yet known about its modifiability by the environment during the lifetime of the organism.

The presence of such a particular kind of a phase-setting system has been responsible for misleading scholars in the field of rhythms to the interpretation that the clocks are almost always somewhat inaccurate. In constancy of fields of factors that, when varying, normally can serve as *Zeitgeber*, the rhythm drifts slowly across the clock hours of the day. The internal timer school has seized upon, and emphasized, this phenomenon, termed free running of the rhythms, as "proof" that the timer is in the organism rather than in the environment. Every individual organism seems to free-run with its own usually variable period. The specific clocks timing them are postulated to be running with exactly the same odd and variable periods. Indeed, the term circadian itself was coined by F. Halberg of the University of Minnesota in 1959 to emphasize the usual non-24-hour character of the rhythms in the unvarying fields of all *Zeitgeber*. However, the clocks timing free-running rhythms were soon to become gratuitously termed circadian clocks because of the dominant hypothesis about the clocks' nature. In reality, it is known that rhythms can become circadian, but it is only hypothesis that the clocks are too.

SOLAR-DAY AND LUNAR-DAY RHYTHMS

Circadian rhythms clearly fall into two families, the solar-day and the lunar-day, which are not distinguished by the term circadian. Yet, the two kinds are extremely easily separated in their environmental response. The first kind, a solar-day circadian rhythm, can be phase set and phase entrained to a 24-hour light-dark cycle. These include the numerous and diverse rhythmic variations that normally comprise the endodiurnal rhythmic complex. These usually free run under conditions in which all the ordinary *Zeitgeber* that can set and synchronize them are held constant. The periods can be either longer or shorter than 24 hours (Figure 1-5). They may keep a relatively constant period over extended intervals, or may abruptly or systematically alter their period at any time with or without explanation. Held under the same conditions, or even when enclosed in the same room, each individual may be displaying its own period, slightly different from that of any others (Figure 2-19). This has been the major observation interpreted to demand private internal timers. The periods may be altered very slightly by differences in the levels of the ambient constant temperature, or in response to changes in the level of the constant illumination. Whereas the period seems to be uninfluenced by almost all drugs and other chemicals, it seems to be quantitatively alterable to a small degree by administering D_2O (heavy water) (Figures 2-20; 2-22) or cycloheximide (Figure 2-16) to the organism.

The second kind, the bimodal lunar-day or tidal rhythms, is most conspicuous in organisms living along the ocean fronts. The rhythms are normally adaptively set to the tides. The time of tide of maximum activity differs among species, with each possessing its individually favored time. These times may, even within the same species upon the same beach, be adjusted to covering and uncovering by water of a specific site on the beach as the tide ebbs and flows. Away from the tides, in the laboratory, these rhythms may persist with their characteristic beach-adapted patterns even for extended times in controlled conditions. In fiddler crabs, they soon stabilize and persist with an accurate lunar-day period under those conditions that favor the maintenance of the 24-hour period of solar-day cycles (Figure 7-20). These conditions are complete darkness or in the natural or an artificial 24-hour light-dark cycle. Under conditions of continuous light that favors a free-running solar-day cycle longer than 24 hours in the crabs, the lunar-tidal cycles comparably free run with a period longer than the lunar day (Figure 1-1).

The explanation of this partially interlocked behavior between the solar-day and lunar-day cycles of the fiddler crabs appears to lie with the effects of light as a *Zeitgeber*. The crab tidal cycles of running activity are phase labile, yet they normally occur day by day about 50 minutes later relative to the solar day ones. Light changes readily phase shift the solar-day cycles, with changes that sweep the lunar-tidal cycles along with them. Many years ago it was shown that although the tidal cycles of crabs could run with their own tidal periods they could be phase shifted by resetting the solar-day cycles. This was dramatically confirmed by H. M. Webb and the author in an experiment in which crabs were subjected to a light cycle of lunar-day frequency. While the solar-day cycles were delayed 50 minutes each day in response to the cycles of light change, the daily reset tidal cycles were being delayed about 100 minutes a day.

It seems readily apparent that the fiddler crabs have two clocks, solar-day and lunar-day ones, of considerable accuracy. This is true despite the theoretical difficulties faced by the internal timer advocates to explain how two rhythms so similar in frequency did not become mutually entrained. For a long time they were hypercritical concerning all evidence that purported to show the two in the same organism at

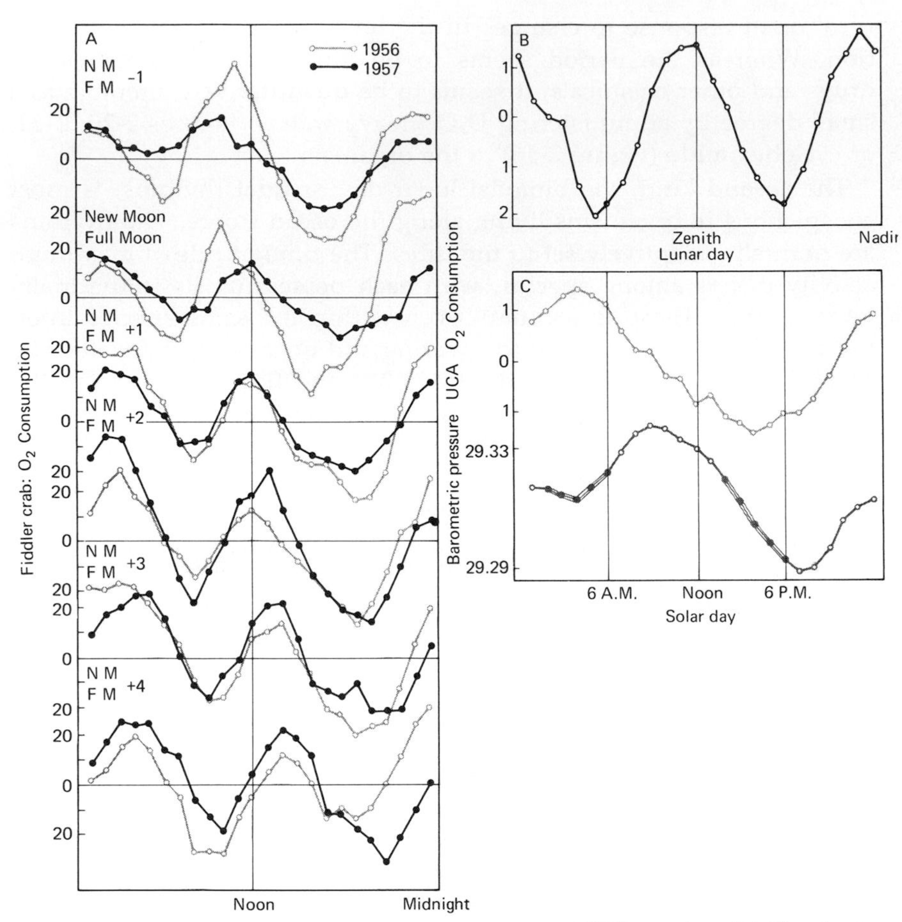

Figure 7-20 On the left are shown the diurnal patterns of fiddler-crab metabolism in constant conditions showing how they change their form systematically with day of the semimonth and how these patterns are reproducible from one year to another. This semimonthly recurrence of daily patterns results from the simultaneous existence in the crabs of the lunar-day and solar-day rhythmic components illustrated on the right. Also illustrated is the mean form of the solar-day tide of the atmosphere for the month of June, with thickened portions marking the times of day-to-day correlations between crab metabolism and barometric pressure. From F. A. Brown, Jr., *Cold Spring Harbor Symp. Quant. Biol.* **25,** 57–71 (1960).

the same time. However, the rhythms timed by both of these clocks are phase labile relative to these clocks and dependent upon common metabolic machinery within the organism which compels them to be phase reset together, relative to their respective basic timers, and yet gradually able to alter their phases relative to one another in response to the tides.

AUTOPHASING

Timing Free-Running Rhythms by Exogenous Clocks

What is known of the mechanism that participates within the organism in the phase responses of solar-day rhythms to *Zeitgeber* rhythms? Let us consider light first, the dominant, and most investigated environmental factor which determines the phase settings of the inherited daily patterns. The solar-day circadian complex includes very importantly a phase-response system. This system can be illustrated by a curve that shows the relative amounts and the direction of a phase shift of the circadian cycle in response to administration of a light stimulus at various times during the daily cycle. Figure 7-21 shows, diagrammatically, the general form of a phase-response curve and its approximate relationship to the activity for a nocturnal animal. The rising curve reaching its peak close to the time of onset of activity describes an increasing delay produced by the light, to a maximum near the time of the onset. The delaying effect of light then gradually diminishes to no light-phasing response, then continues onward to

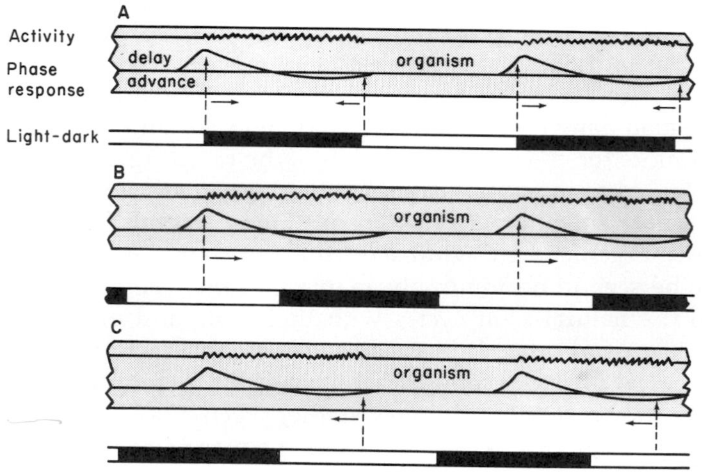

Figure 7-21 (A) The relationship between the activity rhythm of a nocturnal mammal and its closely associated phase-response curve, together with the relationship of these to the environmental light-dark cycles. The animal's rhythm has become fully adjusted to the light. (B) The relationship immediately after an abrupt phase delay in the environmental light cycle such as would result from rapid displacement westward a quarter way round the earth. (C) The relationship immediately after an abrupt phase advance in the light cycle, as would occur after comparable eastward displacement. Horizontal arrows indicate the expected phasing influences of light at times in the biological cycles shown by the vertical arrows. From F. A. Brown, Jr., *Amer. Sci.* **60,** 756–766 (1972).

reach its lowest point, a point at which light exhibits a maximum in phase advance. This typically occurs over the time of termination of activity in the cycle.

This form of the phase-response-activity complex is a highly efficient one for readjusting the cycle phases when the organism is moved eastward or westward to a time zone in which the time of light and darkness are different. Such phase-response curves have been described for many organisms—plant, microorganism, and animal. Indeed, they appear to be invariable concomitants of the phase-labile cycles and used for phase adjustments. It is equally evident that once the cycle has become phase adjusted, the phase-response system no longer puts any shifting stress on the organism, in terms of the external timing hypotheses. On the internal timing hypothesis, the phase-response cycle operates usually day after day to keep the independent internal oscillator system set to the local clock hour.

For any given amount of phase change in the light-dark cycles, the rate of the phase shift is known to vary within limits with the brightness of the illumination. Also, as is evident from the asymmetry between the phase-delay and phase-advance limbs of the curve, the rate of phase shifting of the circadian cycle differs between west to east and east to west geographic translocations, as well as even differing from individual to individual.

This system for resetting the phase of the rhythms which gives usually small daily nudges to the circadian cycles and requires up to a week or more for essential completion of the resetting process is splendidly adapted to the complexly integrated organism. The gradual, systematic phase shifting permits the continuing smooth functioning of a complexly organized individual during the course of the shifting. It can also be seen to be admirably adapted for the resetting by the organism to the natural light cycles with their long and graded twilights during dawn and dusk. It is a readily observable fact that not only do the biological cycles become clearly set to the natural environmental ones, but they resume a relatively stable, precise 24-hour frequency of their rhythms at least as accurate as to a square wave imposed 12:12 light-dark cycle in the laboratory.

What can be the predicted consequences of placing an organism with such a phase-response-activity complex in an environment devoid of any *Zeitgeber* rhythms that ordinarily reset the rhythm phases? More specifically, first, what can our nocturnal organism be predicted to do when held in an environment in unchanging light? The phase-response complex would effect a rhythm in the perceived light in a phase relation to which the organism was not adjusted. It has been

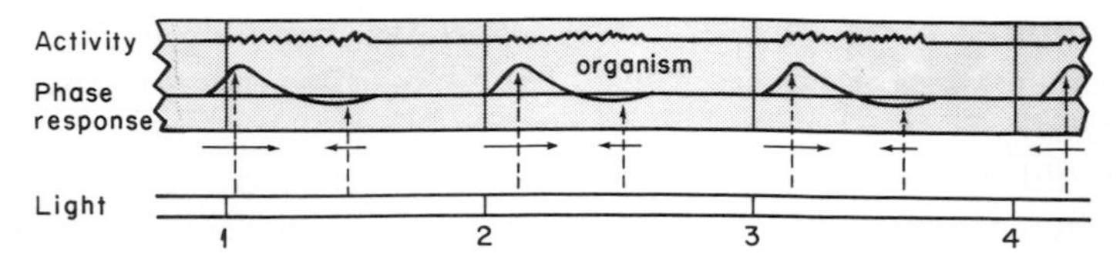

Figure 7-22 The expected phase-shifting behavior of a nocturnal organism in constant light. There are repeated daily phase shifts, or autophasing, in response to the light, which is normally a *Zeitgeber,* but is here held constant. The numerals mark successive 24-hour periods. Horizontal arrows indicate the expected phasing influences of light at times in the biological cycles shown by the vertical arrows. From F. A. Brown, Jr., *Amer. Sci.* **60,** 756–766 (1972).

noted that light over the time of onset of activity delays the whole circadian cycle a little (Figure 7-22). Light later, over the time of termination of the cycle advances it, but usually to a lesser extent since the phase-response curve is asymmetrical. The algebraic sum of the two daily phase shifts is a small residual delay. This, occurring cycle after cycle in the same way and extent, can be expected to generate an observed rhythmic free-running period slightly longer than 24 hours.

Phase-response curve forms vary among individuals. Not only the responsiveness of the individuals in the phase-delay and phase-advance portions of their cycles, but the relative responsivenesses between the two show individual differences. Consequently, just what is expected is observed—that the periods of free-running rhythms vary among individuals held under the same conditions of constancy of normal environmental phase-setting factors.

Diurnal animals also have phase-response systems for light, although less is known about them. In birds for example, the phase-advance portion of the phase response curve advancing the animals activity into the earliest light hours of the day is usually the stronger, with phase delay ordinarily weaker. For such a diurnal animal, the free-running cycles are, as expected, commonly shorter than 24 hours. Crepuscular organisms have not been investigated in this regard, but it seems reasonable to presume that these, too, would show the same individual variabilities among their phase-response curves. Hence, free-running cycles with a spectrum of frequencies would be apparent in conditions of constancy of all normal *Zeitgeber.*

Light, though the dominant, is not the only *Zeitgeber.* Temperature cycles, for example, often serve as *Zeitgeber,* as can a number of other periodic environmental factors. Phase-response curves for temperature have been described for a few species. Since temperature is generally a much weaker *Zeitgeber* than light, the commonly reported either precise mean 24-hour cycles persisting in darkness or smaller deviations from 24 hours commonly observed in darkness can be postulated to result from smaller, or no, response of the circadian cycles to the weaker *Zeitgeber*.

It should be reemphasized that the phase-response relationship depicted by the curve is an intrinsic portion of the circadian cycle. It phase shifts along with the activity. It, therefore, is a component of the subjective day-night cycles during phase shifting to altered light cycles and during free-running in constant conditions.

The term *autophasing* was coined by the author in 1959 for the postulated phase shifting of the circadian cycles under conditions of constancy of all *Zeitgeber* fields. At an earlier date, he had also indicated that it is not possible to distinguish between a rhythm with an intrinsic period slightly different from 24 hours and one in which that difference from 24 hours has resulted from regularly, periodic phase shifts.

One of the early reasons for postulating autophasing resulted from a search of the explanation of the effects of different levels of illumination on the free-running periods, a phenomenon first reported by M. Johnson at Harvard University in 1939 for the deer mouse. He found that the free-running period, longer than 24 hours, became longer as the illumination increased. This was confirmed for many, but not all, nocturnal vertebrates. On the other hand, diurnal vertebrates tend to show free-running cycles shorter than 24 hours and the brighter the illumination the shorter the period. These facts comprise part of Ashoff's Rule. These foregoing facts were just what one might expect if light were acting in a phase-setting role. If at one illumination intensity the algebraic unbalance of phase delay and phase advance were producing the free-running by a phase shifting to light, an increase in light level would yield larger delays and advances and also the difference between them would be greater. Thus, the observed results from the differing light levels would find a simple, straightforward explanation.

Also readily explained by the autophasing concept was the observation of J. Aschoff and R. Wever at the Max Planck Institute in Erlang-Andechs, Germany, that the phase relationship of activity of a finch to a 12:12 light-dark cycle was related to the measured period of the free-running cycles in the bird. If one makes the reasonable assumption that the relationship of the birds' phase-response system to the light cycles reflected the individual form differences in the phase-response curves, the most advanced bird in the 12:12 hour light-dark cycles would have the shortest free-running cycles and the most delayed, the longest. The authors of the study, proponents of internal clocks, arbi-

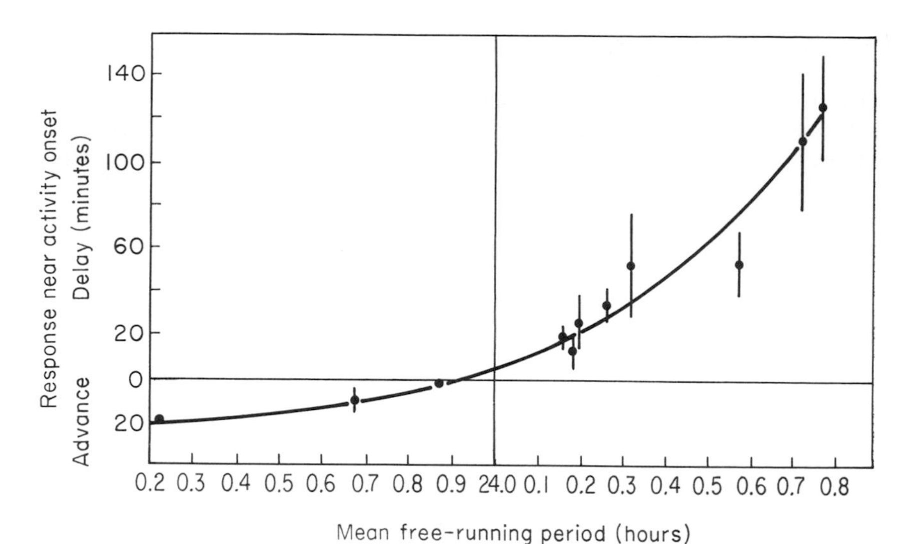

Figure 7-23 Relationship between the measured free-running period and the amount and direction of phase shift near onset for all 11 animals. Height and directions based upon curves drawn by inspection; this curve was calculated by the method of least squares; standard error of estimate is given where it was meaningful to calculate. From J. J. Natalini, *Physiol. Zool.* **45,** 153–166 (1972).

trarily attributed an explanation in terms of intrinsic oscillators. The early bird, when permitted to free run, exhibited the shortest period, and vice versa.

A more direct support for the concept of autophasing came from a later study, on kangaroo rats, by J. Natalini at Northwestern University. He learned that there was a direct relationship between the strength of the phase-delay response close to the time of activity onset in these nocturnal animals and length of their free-running periods (Figure 7-23). He studied eleven individuals, three showed phase advances at this time and had free-running cycles shorter than 24 hours, and eight showed phase delays and periods greater than 24 hours. Plotting the relation, the best smooth curve went essentially through the point of no free-running, or 24 hours, at the point of neither phase delay or advance at that time.

D. Avery, of the University of Virginia, having recently found highly significant correlations between free-running periods of rats in light with the transient cycle lengths following reversal of lighting at the same intensity, concluded that the rats' free-running periods were generated by the autophasing of cycles very close to 24 hours in length.

Factors Modifying Circadian Periods

In now classical studies, E. Bünning had discovered in 1935 that the length of the free-running period in bean sleep movements involved a genetic element for its determination. Much more recently, R. Konopka and S. Benzer have learned that the free-running periods in *Drosophila* could be affected by single gene changes. Comparable evidence has been found for genetic determination of free-running periods in *Neurospora* by J. Feldman and M. Hoyle, at the State University of New York at Albany, and of *Chlamydomonas* by V. Bruce of Princeton. The genetic determination of the length of the free-running periods for any given environmental conditions is readily accounted for by the autophasing concept. The genes could be altering slightly the forms of the phase-response curves among individuals in a population just as surely as genes determine the great differences in them between, say, nocturnal and diurnal organisms.

The circadian free-running period commonly is variable within an individual. The frequency may change sharply and abruptly, or systematically over a period of days or weeks, or longer, often without obvious cause. Semimonthly, monthly, and annual variations may be evident in observations on the period change, as well as variations with age of the individual. The latter was reported recently by C. Pittendrigh and S. Daan of Stanford University. Even more, an ongoing free-running rhythm may cease completely, to start up again, apparently spontaneously, at a later time right on its free-running time schedule. Therefore, a process within the body can apparently become dissociated and later reassociated with the clock-dependent rhythms. A rhythm in crabs may be depressed in amplitude in light, a depression that persists in continuous darkness. All these behaviors need not involve directly the basic clocks at all, only the clock-timed behavioral patterns as they autophase.

A very extraordinary property of the clock-timed rhythms is the often complete or nearly complete temperature independence of their periods. Q_{10} values are very seldom found beyond the range, 0.8 to 1.2, instead of values of 2 to 3 or higher as expected for biological or biochemical activities in general. It is significant that those free-running rhythms in organisms for which complete temperature independence of period has been reported or implied were indistinguishable from 24 hours. Most free-running rhythms that deviate from 24-hours show values slightly different from 1.00. While it may be possible to postulate a temperature-compensated biophysicochemical system able to do this, and, in fact, theoretical models have been proposed, it seems improbable to me that this was done. If it were accomplished, natural selection might have been expected to have rendered all organisms, including poikilotherms, with their physiology and behavior essentially uninfluenced in rates by temperature change. After all, these have their clocks temperature-independent or compensated. A powerful natural selective pressure would have been upon them to do this. But it was not done. Rather it seems more probable that only the pacemakers timing the rhythms are fully temperature independent, with the small commonly observed temperature-dependence of the rhythms occurring during the phase-resetting process when the processes are recurring free from the clock.

The solution to this problem is confounded in that temperature is both a *Zeitgeber* with its own phase-response curve, and also exerts a strong kinetic effect. If one makes a reasonable assumption that the clocks are completely temperature-independent in their periods, the small autophasing effects of temperature (or other phasing factors) would provide almost never more than a 3-or 4-hour (12 or 17%) deviation from the 24-hour period, usually far less. The temperature coefficient for this small fraction of the total cycle could readily be within the usual range of 2 to 3 for a biochemical phenomenon, which provides then for the whole cycle, including the large temperature-independent portion, the very small coefficients usually observed.

Elaborations of the concept of autophasing can be anticipated which would explain equally well all other observations known and probably still to be discovered in solar-day circadian rhythms. And all rhythm properties would have been accounted for without any compulsion to postulate a temperature-compensated, fully autonomous, clock system for organisms.

The Significance of Tau

The free-running period, given the Greek letter τ, is a measurable value. This period is believed by supporters of the internal timer hypothesis to represent the actual period of the clock oscillator when running freely under the experimental conditions of constancy of all *Zeitgeber* fields. On the other hand, it is the difference between τ and the period of the day, or lunar day, which is the significant one for the external timer school. Differences between τ's are believed to indicate the differing effects of autophasing. In these latter terms, not only are effects of illumination, temperature, chemicals, and genotype accountable, but also such parameters as the ranges of entrainment to alterations in periods of imposed artificial *Zeitgeber*. Relative co-ordination—in which an ongoing rhythm, while not entrained to an imposed period field, has its period altered by this field—may also be explained. These can, theoretically, be accounted for by known properties of the phase-response system.

The τ's of rhythms are not only variable, but are only within predictable limits. Each must be determined experimentally even within a single species, and under the specific environmental conditions. Even thereafter, the value may change essentially unpredictably. Hence, τ is not itself a particularly useful characteristic. Its usefulness is in its interpretation for the solution of the clock problem. In investigating circadian patterns in man, such researchers as F. Halberg of the University of Minnesota and A. Reinberg of Paris deal chiefly with 24-hour, "entrained" cycles, and thereby avoid the basic questions raised as a consequence of the two alternative theories of time measurement. However, the detailed 24-hour patterns of variation in an individual would be expected to vary with the clock hour of the day to which the cycle was adjusted, as a consequence of influences of subtle geophysical daily patterns of the organism, and with its responses to *Zeitgeber* parameters.

Many scientists find it difficult to attribute to a biological system, by itself, the relative precision of the circadian period which will continue for days or months, even indefinitely, without any recourse to environmental cycles. Such precision may continue even within reproducing systems of organisms with their repeated bodily reorganizations. Such precision seems nonbiological. Much easier for these persons would be the concept of a continuing reference frame of rhythms timed by the inexorable, periodic motions of the earth, its moon, sun, and stars relative to one another. These would be mediated to the organisms through biological effects of the atmospheric periodicities that these motions engender.

In summary, it is evident that the free-running periods need not be, and probably are not, those of a basic biological clock that times each of them. However, it must be pointed out that the rhythmic changes themselves sometimes appear to be able to serve as clocks of a second order, which can be used in sun, and possibly also moon, navigation by animals (Chapter 5). This is indicated by the fact that homing birds, for example, can under certain circumstances under sunlight be induced to take predictably altered "home" directions following resetting of the circadian cycles by artificial light changes. It is now known, however, that "clocks" and navigation by sun and moon are not essential either to "homing" or to migrations.

GEOPHYSICALLY DEPENDENT RHYTHMS

With the extraordinary sensitivities of living creatures to a host of the environmental electromagnetic parameters, together with the well-known periodic character of these in the atmosphere, generated fluctuations in the organisms can be expected. These have been found in every animal or plant in which they have been sought. Essentially, the

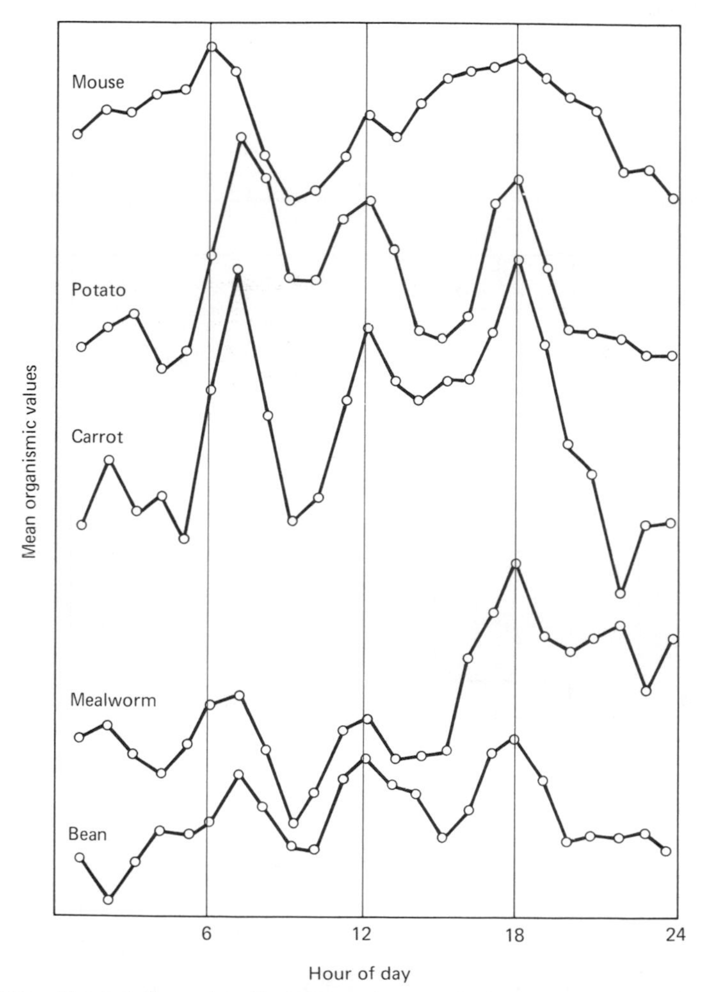

Figure 7-24 Mean daily metabolic patterns in sprouting potatoes, carrot slices, mealworm larvae, and germinating beans, and the pattern of a daily variation in the response of mice to a fivefold increase in background radiation, all in constant conditions. From F. A. Brown, Jr., *Can. J. Bot.* **47,** 287–298 (1969).

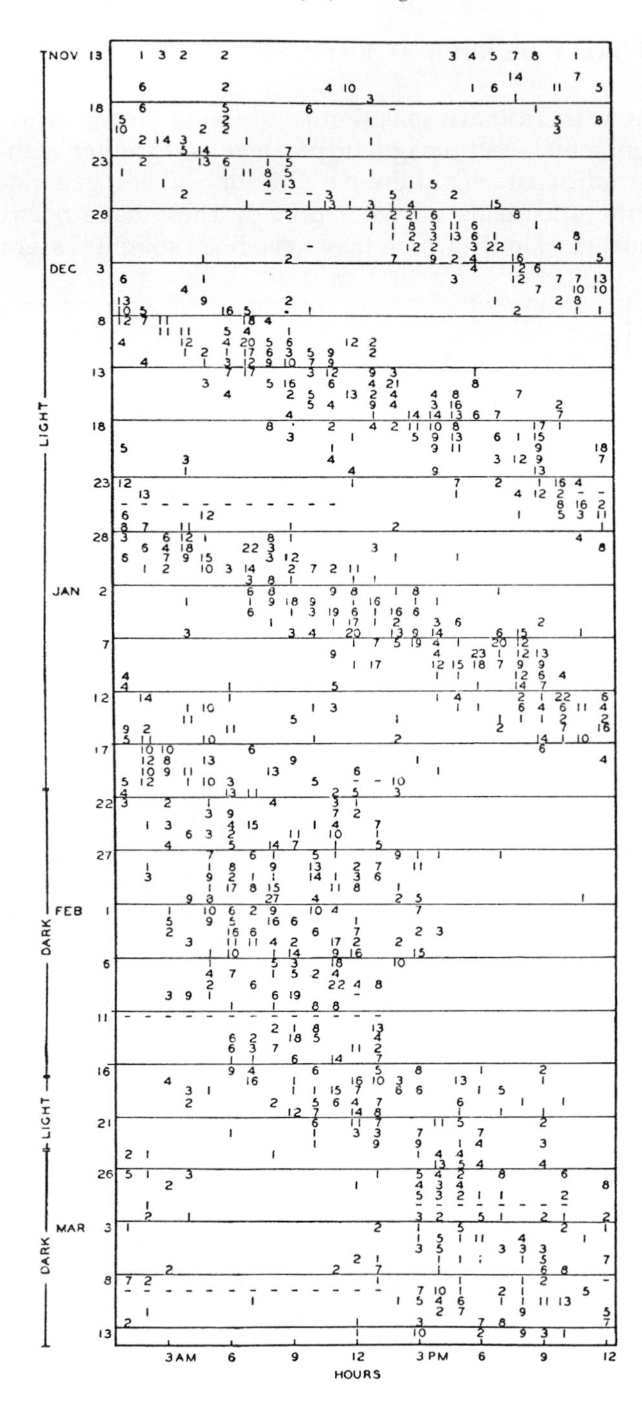
NOV 13
18
23
28
DEC 3
8
13
18
23
28
JAN 2
7
12
17
22
27
FEB 1
6
11
16
21
26
MAR 3
8
13
LIGHT
DARK
LIGHT
DARK
3AM
6
9
12
3 PM
6
9
12
HOURS

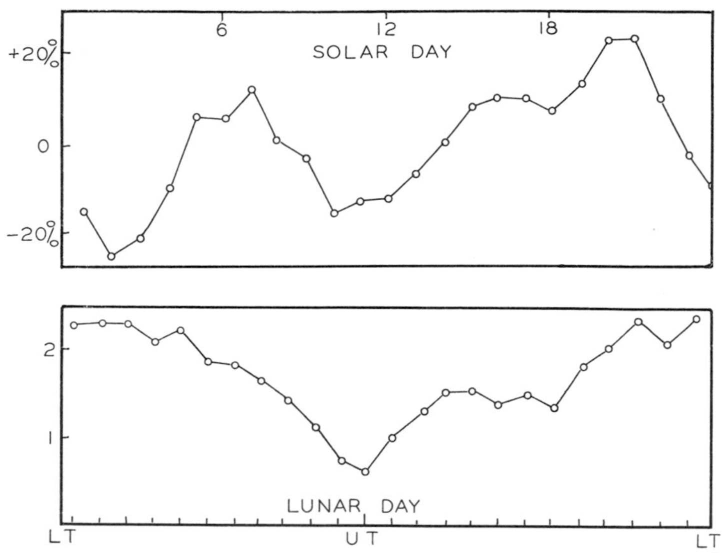

Figure 7-26 The mean solar-day and lunar-day patterns of activity of the white rat of Figure 7-25 during the first 59 days (two synodic months) in continuous dim illumination.

same forms of low amplitude mean patterns of 24-hour change have been described for respiration in sprouting potato plugs, bean sprouts, young chick embryos, mealworm larva, and carrot slices (Figure 7-24) and in spontaneous activity of mice and rats. In many of the foregoing, there was no evidence for the existence of a free-running circadian cycle occurring at the same time within these organisms. In rats and mice, these cycles appear to underlie large amplitude circadian variations from which they could be extracted by determining the mean solar-day (24-hour) variations in spontaneous activity as the circadian cycles free-ran across the day in continuous weak illumination for the rat (Figures 7-25, 7-26), or by noting the 24-hour pattern in fluctuation in response of mice to an experimental fivefold increase in background radiation produced by a ^{137}Cs source (Figure 7-15).

A factor that has complicated the investigation of these geophysically dependent organismic variations has been the existence of a complementarity in the character of organismic response to the pervasive external variations. There seems to be an ability of organisms to alter the sign of their response to the environmental variations, which gives

---

Figure 7-25 Relative amounts of running activity, in arbitrary units, of a male white rat as a function of time of day in continuous low illumination (Nov. 13–Jan. 21), darkness (Jan. 21–Feb. 16), illumination (Feb. 16–Feb. 23), and darkness (Feb. 23–Mar. 13). From F. A. Brown, Jr., J. Shriner, and C. L. Ralph, *Am. J. Physiol.* **184,** 491–496 (1956).

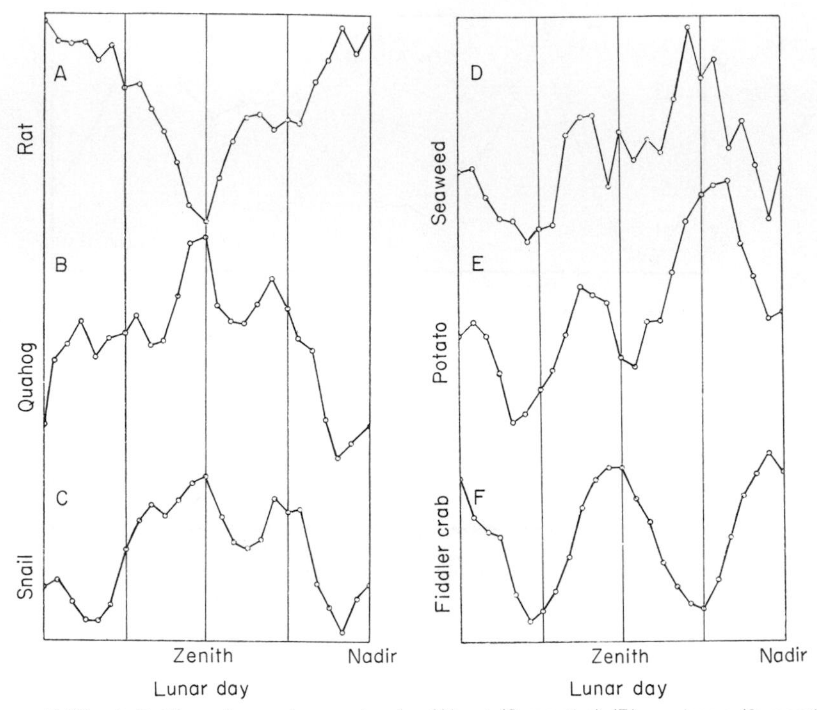

Figure 7-27 A–F. Mean lunar day cycles for (A) rat (2 months) (B) quahogs (8 months) (C) snails (1 month) (D) Fucus (5 summers) (E) potatoes (3 years) and (F) crabs (5 summers). From F. A. Brown, Jr., *Cold Spring Harbor Symp. Quant Biol.* **25,** 57–71 (1960).

rise to either plus or minus correlations with them. It had been increasingly suspected over a number of years that this might be true. Two species could show mirror-imaged mean patterns relative to one another; for wholes or parts of the cycles (Figure 7-27) also, one species could exhibit a clear positive correlation with primary cosmic radiation, while another species could be showing just as definite a negative one (Figure 7-6). A portion of a geophysically correlated pattern could mirror image an earlier observed pattern with time, often even displaying an annual variation in sign, as was described for potatoes. With this state of affairs and usually without any manner of prediction, the geophysically correlated fluctuations, while demonstrably real, did not conform to the demand of ordinary science for reproducibility at will.

Within the past 2 or 3 years, some substantial strides have been made in the firm establishment of the occurrence of the two states. The same species at the same time may be driven to either predominantly plus or minus response states. Slow rotation on tables in the earth's field or rotations of very weak magnetic fields for the organisms can usually cause the organisms to assume one or the other of the two states or mixtures. Slow counterclockwise table rotation leads to one state, clockwise rotation to the other. Organisms on stationary tables are influenced in just the opposite manner by rotation of their magnetic fields by weak bar magnets. Counterclockwise table and clockwise magnet rotations tend to produce one sign, and clockwise table and counterclockwise magnet, the other sign.

Organisms inside a metal-sheathed constant-temperature cabinet respond with opposite sign, even to having a mirror-imaged monthly cycle, to those in the laboratory just outside the cabinet. Evidently, not only do parameters of the atmospheric electromagnetic fields produce substantial alterations within the organisms, but other alterations in these fields are able to determine the sign of the organism's response. It is this sign-reversing capacity that has occurred without apparent cause that led some of us several years ago to discuss the response to periods without a fixed pattern of observed response. The increasing data available make it possible to understand how this is done—complementarity or a flip-flop in responsiveness.

Perhaps, collectively, all organisms over time would show no algebraic residual response, while, in reality, every organism could be receiving periodic information all the time. As with the controversy of free-running clocks versus autophasing against fixed clocks, the investigation of the mysterious clock phenomenon is again leading to further details of the phenomena that have been obfuscating relatively simple issues, and even probably leading to erroneous conclusions and hypotheses.

When it is resolved why very slow rotation of the organism relative to the magnetic field in one direction produces an effect the opposite of rotation in the other direction, another substantial inroad will undoubtedly be made into the beautifully subtle ways in which terrestrial creatures have adapted themselves and their periodisms to their spinning and sun-orbiting planet.

The vast majority of holders of the internal timer hypothesis have been forced by such studies, as all the foregoing and others, to concede that organisms can often have their postulated endogenously clock-timed cycles phase synchronized to the subtle, pervasive, external geophysical ones. While this is a tremendous concession on their part, this is probably only the first step. Any subtle exogenous periodism that can synchronize a biological rhythm can, obviously, also be the primary timer of that rhythm. A subtle-field synchronized cycle can be simply a case in which the subtle field has come to overwhelm the autophasing phenomenon; in a free-running cycle, on the other hand, autophasing has assumed the dominant role. The cycle, fully labile with regard to its phase relations to the geophysical cycle, remains throughout it all timed by that same geophysical cycle.

There is adequate reason to presume that the circatidal cycles obey a fully comparable relationship to environmental lunar-day cycles, either adhering faithfully to the external periods or free-running relative to them. Free-running menstrual cycles, evidence suggests, are underlaid by a variation with a period of the natural synodic month. And relatively large annual geophysically correlated variations appear to be present in every organism in which they have been appropriately sought. All can be the basic, completely temperature-independent timers for their respective circa periodisms, with the circa cycles autophasing or, particularly in the case of the longer monthly and annual periods, possibly simply drifting gradually relative to the periodic subtle input.

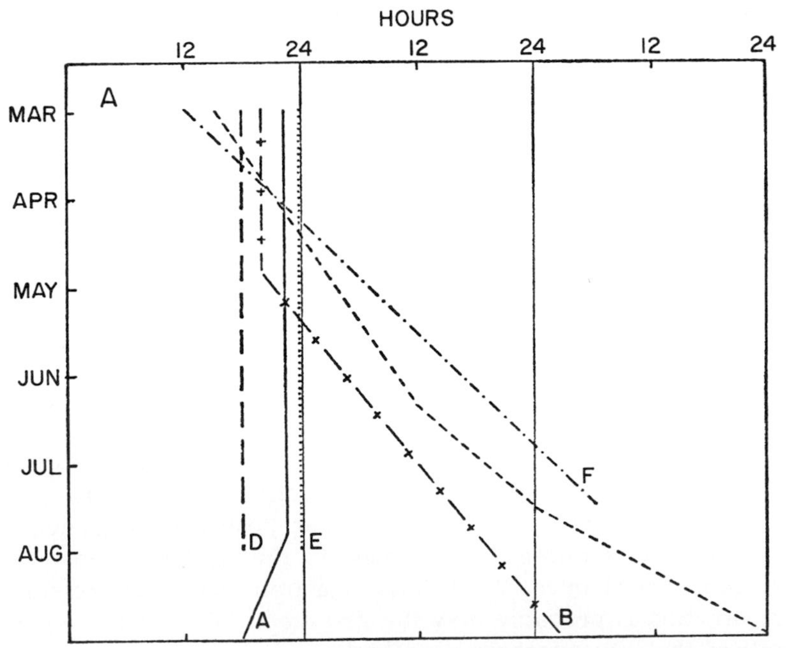

Figure 7-28 Time of day of estimated center of period of maximum activity and its changes for each of six mice over a 6-month period. From E. D. Terracini and F. A. Brown, Jr., *Physiol. Zool.* **35,** 27–37 (1962).

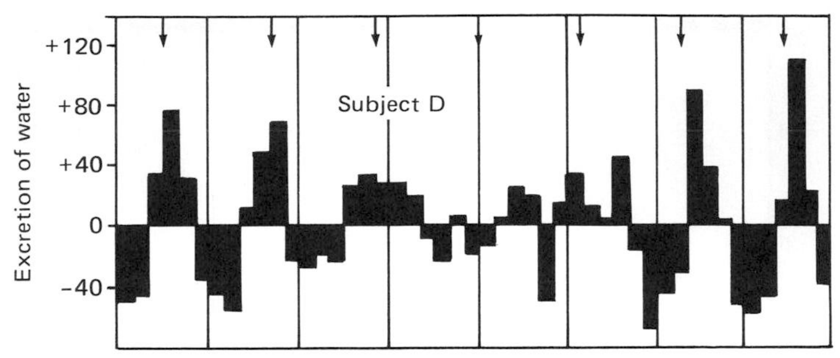

Figure 7-29 Urine flow, percentage deviation from mean, in a subject during eight 21-hour "days." Arrows indicate midday by real time. From P. R. Lewis and M. C. Lobban, *Q. J. Exp. Physiol. Cogn. Med. Sci.* **42,** 371–386 (1957).

PRECISION OF SOLAR-DAY CYCLES

Another argument for external timing of the biological clocks is readily seen in the mean precision of the rhythms they time. It was stated by E. Bünning about a half dozen years ago that if periodisms of 24 hours were observed under conditions in which no obvious factors were synchronizing them to this period, then this would be clear evidence for external timing. There are ever-increasing numbers of reported instances where this is the case. A male white rat had cycles indistinguishable from 24 hours during two periods in darkness, with activity at two quite different times of day (Figure 7-25). These two periods were interspersed with longer runs in constant dim light in which a $25\frac{1}{4}$-hour free-running period was evident. The records of 6 mice in independent recorders in the same continuous, dim-light field were informative (Figure 7-28). Two showed free-running rhythms from the start. Two others held to 24 hours for 2 and 5 months, thereafter abruptly becoming free-running. The final two retained 24-hour periods throughout the 5-month interval they were observed. P. Lewis and M. Lobban, working in summer in Spitzbergen, found evidence for a 24-hour periodicity in man to underly a larger amplitude imposed concurrent 21-hour cycle in water excretion (Figure 7-29).

Others working with populations of Gila monsters and kangaroo rats observed statistically precise 24-hour periods. Persistent 24-hour cycles appear to display also a minimum in their standard deviation when compared with other frequencies. Briefly, the ability of organisms to maintain an accurate mean 24-hour cycle far exceeds the expectation from a random series of circa periods spanning the 24-hour

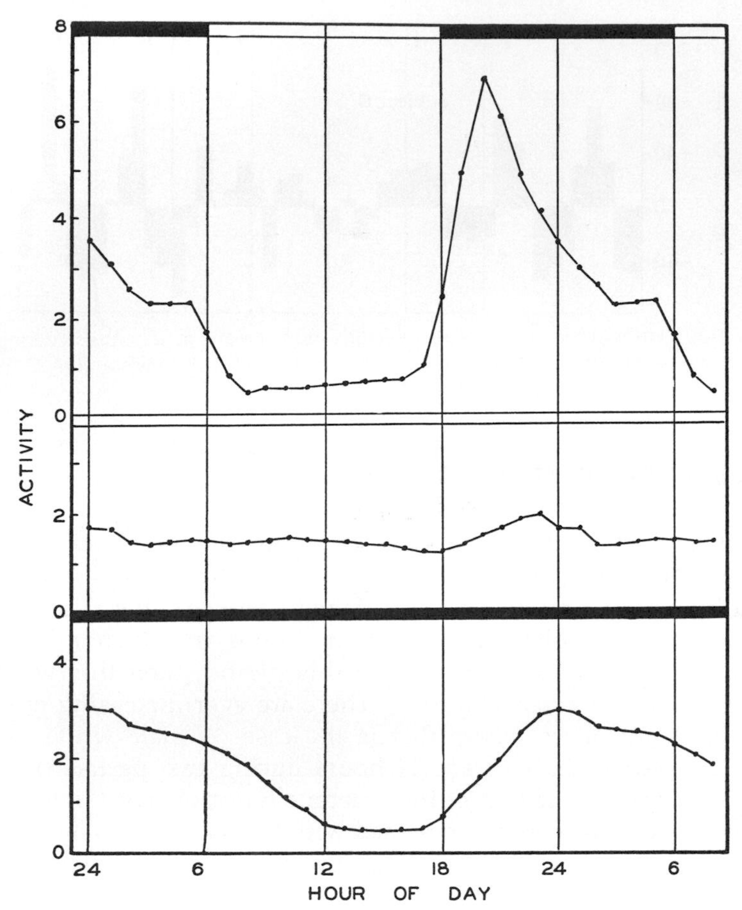

Figure 7-30 Average hourly running activity for four hamsters during (A) 6 months in 12:12 light-dark cycle; (B) 3 months in light; and (C) 3 months in darkness. From F. A. Brown, Jr., and C. S. Chow, umpublished.

period. More recently, a covert low amplitude 24-hour mean variation underlying 12 hamsters all free-running for 3 months in continuous dim illumination was found by C. Chow and F. Brown, Jr., in darkness to entrain, within a few days, the free-running cycles of all the hamsters. The entrained overt 24-hour periodicities had essentially the same detailed mean form and phase as the 24-hour pattern observed in constant light (Figure 7-30). This form of synchronized cycle in darkness was quite different from that of 24-hour cycles entrained to a light-dark one. L. G. Johnson's study at Northwestern University of rhythms of chick embryos during a critical 5-day period extending through the differentiation within them of a completely functional

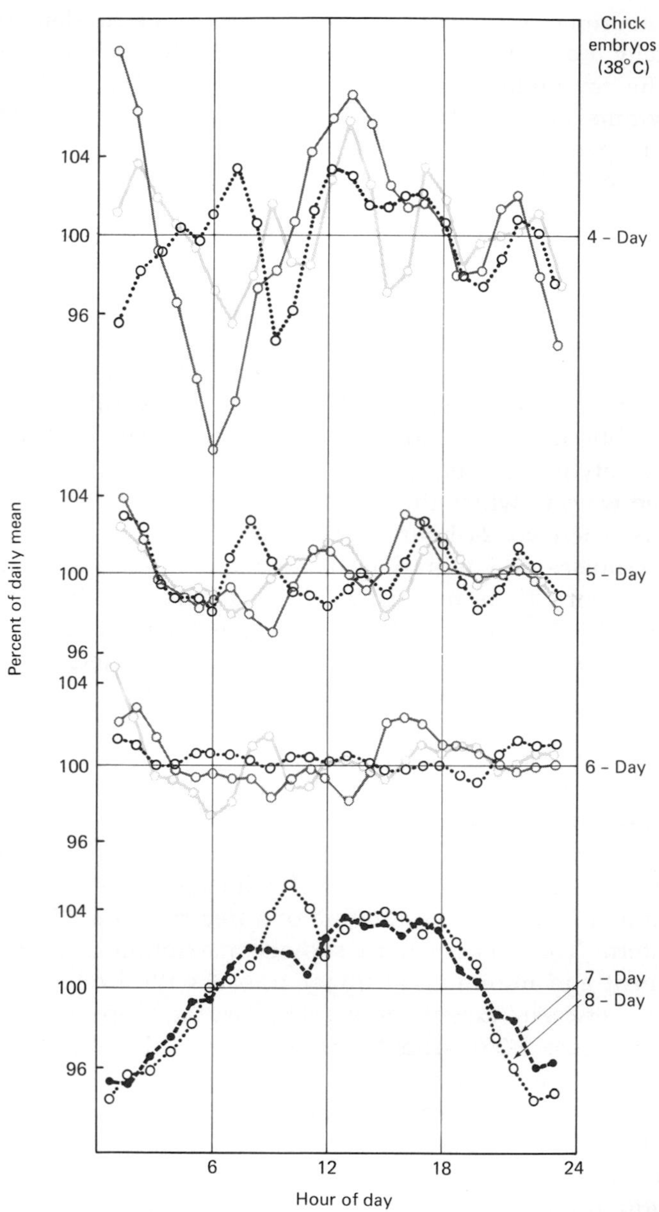

Figure 7-31 Mean daily metabolic patterns of chick embryos during early development in what biologists customarily consider to be constant conditions. For 4 to 6 days of incubation, the dotted curves are for spring, the faint curve for fall, and the solid curve for winter. Of the bottom two patterns, both for spring, the solid curve is for day 7, the dotted for day 8. All are 3-hour moving means. Calculated from original data of L. G. Johnson, *Biol. Bull.* (*Woods Hole, Mass.*) **131,** 308–322 (1966).

sensory-neuromotor system suggests a shift from a fundamental environmentally induced pattern to a 24-hour genetic one set to the proper time in the day-night cycles for the diurnal chicks (Figure 7-31). Circadian rhythms have been found by T. Bryant of the University of Georgia to persist even in dry seeds in which virtually all chemical reactions are in abeyance.

PROPENSITY FOR LUNAR PERIODISMS

The persistent 24-hour rhythms in constant conditions are not as persuasive arguments for natural external timing as the accurately timed lunar-day or tidal period, since these would reflect much weaker subtle-field variations than for the solar-day ones. Since human society is a 24-hour periodic one, there are innumerable man-influenced ways in which the 24-hour period could be conveyed to the organisms. There are 24-hour cycles in line voltage of our power lines. There are noises and vibrations from human activities, odors, and many other periodic factors. However, man does not live by a lunar-day or 24.8-hour schedule. If these were present in constant conditions, then truly natural subtle atmospheric parameters would be involved. Again, the precision of this cycle in early studies of fiddler crabs was impressive. The later investigations by M. Webb, F. Barnwell, E. Naylor, and J. Palmer, referred to in Chapter 3, have confirmed the great mean precision of the tidal cycle that could often persist in constant darkness or in 24-hour light-dark cycles.

However, even more impressive has been a persistent lunar-day fluctuation in male rats. The form of a lunar-day variation was first disclosed in a 2-month investigation of a free-running, 25¼-hour circadian pattern. The lunar pattern showed maximum activity at lower lunar transit and minimum at upper transit with a relatively smooth gradation between, except for a secondary and lesser minimum at moon set (Figure 7-26). About 3 years later, E. D. Terracini and the author, working with another male white rat, noted that for the whole 2-month period of observation a free-running circadian rhythm was present that was indistinguishable from 24.8 hours. When the cycle for the study was quantified, the detailed cycle form and its relations to lunar midnight and noon, and even to moonset were amazingly, very close to those of that cycle which had been present concurrently with the 25¼-hour rhythm in the first rat. The lunar-day cycle had apparently both entrained and given its form to a cycle normally synchronized to the solar day (Figure 7-32).

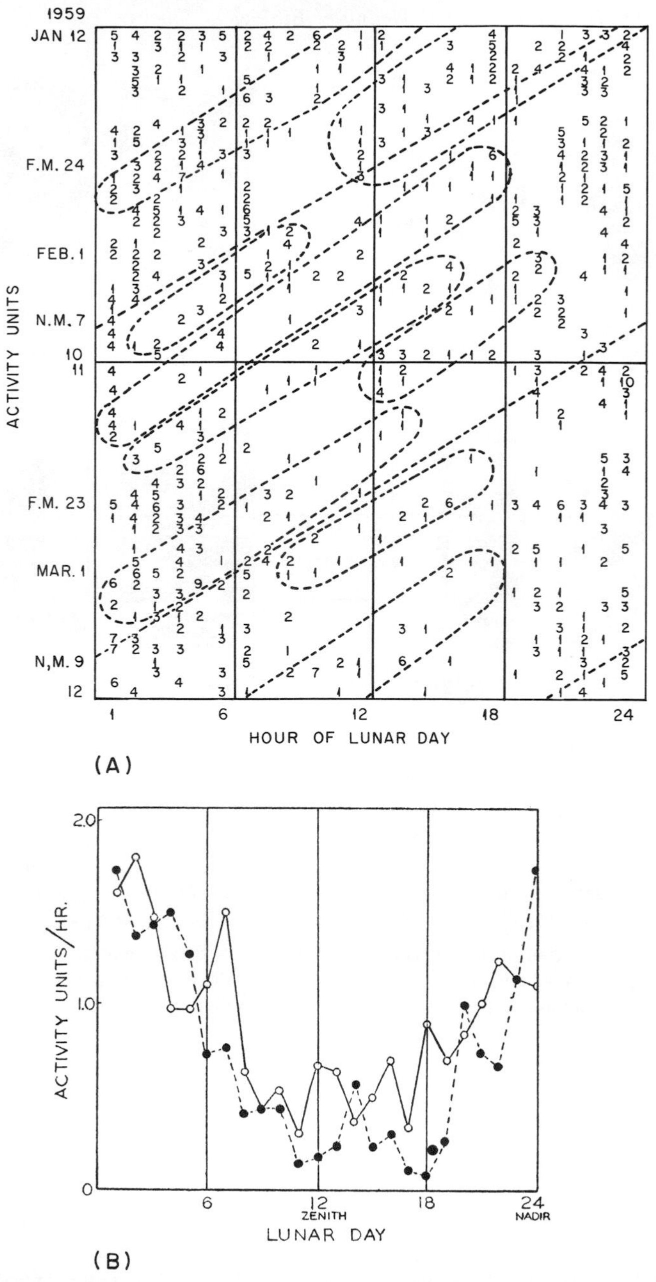

Figure 7-32 (A) Chart including activity of male white rat kept in constant illumination and temperature for each lunar-day hour during 60-day period. (B) Average lunar-day cycle of running of the rat for 30-day period, January 12 through February 10 (solid line); for 30-day period, February 11 through March 12 (broken line). From F. A. Brown, Jr. and E. D. Terracini, *Proc. Soc. Exp. Biol. Med.* **101,** 457–460 (1959).

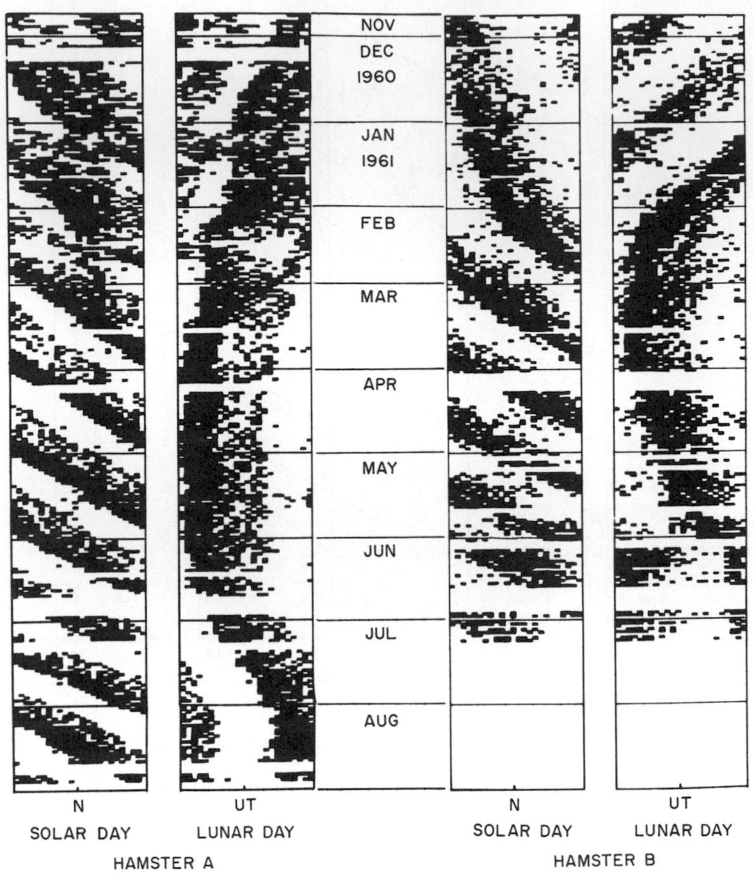

Figure 7-33 Solar-day and lunar-day plots for each of two hamsters held in continuous dim illumination. Note the tendencies of both animals to assume the lunar period for longer or shorter intervals. From F. A. Brown, Jr., *Proc. Soc. Exp. Biol. Med.* **120,** 792–797 (1965).

Similarly, two male hamsters steadily recorded in continuous dim light, one for about 7 and the other for about 9 months, both showed—independently of one another—an unusual propensity of the free-running circadian rhythm to be trapped again and again into a 24.8-hour period (Figure 7-33). Planarian worms, tested in the morning 5 or 6 times a week for the strength of phototaxis showed each year, for five consecutive years (Figure 7-34), a monthly variation with minimum over the day of full moon, and maximum over new moon. There was a superimposed annual modulation. Beans placed in water over a 12-month period displayed a quarterly lunar variation in their rate of water uptake during their first 4 hours with maxima very close

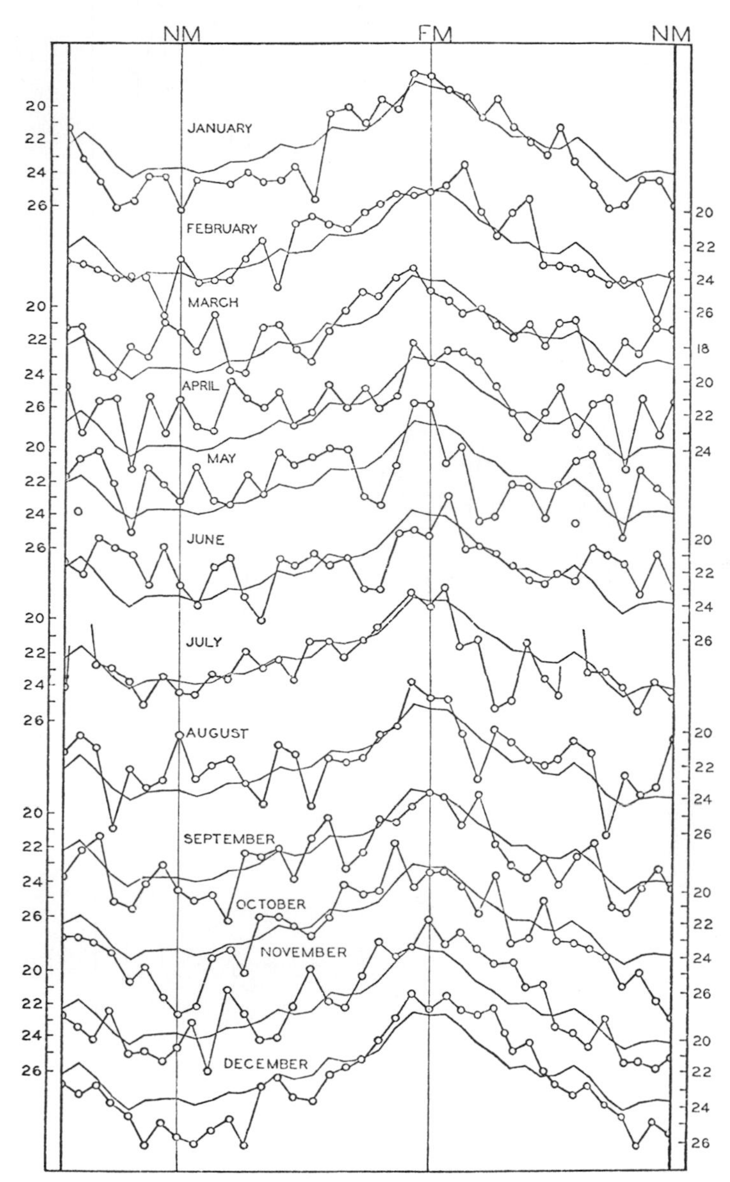

Figure 7-34 Average results for each month of the year obtained from a 5-year continuous investigation of strength of turning of planarian worms away from a light on their right. 0° would be no response; ordinate values are degrees of turning away from the light. The 12 unbroken curves are repetitions of the average monthly rhythm for the year.

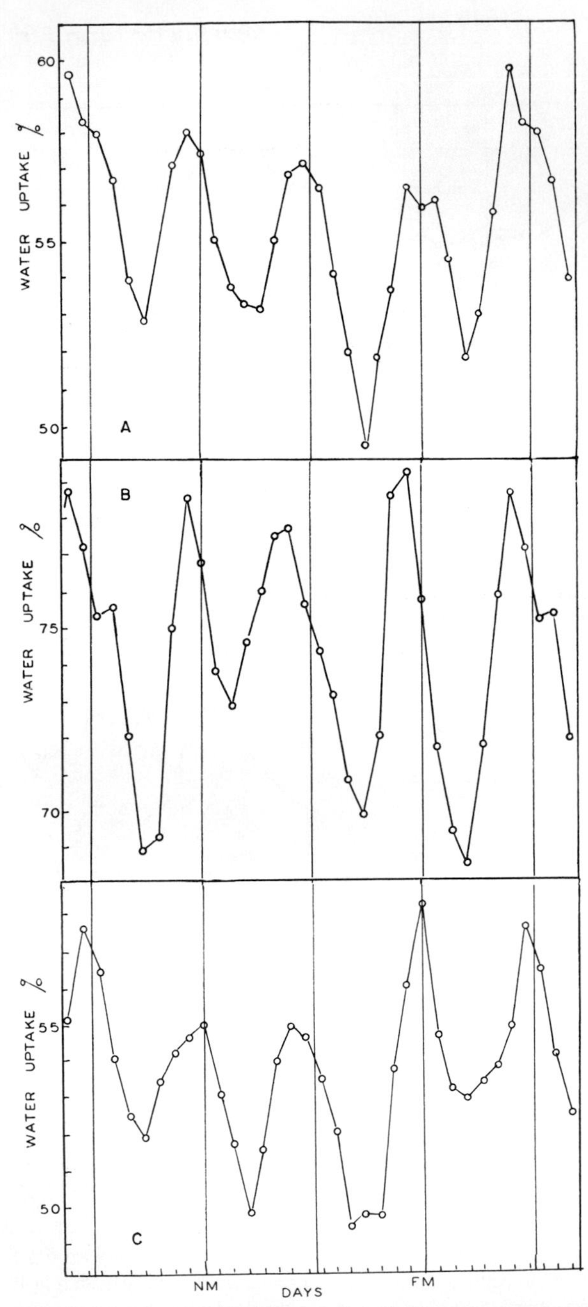

Figure 7-35 (A) Mean synodic monthly pattern for Series over period May 15 through August 18, 1972, in Evanston, Ill.; (B) for Series at Woods Hole, Mass.; and (C) for Series for the period September 25, 1972, through January 5, 1973, in Evanston, Ill. All are three-day moving means. Modified from F. A. Brown, Jr. and C. S. Chow, *Biol. Bull.* (*Woods Hole, Mass.*) **145,** 265–278 (1973).

to the lunar quarters. Only water absorption by physical processes is believed to be occurring. The similarity between spring and fall in Evanston, Illinois, and summer in Woods Hole, Massachusetts are notable (Figure 7-35). J. Aschoff had reported that the free-running body temperature circadian rhythm of a man without timing cues ran with a 24.8-hour rhythm while his sleep-waking cycle possessed a 32-to 33-hour period (Figure 7-19). T. Hoshizaki at the Space Laboratory of the University of California, Los Angeles, recently has found, simultaneously, both 24-hour and 24.8-hour periods in the chimpanzee in constant conditions.

Wever, long an avid supporter of internal timers, has very recently reported widespread occurrence of a 24.8-hour periodism in man—a period, he conceded, that was indistinguishable from an exogenous lunar-day frequency.

The menstrual cycle of the human female between 15 and 40 years of age averages close to 29.5 days, the average synodic monthly period

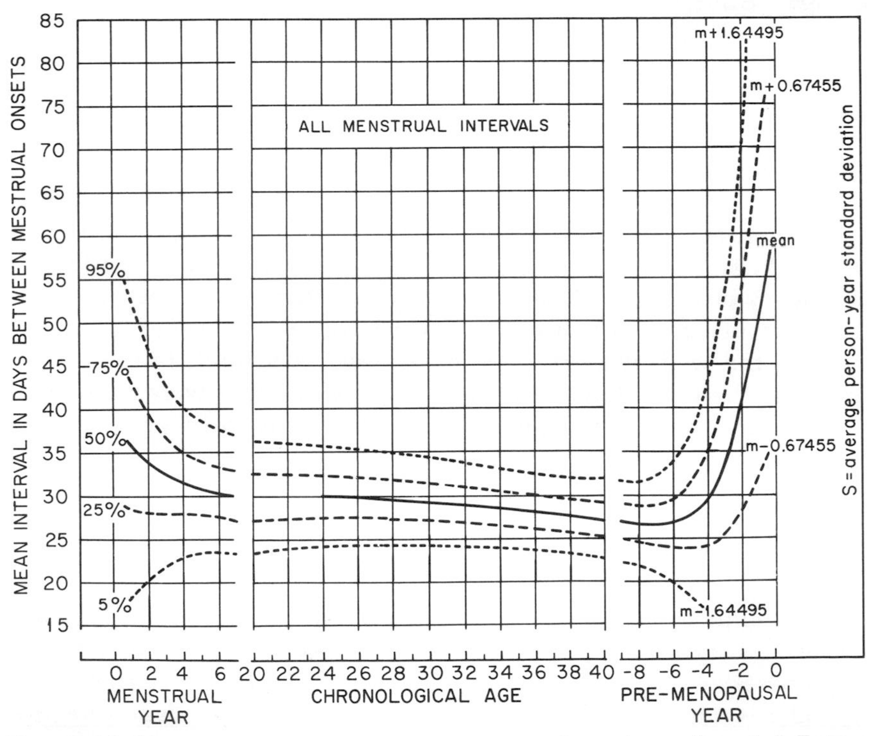

Figure 7-36 Human menstrual intervals in three zones of experience. From A. A. Treloar, *in* "Natural Family Planning" (W. A. Urricchio and M. K. Williams, eds.), pp. 64–71. Human Life Foundation, Washington, D. C., 1973.

(Figure 7-36). Investigating about three-quarters of a million births in two large New York City hospitals, W. Menaker and A. Menaker reported a synodic monthly variation in frequencies of the births. This last suggests that a monthly cycle underlies the free-running menstrual and ovulation cycles in the human. E. Dewan, dealing with a group of women with a wide range of menstrual period lengths, discovered that providing nocturnal illumination on days 14–17 following the initiation of menstruation induced the cycles of all the women to converge toward a common period of 29.5 days. This period length, that of the natural month, was not contributed by the procedure. Days 14–17 for the nocturnal illumination had been suggested to him by reports that some primates in nature menstruate at new moon and ovulate at full moon. Also, monthly variation was found to occur in hamsters for each of two consecutive years in 12 : 12 light-dark cycles (Figure 7-37).

ANNUAL RHYTHMS

While circannual cycles are being reported in constant conditions, persistent annual cycles in a variety of organisms and phenomena have been described. These very commonly possess their maxima in July-August and minima in October-December (Figure 7-38). They have been reported for O_2 consumption in potatoes and bean sprouts, orientation of planarians to a very weak gamma radiation source, nitrate reduction rate in an alga, food uptake in woodchucks, and water uptake in bean seeds. Most impressive of these were, perhaps, the bean sprouts that were investigated during the years four, five, and six, that the seeds were retained in constant conditions in the laboratory, and the woodchucks with which futile attempts were made with temperature changes and photoperiod alterations to change the cycle form.

EXTERNAL TIMING AS A SCIENTIFIC HYPOTHESIS

Any sound scientific hypothesis must fulfill five conditions: (a) relevance, (b) explanatory power, (c) compatibility with established hypotheses, (d) simplicity, and (e) testability. The external timer hypothesis meets all five of these criteria. The first two were met through the general acceptance of the phase-lability and response systems of organisms. The third became anchored in the demon-

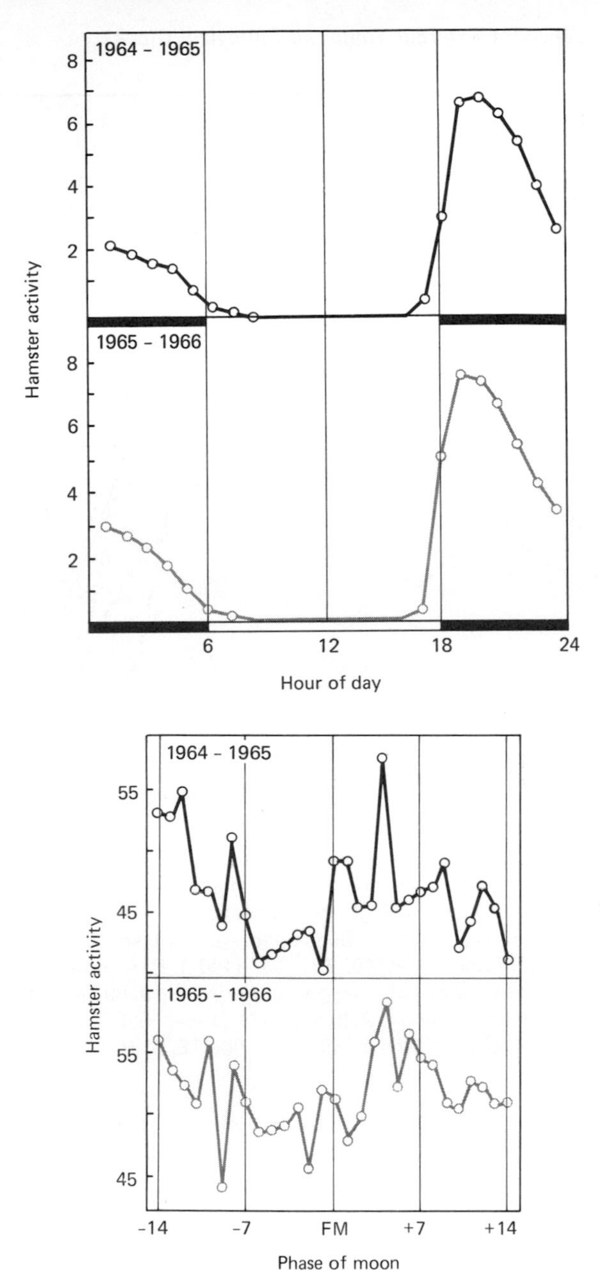

Figure 7-37 Daily and monthly variations in activity of hamsters maintained in alternating 12-hour periods of light and darkness. The solid black horizontal bars signify the times of darkness. On the left are the mean daily activity patterns for two hamsters from 1964–1965 and of two different hamsters during the next 12 months. Shown below are the mean daily amounts of activity as they vary with day of the synodic month for each of the two years. From F. A. Brown, Jr. and Y. H. Park, *Proc. Soc. Exp. Biol. Med.* **125,** 712–715 (1967).

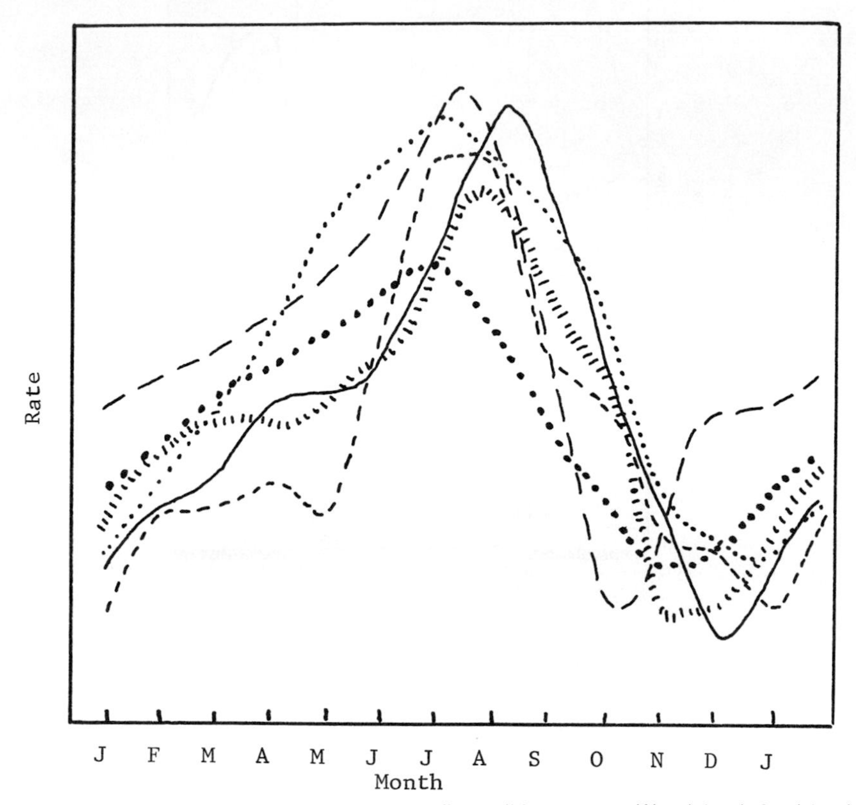

Figure 7-38 Annual variations in "constant" conditions. · · ·, Woodchuck food intake. ●●●, Potato O_2 consumption. ///, Algal nitrate reduction. ———, Planarian gamma response. ___, Bean O_2 consumption. ---, Bean-seed water uptake. (Assembled from various sources: D. E. Davis, *Physiol. Zool.* **40,** 391–402 (1967); F. A. Brown, Jr., *Scientia* **103,** 245–260 (1968); E. Kessler and F. C. Czygan, *Experientia* **19,** 89–90 (1963); F. A. Brown, Jr., and Y. H. Park, *Nature* (*London*) **202,** 469–471 (1964); E. F. Lutsch, Ph. D. Dissertation, Northwestern University, Evanston, Illinois (1962); F. A. Brown, Jr. and C. S. Chow, unpublished.)

strations of absence of "constant conditions" for organisms and responsiveness to a host of parameters of the ambient electromagnetic fields. Simplicity, a qualitative property, is more difficult to evaluate. However, pitting against one another the two concepts available for the timer—straighforward demonstrable responsiveness to environmental rhythms, on the one hand, and a hypothetical, still undisclosed, cellular mechanism generating regular, temperature-compensated periods within each organism, on the other hand—the greater simplicity once judged for the internal timer hypothesis has perhaps now yielded its position to a greater simplicity of the external.

Finally, the external timer hypothesis has been claimed by a few investigators to be unscientific on the grounds of displaying a lack of testability, while they claim testability for the internal timer. In fact, however, the external timer view appears to be the more easily establishable, while simultaneously being the most difficult to exclude. Witness the predicted systematic discoveries made over the recent years of the organismic-environmental interactions and the disappearance of the concept of constancy of conditions. Also note the many evidences for environmental determination, in large measure, of cycle timing and cycle forms. Note also the tests designed by Natalini and by Avery, both of which indicated free running periods to be generated from close to 24-hour ones by autophasing.

Indeed, major concepts of biology have commonly arisen from observation and induction, rather than by deduction from what is known. Efforts in postulation of hypotheses and deducing tests for them may never lead to the correct answers if the hypotheses are rooted in established ones which are not relevant to the problem at hand. The clocks of life appear to demand an admission of ignorance about a lower level of organization of life. The problem should be defined first as involving either (a) an exclusively organismic or (b) an organismic-environmental clock. The hypotheses to be formulated, experiments to be designed, and the tests to be made are quite different for the two concepts. If organismic, the tests would probably parallel and be compatible with the nonclock reductionism prevalent today in biology. If organismic-environmental, the tests should be quite different and center on demonstration of the field factors involved, the mechanisms of their reception and perception, and the integration of their effects, with other inputs, into the total economy of the individual.

GENERAL CONCLUSIONS

New discoveries in geology led a few years ago to a revolution in the geologists' view of the nature of the earth from the long-held fixistic with stationary continents to a mobilistic with drifting tectonic plates. Commensurate discoveries in biology that have been described here will in all likelihood soon lead to a comparable revolution in the biologists' view of the timing of biological rhythms. From the earliest recognition of clock properties of organisms, the two current views have been voiced. In timing its rhythms, the organisms were either (1) independent of, or (2) dependent upon, the environmental rhythms.

During the discoveries over the years of many of the known extraordinary properties of the rhythms, it seemed at one time the best hypothesis to postulate that the organism had evolved an intrinsic, largely temperature and drug-compensated clock system. It seemed the simplest and most straightforward one and, hence, followed Occam's rule in the selection of a hypothesis.

This hypothesis has, however, turned out to be a relatively sterile one in disclosing the clock mechanism. And it has, of course, depended upon negative evidence for its continuation. No external source of information has yet been proved to be needed. However, a single, truly positive experiment could demolish the hypothesis. The hypothesis has been reinforced by the formulation of refinements in the hypothesis for the explanation of many of the observed characteristics of the rhythms and, heuristically, by the design of theoretical physical or mathematical models. All these can be no stronger than the basic hypothesis that an intrinsic clock-propertied oscillatory complex exists within each and every organism, indeed, in every organ, tissue, and cell of it. It is small wonder, therefore, that with a hypothesis so insecure, and unsupported by any direct evidence in favor of the existence of such a clock at all, the proponents of internal timing almost never mention the alternative hypothesis of external timing, except occasionally to attack evidences for this opposing view. Such attacks have centered upon statistical methods, experimental procedures, and, incredibly, on ambiguities of interpretations.

Support for the case for internal timing of the biological clocks has seemed, in ever widening measure, to involve suppression of all evidence for the alternative external one. Intuitively, or rationally, advocates of internal timing seem to sense increasingly the essential futility of their view which cannot be established short of finding the mechanism of a clock that can and does operate with all the astounding observable biological properties and over such extended intervals without any periodic input. In view of the complete absence of any progress on this score during many years of work by very able physiologists and biochemists, this seems to be a highly unlikely event.

On the other hand, the progress being made in the discovery of facts that strengthen the possibility of external timing has been great and is accelerating rapidly. More than 10 years of work in the author's laboratory which established that rhythmic information from the physical environment was still reaching organisms in the most constant environments with which physiologists dealt was finally, but only begrudgingly, acknowledged. This together with the frequency, intuitively well beyond chance, that organisms, or populations of them, could retain in constancy of all obvious environmental factors accurate mean geophysical periods finally compelled the internal timer advocate to concede the not uncommon operation of subtle *Zeitgeber* or phase setters.

Currently, the study of responses of organisms to extremely weak electromagnetic parameters of a host of kinds and quantum energies has proven that the yesteryear concept of organisms in constant conditions, that prevailed when the concept of internal timing seemed compelled, has been fully outmoded. Spanning the gamut of the electromagnetic spectrum, including parameters of strengths and vector directions, organisms have been demonstrated to display such high degrees of sensitivities that they can perceive, through such fields, the passage of the environmental cycles, recognize different sites on the planet, and even recognize the presence of neighbors. The clocks of life have never been shown to run with their remarkable regularities in truly constant conditions. Would W. Pfeffer have adopted, as he did reluctantly in 1915, the internal clock hypothesis had he known all the recent findings?

And now if one simply assumed, no longer unreasonable in terms of all the foregoing, that the clocks of organisms depend upon a continuous interaction of organisms with their subtle environment, then all the rest of the known properties of the observed clock-timed rhythms fall splendidly into place. Everything else can conform to the conventional rules of physics and chemistry. First, there is the development of inherited recurring patterns linked to one or another of the geophysical cycles. Second, there is the development of a phase-response system and adaptive resettability of the rhythms by relevant environmental stimuli, including dominantly the light and temperature cycles. Third, with the phase lability of the rhythms and their peculiar phase-response-activity complex, free-running cycles slightly modified in period by differing light and temperature levels, as well as influences of genetics and some chemicals, are rationally explained as effects on autophasing or of systematic phase drifting relative to the natural exogenous cycles. Indeed, from this start, it seems quite probable that every other property of the rhythms, known or still unknown, can be accounted for by the appropriate elaboration of the external timer hypothesis. Such properties would even include the jet syndrome, phase dissociation, geophysically dependent variations, and time training.

Furthermore, in terms of external timing, the clocks of living creatures will more readily become part of a larger problem, including homing and navigation of organisms. The recognition of, and reaction to, spatial coordinates, as well as to temporal ones, will be thrown into a common pool for solution. Clocks once believed necessary for the phenomena are found not to be so. Space and time appear to have converged for organisms into a continuum, as in nature. To think any longer of a sense of time in total isolation from the remainder seems to resemble more a limited intellectual game than scientific exploration for the truth wherever it lies.

Before this problem of internal versus external timing is ultimately resolved, however, there will probably be a large play on terms and semantics. Every attempt will be made to hold to every still tenable vestige of the internal timer hypothesis; as the change over from the currently dominating internal to the external hypothesis occurs, one will see again and again spurious references to specific researchers who finally settled the matter. The truth will be, however, that a host of individuals and experimental results over many years will have collectively compelled a historical transition from one paradigm to a new one. It will be that a mass of sound descriptive new information leading to an alternative, more secure and widely embracing hypothesis has finally been recognized. Major advances in science are often characterized by an increasing unification and simplification.

If autonomous, clock-propertied oscillators timing the circa periods should ultimately be proved to exist within protoplasm, the numerous discoveries that have been made during the search for means of environmental timing would still remain as a permanent contribution to biology.

SUMMARY

Organisms behave as if they were accurate, or moderately accurate, clocks or tremendous batteries of clocks whose rhythmic cycles are normally integrated into a characteristic phase-map complex. The periods of the clock-timed rhythms appear to be heavily compensated for, or independent of, temperature and to virtually every chemical substance that can influence reaction rates in organisms.

Research over the past 15 years, rapidly increasing in volume and numbers of laboratories concerned, has disclosed that organisms have never been in constant conditions during any laboratory study of rhythms. Organisms are steadily apprised of weather changes and all atmospheric periodicities. This information can be conveyed steadily to organisms in constant conditions by way of magnetic, electric, and extremely low frequency electromagnetic fields, as well as those of high energy background radiation. With these capacities, the organism can resolve the differences within these fields occurring with not

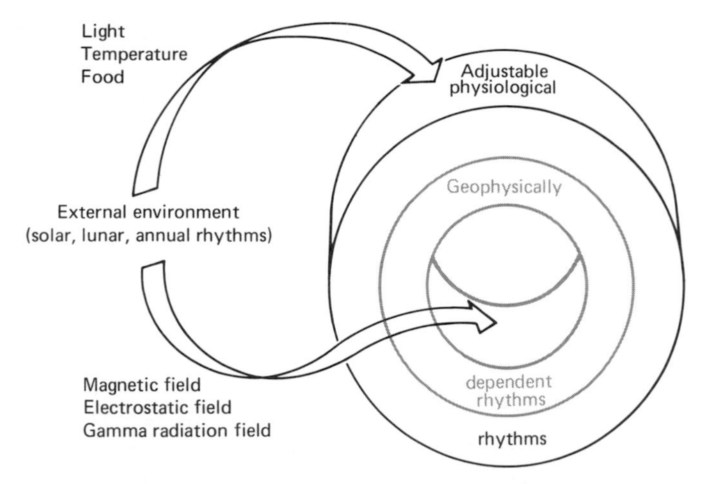

Figure 7-39 Diagrammatic representation of the duplex nature of biological rhythms. One, represented by the outer ring, includes the circadian, tidal, and other rhythmic patterns adapting the organism to the specific rhythmic patterns of its environment. The other, represented by the inner ring, describes underlying, deep-seated, rhythmic fluctuations, direct responses to subtle, rhythmic forces of the physical environment. From F. A. Brown, Jr., *Adv. Astronaut. Sci.* **17,** 29–39 (1964).

only time, but also probably obtain information as to their positions and orientations in space and the map sense.

All the clock-timed rhythms can be readily accounted for by postulating that the rhythms are recurring patterns genetically generated to fit the periods of clocks which are, in turn, timed accurately by rhythmic changes in the organisms' ambient physical environment (Figure 7-39).

The inherited rhythms, once developed in the individual, adaptively become capable of a labile phase relationship with the publicly timed clock cycles. They can be phase displaced to any degree in response to light cycles or other *Zeitgeber.* Such changes follow geographical translocations, or altered artificial light-dark schedules in the laboratory.

To accomplish the phase lability a phase-response system comprises a component of the inherited circadian cycle. Under conditions of constancy of all *Zeitgeber* fields and with such a phase-response system, the cycles would be expected ordinarily to self-phase shift, or *autophase,* cycle after cycle. This is postulated to be the explanation of free-running rhythms deviating from any natural geophysical period. It also can readily explain all the well-known free-running properties, including such responses as those to differing light levels, temperature levels, D_2O, cycloheximide, and genotype.

Lunar-tidal and day, semimonthly and monthly, and annual clock-timed inherited rhythms are also presumed, on the basis of the evidence available, to be similarly inherited rhythms timed by, or phase shifting relative to, the accurate environmental rhythms.

In short, all known properties of the clock-timed rhythms can be accounted for without any requirement for the postulated, but still undisclosed autonomous, temperature-compensated internal, clock-timing system. All issues are restored to the well-known, conventional rules of physics and chemistry.

Finally, the holistic approach to biological clocks supported in this chapter is the only one that can account for all the reported properties of the rhythmic systems of organisms, including geophysically dependent cycles and subtle-field entrainment. The reductionist attack can, at best, deal with only a limited portion of them.

Selected Readings

Avery, D. L. (1974). The relationship of light-dark reversal transients and free-running rhythms. *Int. J. Chronobiol.* **2;** 223–232.

Barnwell, F. H. (1966). Daily and tidal patterns of activity in individual fiddler crabs (Genus *Uca*) from the Woods Hole region. *Biol. Bull.* (*Woods Hole, Mass.*) **130;** 1–17.

Bennett, M. F. (1963) The phasing of the cycle of motor activity in the fiddler crab, *Uca pugnax*. *Z. vergl. Physiol.* **47;** 431–437.

Bennett, M. F. (1974). "Living Clocks in the Animal World." Thomas, Springfield, Illinois.

Bernhardt, P. A. (1974) Separation of lunar and solar periodic effects in data. *J. Geophys. Res.* **79** (28); 4343–4349.

Brown, F. A., Jr. (1954). Biological clocks and the fiddler crab. *Sci. Am.* **190;** 34–37.

Brown, F. A., Jr. (1959). Living clocks. *Science* **130;** 1535–1544.

Brown, F. A., Jr. (1960). Response to pervasive geophysical factors and the biological clock problem. *Cold Spring Harbor Symp. Quant. Biol.* **25;** 57–71.

Brown, F. A., Jr. (1962). Responses of the planarian, *Dugesia*, and the protozoan, *Paramecium*, to very weak horizontal magnetic fields. *Biol. Bull.* (*Woods Hole, Mass.*) **123;** 264–281.

Brown, F. A., Jr. (1962). Response of the planarian, *Dugesia*, to very weak horizontal electrostatic fields. *Biol. Bull.* (*Woods Hole, Mass.*) **123;** 282–294.

Brown, F. A., Jr. (1963). An orientational response to weak gamma radiation. *Biol. Bull.* (*Woods Hole, Mass.*) **125;** 206–225.

Brown, F. A. Jr. (1969). A hypothesis for extrinsic timing of circadian rhythms. *Can. J. Bot.* **47;** 287–298.

Brown, F. A., Jr. (1971). Some orientational influences of non-visual terrestrial electromagnetic fields. *Ann. N.Y. Acad. Sci.* **188;** 224–241.

Brown, F. A., Jr. (1972). The "clocks" timing biological rhythms. *Am. Sci.* **60;** 756–766.

Brown, F. A., Jr., and Chow, C. S. (1973). Interorganismic and environmental influences through extremely weak electromagnetic fields. *Biol. Bull.* (*Woods Hole, Mass.*) **144;** 437–461.

Brown, F. A., Jr., Hastings, J. W., and Palmer, J. D. (1970). "The Biological Clock: Two Views." Academic Press, New York.

Bünning, E. (1935). Zur Kenntnis der erblichen Tagesperiodizitat bei den Primärblattern von *Phaseolus multiforus. Jahrb. Bot.* **81;** 411–418.

Bünning, E. (1973). "The Physiological Clock." Springer-Verlag, Berlin and New York.

Chapman, S., and Bartels, J. (1940). "Geomagnetism," 2 vols. Oxford Univ. Press (Clarendon), London and New York.

Cole, L. (1957). Biological clock in the unicorn. *Science* **125;** 874–876.

DeCoursey, P. J. (1960). Daily light sensitivity rhythm in a rodent. *Science* **131;** 33–35.

Enright, J. (1965). The search for rhythmicity in biological time-series. *J. Theor. Biol.* **8;** 426–468.

Halberg, F. (1959). Physiologic 24-hour periodicity: general and procedural conditions with reference to the adrenal cycle. *Z. Vitam.-Horm.-, Fermentforsch.* **10;** 225–296.

Harker, J. E. (1963). "The Physiology of Diurnal Rhythms." Cambridge Univ. Press, London and New York.

Johnson, M. S. (1939). Effect of continuous light on periodic activity of white-footed mice. (Peromyscus). *J. Exper. Zool.* **82;** 315–328.

Kalmijn, A. J. (1971). The electric sense of sharks and rays. *J. Exp. Biol.* **55;** 371–383.

Konopka, R. J., and Benzer, S. (1971). Clock mutants of *Drosophila melanogaster. Proc. Natl. Acad. Sci. U.S.A.* **68;** 2112–2116.

Lindauer, M., and Martin, H. (1968). Die Schwereorientierung der Bienen unter dem Einfluss des Erdmagnetfeldes. *Z. Vergl. Physiol.* **60;** 219–243.

Lissman, M. S. and Machin, K. E. (1958). The mechanism of object location in *Gynarchus nilotus* and similar fish. *J. Exper. Biol.* **35;** 451–486.

Menaker, W., and Menaker, A. (1959). Lunar periodicity in human reproduction: A likely unit of biological time. *Am. J. Obstet. Gynecol.* **77;** 905–917.

Pfeffer, W. (1915). Beiträge zur Kenntnis der Entstehung der Schlafbewegungen. *Abh. Säechs. Akad. Wiss. Leipzig, Math.-Phys. Kl.* **34;** 1–154.

Rommel, S. A., Jr., and McCleave, J. D. (1974). Oceanic electric fields: Perception by American eels? *Science* **176;** 1233–1235.

Solberger, A. (1965). "Biological Rhythm Research." Amer. Elsevier, New York.

Walcott, C., and Green, R. P. (1974). Orientation of homing pigeons altered by a change in the direction of an applied magnetic field. *Science* **184;** 180–182.

Webb, H. M. (1950). Diurnal variations of response to light in the fiddler crab, *Uca. Physiol. Zool.* **23;** 316–337.

Wiltschko, W., and Wiltschko, R. (1972). Magnetic compass of European robins. *Science* **176;** 62–64.

8

Models and Mechanisms for Endogenous Timekeeping

LELAND N. EDMUNDS, JR.

INTRODUCTION

The subject matter within the purview of a biologist differs from that in the domain of the physicist chiefly on the basis of the degree of complexity of its organization. Although the physicist may be quite content in analyzing the organized system into its component parts, the biologist—particularly the evolutionary biologist—is intrigued by the natural origin of such organization. And whereas the physicist sees the passage of time marked by a loss of organization (an increase in entropy), the biologist is confronted with a continuously reproducing and evolving set of highly organized living systems. As C. S. Pittendrigh has pointed out in an engaging essay, this observed organization in living forms is strongly history-dependent since it arose through the twin processes of natural selection and adaptation. Within the set of living systems, an organism is said to be adapted if it has thrived by means of differential reproductive success, and its adaptation is reflected in its total organization. Biological problems, therefore, pivot upon the complexities of biological organization.

To any student of biology who has studied cell structure and metabolism and who has even cursorily examined the physiology of more complex tissues, organs, and organ systems, it is almost self-evident that the spatial organization and the functioning of living organisms are inextricably intertwined. Of equal importance, however, is the temporal dimension of biological organization. F. Halberg of the University of Minnesota particularly has stressed that at the the physiological level, not only must the "right" amount of the "right" substance be at the "right" place, but also this must occur at the "right" time. Furthermore, it is often necessary for the organism to be in the right place at the right time with respect to food supply, sexual partners, or optimal physical environmental conditions. Indeed, the environment is highly periodic in many of its variations, so that it is often advantageous, if not essential, for the organism to adapt to these periodicities.

Functional Biochronometry

That organisms can and do measure time in some manner should now be abundantly clear from the quite diverse phenomena surveyed in the preceding chapters. These phenomena, explicitly demonstrating functional biochronometry, fall rather neatly into four different categories: (1) *persistent rhythms* having daily (circadian), tidal, lunar monthly, and perhaps, yearly (circannual) periods (Chapters 2 and 3); (2) the *Zeitgedächtnis,* or time sense of honeybees (Fig. 2-28); (3) seasonal *photoperiodism,* where many plants and animals perform a certain function at a quite specific time of the year by what may well be essentially a daily measurement of the length of the day (or night) (Chapter 6); and (4) *celestial orientation* and *navigation,* in which the sun, moon, or stars are used as direction givers, which implicate a timing system to compensate for the continuously, but predictably, shifting positions of these celestial bodies (Chapter 5). One can speculate that the last three categories of timekeeping are probably relatively recent developments in the history of organisms, which perhaps represent sophisticated variations on the more ancient evolutionary theme of circadian chronometry. If so, then an understanding of the nature of circadian (and circatidal) timekeeping mechanisms is crucial to the elucidation of the higher-level phenomena.

It should be noted that these four categories of time measurement do not include so-called physiological periodic events, such as heartbeat, alpha brain waves, epileptic seizures, or the emergence of 17-year cicadas, to name a few. Although these types of rhythms may, indeed, be generated by endogenous mechanisms, they lack external correlates and constitute clearly "private" phenomena, dependent to a large extent on the general physiological characteristics and state of the organism. Rather, the rhythms to be discussed here are deliberately limited to those that typically have periods approximately equal to those of the various geophysical periodicities resulting from the movements of the sun and moon relative to the earth and that persist upon removal of as many of the various synchronizing or entraining agents (such as light and temperature cycles) as is possible.

The central thesis throughout this text, then, has been that there is a selective premium for such temporal adaptation, especially to solar and lunar periodicities, and that these adaptive features can be and have been attained by organisms through their possession of some sort of timing mechanism—in particular, an endogenous "biological clock" which is responsive to and which can be synchronized, reset, and otherwise modulated by those environmental periodicities (such as light and temperature cycles) that the organism has encountered, and to which it has adapted, throughout its evolutionary history. The enigma, of course, is the nature of the clock(s).

Exogenous and Endogenous Timing Hypotheses

The basic tenets of both the exogenous and endogenous "schools" of thought has been briefly outlined in Chapter 1. It is fascinating (as, no doubt, future historians of chronobiology will agree) that from a purely formal standpoint all the definitive characteristics of well-formed circadian clocks—ubiquity, approximate 24-hour period, entrainability, persistence, phase shiftability, and temperature-compensation—are accounted for by either hypothesis. The situation in a nutshell is perhaps best illustrated in the now-classic verbal exchange between C. S. Pittendrigh (a leading proponent of endogenous timing) and F. A. Brown (the foremost advocate of exogenous timing) during the discussion following the presentation of the former's paper at the Cold Spring Harbor Symposium on "Biological Clocks" in 1960:

> **Brown:** ". . . No one can doubt that an inherited clock-system is present in organisms. But in insisting upon a self-timed, or fully autonomous, living clock, there always lurks the possibility that we are pursuing a ghost."
>
> **Pittendrigh:** ". . . Apparently we all—Dr. Brown now included—agree 'that an inherited clock-system is present in organisms.' The remaining areas of dispute concern the issues in his last sentence. The question of the ghost is simple—either it is an aspect of living organization, or an unknown geophysical variable. My taste in ghosts suggests the latter. . . . We both will have fun in any case."

One of the pleasures, perhaps, for a student of biological clocks is that he is confronted with two such opposing theories; likewise, such a situation presents an opportunity for the beginning investigator to "straighten out the mess." But how does one proceed empirically to attempt to discriminate between the two alternative theories? Unfortunately, the most direct approach is not yet financially feasible. If an organism could be sent in a space capsule outside of our solar system (or at least beyond the orbit of Mars), presumably beyond the influence of any 24-hour geophysical periodicity generated by the rotation of the earth about its axis, and if this organism continued to exhibit a circadian rhythmicity (as monitored by telemetry), then surely this would constitute the strongest type of evidence for a fully autonomous, self-sustaining oscillation. (Converse results, of course, would be ambiguous.)

But again, where does this leave us with regard to distinguishing the alternative theories? The mechanism of the putative endogenous clock, the nature of the hypothesized autophasing machinery, and the postulated ultimate subtle geophysical cues are still to be identified. Furthermore, Occam's razor and the law of parsimony cannot be wielded to prevent the multiplication of unnecessary entities since both theories are exotic to the uninitiated and were rejected out-of-hand by many scientists as little as 10 years ago. Nevertheless, valid options do remain. One is entitled to ask several pointed questions: What is the *predictive capability* of each contesting theory? Are these predictions *uniquely testable?* To what extent have these predictions been confirmed or disconfirmed by *empirical results?* To what degree has each theory generated excitement within the field and led to further lines of experimental questioning? And, to my mind most importantly, what headway has been made by the proponents of either theory toward understanding the mechanism of the underlying clock at a *phenomenological level* (as contrasted with a purely formal model)? The answers to these queries, therefore, are important in the assessment of a given theory or approach.

The goal of this final chapter, then, is to marshall the evidence to a large extent surveyed in Chapter 2 for a truly *autonomous, endogenous, self-sustaining oscillation*—a "biological clock" in the most restrictive sense—by means of which temporal organization could be maintained with more or less appropriate phase relationships to the physical and biotic environment, even if normal external stimuli were late or occasionally failed to appear or to be perceived by the organism. Models and mechanisms for circadian timekeeping will be emphasized. The observer must decide for himself the relative merits of this "endogenous" approach and the "extrinsic timing hypothesis" detailed in Chapter 7.

A word of warning to the uninitiated, however, is in order: an attempt has been made to avoid the black-and-white fallacy of portraying the field as a battleground between two opposing factions; nothing could be farther from the truth. Although historically there has been controversy along these lines, especially in the late 1950s and early 1960s, in general, this approach has yielded diminishing returns, and the graduate students of the earlier workers in particular, as well as relative newcomers to the field, no longer are turned on by (much less become uptight about) intrinsic-extrinsic "ghost-hunting." Workers have now become quite pragmatic. As M. Menaker at the University of Texas in Austin has put it, "At the moment continued flogging of the exogenous-endogenous horse is not likely to yield even a muffled heuristic whinny." By either broad theory there must be a wondrous amount of biological machinery involved within the cell whose detailed mechanism is virtually completely unknown. First, the types of lower-level hypotheses that have been and are being tested will be discussed in an attempt to elucidate first the formal and then the biochemical and molecular nature of this clockwork. It is really of only secondary importance whether it ultimately proves to be an autonomously self-sustaining oscillator or some sort of autophasing machinery dependent on an external subtle geophysical variable. My guess is that the two seemingly divergent approaches will tend to converge even more than they already have.

FORMAL MODELS FOR AN AUTONOMOUS CLOCK

When one considers the historical importance of *Drosophila* in the biological sciences, perhaps it is not so surprising that this lowly fly has afforded one of the most fruitful clock systems and has led to the formulation of several explicit, formal models for endogenous circadian clocks and their entrainment by light and temperature cycles. It is instructive to examine this system in some detail, not only because of the relatively high predictive power of such models, but also because the development of these models represented the second phase in the usual sequence of stages characterizing the growth of a new scientific field: first, the *descriptive phase* in which the phenomena (biological rhythms) were documented throughout the 1950s; second, the *modeling phase*, in which the effects of light and temperature perturbations on rhythms were formally described (commonly referred to as "black-box analysis") in the 1960s; and now, the stage in which attempts are being made to elucidate the mechanism of the putative underlying oscillator using biochemical and molecular species.

Phase Shifts and Transients

We have already seen that emergence (eclosion) activity in *Drosophila* pupal populations is gated to a specific recurring time interval during successive days in either light-dark cycles or in subsequent continuous darkness and that the period of this rhythmicity is remarkably insensitive to temperature over the normal physiological range (Figure 2-24b). Furthermore, C. S. Pittendrigh and co-workers at Princeton University discovered that the phase of this so-called free-running rhythm can be shifted by single, aperiodic light signals, as well as by the shifting of light cycles. The magnitude and the sign of the phase shift ($\Delta \phi$) are a function not only of the intensity and duration of the perturbing signal, but, more importantly, the phase of the circadian rhythm (that is, subjective circadian time in continuous darkness) at which the signal is given. Transients, involving the temporary shortening or lengthening of one or more successive, interim cycles, typically precede the attainment of a new steady-state equilibrium.

These findings are illustrated for *Drosophila* in Figure 8-1 (cf. Fig. 2-25). Single 12-hour signals were applied at different subjective times in different experiments. The plotted points (Panel 1-B) represent mean times of eclosion activity over a time span of 4 days; there are 13 rows of such points corresponding to 13 different cultures, each of which had been reared previously in a light cycle alternating 12 hours of light with 12 hours of darkness. At the beginning of the experiment (hour 24), the eclosion rhythm in all the cultures had identical phases (the median of the eclosion peak is taken as the phase reference point); normally, the light would have come on again, and the flies that had sufficiently developed would have eclosed 2 to 3 hours thereafter. Instead, all cultures were placed in continuous darkness except for the 12-hour light signal that each received. This signal was given at successively later intervals (2 hours) in the 13 cultures so as to scan across the entire subjective circadian cycle of the rhythm. The results presented, therefore, constitute a composite plot.

It is clear that the light signals generate phase shifts in the eclosion peaks whose magnitude and sign vary, depending on the phase of the cycle at which the stimulus is given. Notice that the first 8 cultures (denoted by open circles) initially exhibit a lengthening, or delay, of the free-running period ($-\Delta \phi$), while the remaining 5 cultures (solid

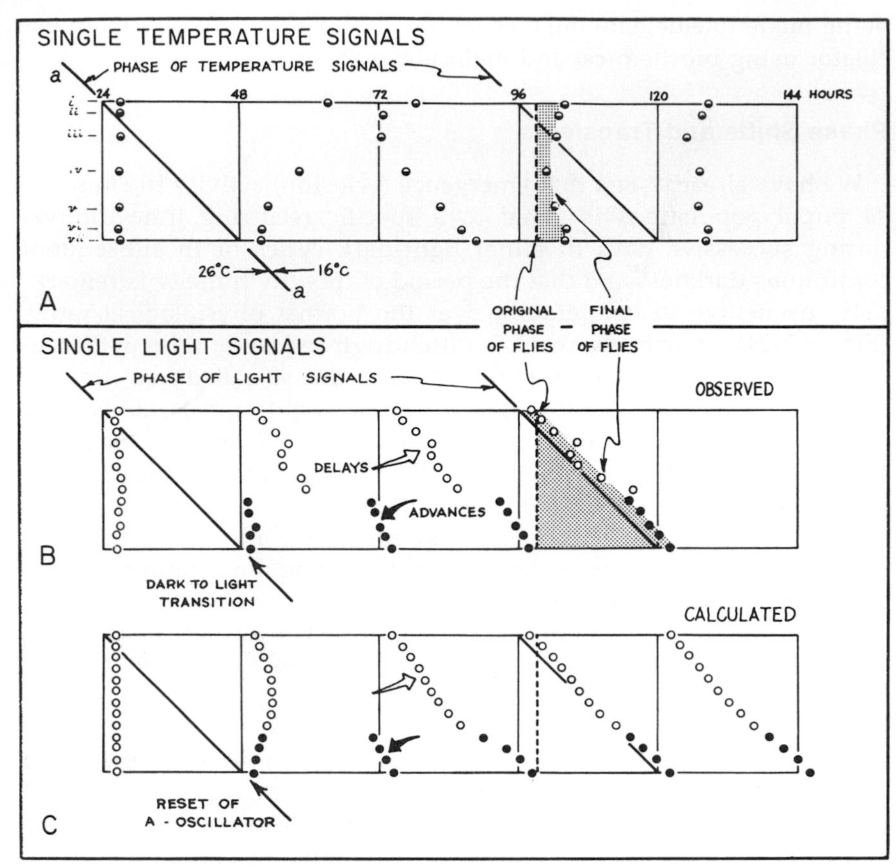

Figure 8-1 Light- and temperature-induced transients in the eclosion rhythm of *Drosophila pseudoobscura.* (A) Effects of single temperature signals. Each horizontal row of points represents medians of eclosion peaks in individual cultures. The cultures had previously been in a *LD 12:*12 light cycle; the time shown represents hours elapsed since the last dawn of this cycle. The 7 cultures were in constant darkness throughout the experiment. The temperature was dropped from 26°C to 16°C at successively later times in the different cultures as represented by the diagonal line between hours 24 and 48; this line is redrawn for reference between hours 96 and 120. (B) Effects of single 12-hour light perturbations given at successively later times in the 13 different cultures shown, as is indicated by the first heavy diagonal line (repeated later for comparison only). (C) The predicted behavior for the light-induced transients as derived from the mathematical formulation of the dual oscillator scheme. [C. S. Pittendrigh, *Cold Spring Harbor Symp. Quant. Biol.* **25,** 159–184 (1960).]

circles) show a shortening, or advance ($+\Delta\phi$), of their periods. These perturbations are only temporary, however, and are appropriately called *transients*. By the fourth experimental cycle (i.e., elapsed 24-hour timespan), the phase shifts are complete and a new steady-state equilibrium has been attained. The absence of data after hour 120 is due to the completion of eclosion in all the experimental cultures. Note that by the end of the experiment the medians of eclosion of the 13 cultures are no longer aligned at the same point in real time but, instead, are arranged along a diagonal line approximately parallel to that denoting the onset of the initial perturbing light signals. Indeed, eclosion now occurs in each case about 2 to 3 hours after the time the light signal would have fallen had more than one light exposure been given. Much shorter light signals have been utilized (e.g., 15-minute pulses) to achieve the same effect. Thus, a single pulse of light, which can convey essentially no timekeeping information in itself to the organism, can generate the appropriate phase shifts necessary for the rhythm to continue to maintain the same phase angle relationship with the signal as with the normally experienced light cycle.

From the results of the *Drosophila* phase-shifting experiment, one can construct a *phase-response curve* for the organism, which is nothing more than a plot of the magnitude and sign of the phase shift (if any) of a free-running rhythm, induced by a single stimulus such as a light pulse, as a function of the phase of the biological oscillation during which the stimulus was applied. Such a curve for *Drosophila* based on 15-minute signals of white fluorescent light is shown in the upper left-hand panel of Figure 8-2. It may be interpreted as a plot of the *periodically changing sensitivity* to the light signal itself and will become one of the cornerstones for a theory of the formal mechanism of the entrainment of a circadian rhythm by light cycles. In fact, a study of the comparative morphology of the phase response curves obtained for a number of different rhythms in a variety of plants and animals—six of which are shown in Figure 8-2—reveals striking similarities in their qualitative features. To an evolutionary biologist, whether or not he subscribes to the intrinsic or extrinsic timing hypothesis, these homologues (homomorphs?) cannot help but suggest a unity and, perhaps, convergence in mechanism (although a case could be made also for divergence from a basic, primitive clock mechanism).

Finally, let us examine the effects of temperature signals in the *Drosophila* system. In this case, a single step-down in temperature from 26° to 16°C performed at different circadian times in 7 different cultures maintained in constant darkness following prior entrainment by the usual light cycle (Panel 1-A, Figure 8-1). In this particular experiment, although large initial transient phase shifts were observed between hours 48 to 96, the final steady-state phases of the cultures were approximately the same as they had been before the step-down in temperature was applied. The interpretation at the time was that

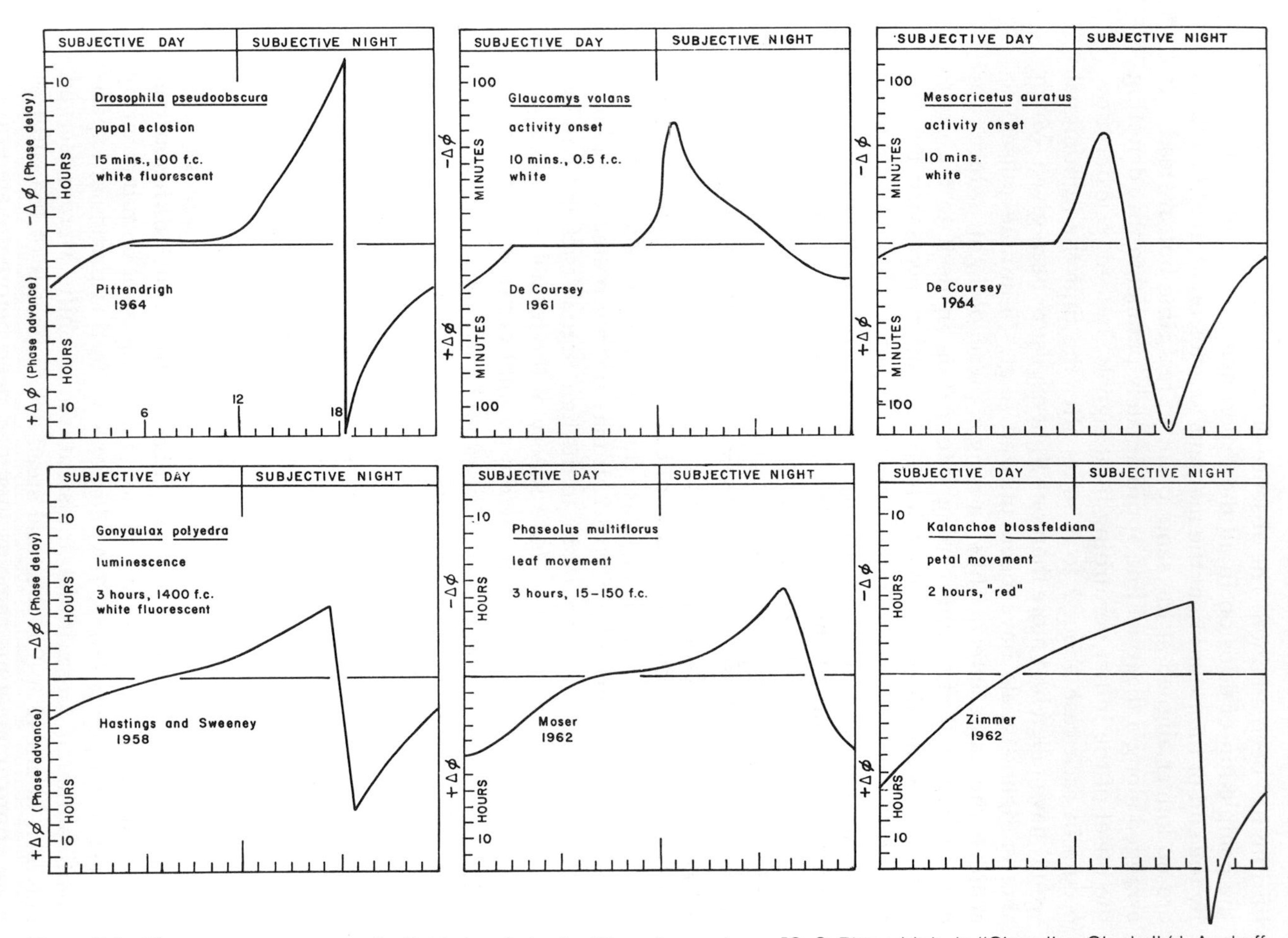

Figure 8-2 Phase-response curves for light signals in six different organisms. [C. S. Pittendrigh, *in* "Circadian Clocks" (J. Aschoff, ed.), pp. 277–297. North-Holland Publ., Amsterdam, 1965.]

single temperature perturbations (at least of this type) are not effective in phase-shifting circadian rhythms; in the light of more recent evidence (see Figure 2-27), however, this conclusion had to be modified.

The Coupled Oscillator Model

On the basis of these results with *Drosophila* and similar findings in many laboratories for a variety of rhythms, a general school of thought began to emerge in the 1950s whose basic tenet was that circadian (and presumably other) clocks have arisen through natural selection and constitute *endogenous, self-sustaining oscillations* (dubbed ESSO's) having their own innate, free-running period that is revealed under conditions of constant illumination (or darkness) and temperature (and presumably any other subtle geophysical factor). These clocks, however, would be subject to entrainment by appropriate light and temperature cycles and could be initiated, reset, and phase shifted by single, short light or temperature perturbations. On the other hand, their period would be remarkably insensitive to changes in temperature within the physiological range, perhaps effected by a series of compensatory reactions within the organism. The underlying oscillator(s) driving the hands of the clock or observed overt rhythms might be expected to have a cellular origin and molecular or biochemical basis, although some workers felt that the ESSO was less a discrete entity than it was a basic attribute of the temporal organization of living systems.

To explain the effect of light and temperature signals—and especially the resulting transients—in the *Drosophila* eclosion system (Figure 8-1), C. S. Pittendrigh and V. G. Bruce proposed a two-oscillator scheme as diagrammed in Figure 8-3. This formal model assumes that at least two distinct, self-sustaining oscillating systems underlie the assayed circadian rhythm of emergence activity (refer to Figure 2-24B). [The association of several transients with precise determination of ultimate phase and the fact that the transients themselves switch from delay to advance at about 18.5 hours after "subjective dawn" (i.e., the time that the lights would normally have come on in a light cycle), as seen in the phase-response curve for *Drosophila* (Figure 8-2), were felt to present great difficulties for any model based on a single oscillator.] One of these (the A-oscillator) was postulated at the time to be the pacemaking, temperature-compensated, ultimate clock, sensitive to light signals and immediately reset (phase shifted) by them. Coupled to it, and driven by it, is the B-oscillator, which was taken to be light insensitive and temperature sensitive, and was assumed to more or less

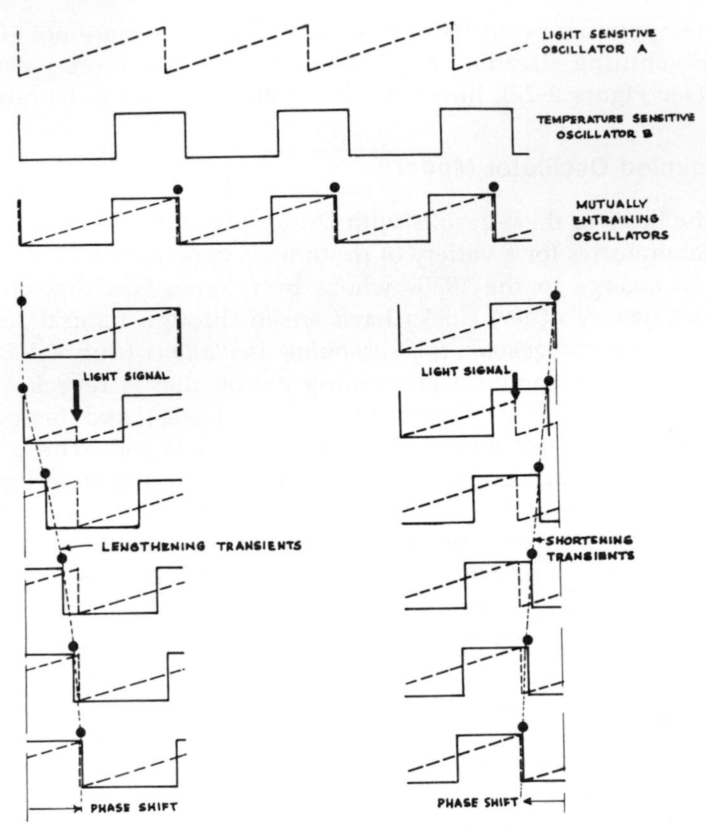

Figure 8-3 Pittendrigh and Bruce's dual oscillator scheme for the *Drosophila* eclosion rhythm. The A and B oscillators are shown both free-running and coupled. When coupled, A entrains and drives B. The black dots represent an arbitrary point in the cycle of the B oscillator, which represents the assayed phase reference point (the median of the eclosion peak). The two lower figures, showing 6 successive cycles, illustrate what is to be expected if a light signal completely and immediately resets the A oscillator and if the B oscillator is then gradually reentrained by the A oscillator. [C. S. Pittendrigh, *in* "Circadian Clocks" (J. Aschoff, ed.), pp. 277–297. North-Holland Publ., Amsterdam, 1965.]

directly underlie the overt persistent rhythm of eclosion. When the phase of the A-oscillator was shifted, it would take several days (i.e., 24-hour cycles) before the B-oscillator could be reentrained by A. The empirically observed, light-induced transients, then, would be a reflection of the driven B-oscillator gradually regaining phase with its pacemaking, light-sensitive driver; transient periods would be either longer or shorter than the innate, free-running period (τ_{FR}). Finally, temperature-induced transients were considered to be only a temporary derangement of the temperature-sensitive B-oscillation, for eventually the temperature-insensitive A-oscillator would regain control of B and the phase and period would then return to normal.

It must be emphasized that this is a formal model only, concerned with the relationship among various internal subsystems of a "black box" as inferred by empirically observing its output following a variety of perturbing inputs. The model does not pretend to describe the molecular details of the putative clock mechanism. Nor is the clock system necessarily limited to only two oscillators—a population of strongly or weakly coupled clocks (e.g., in a tissue) could exist. Finally, the original dual oscillator scheme has had to be revised somewhat in view of the discovery by W. F. Zimmerman during his doctoral research at Princeton University that single temperature pulses and steps can indeed generate steady-state phase shifts (although not as dramatic as those induced by light signals), which implies that temperature can directly (or indirectly by strong feedback from B) affect the A-oscillator (Figure 2-27).

THEORY FOR THE ENTRAINMENT OF A CIRCADIAN RHYTHM BY LIGHT CYCLES

Most important, however, has been the usefulness and predictive power of the coupled-oscillator model: from it has arisen a comprehensive theory on the formal mechanism of the synchronization of a circadian rhythm, such as eclosion by light cycles which is, by several orders of magnitude, more detailed, testable, and, to date, empirically verified than any other existing hypothesis. Once again, the story begins in the laboratory of C. S. Pittendrigh (although it later branches out to other workers).

There are several basic assumptions of this theory. In the first place, it is assumed that an appropriate light signal (e.g., a 15-minute pulse) engenders an almost *instantaneous phase shift of the A-oscillator.* Thus, the phase-response curve for the *Drosophila* eclosion rhythm (Figure 8-2) describes the immediate effect of light in phase-shifting the circadian system, although in actuality it is based on the ultimate net steady-state phase shift of the rhythm (and its underlying B-oscillator) observed some 6 days after the light signal was imposed. Finally, the phase reference point (median of eclosion in the *Drosophila* system), although empirically observed to occur at circadian time (CT) 3.3 (i.e., 3.3 hours after subjective dawn), is *computed* to occur 8.8 hours after the sharp discontinuity, or "breakpoint" on the phase-response curve at CT 18.5. This, in a very superficial sense, might be considered a "trigger" that sets into motion the processes leading to eclosion. [Note that CT 18.5 + 8.8 = CT 27.3 and that if the free-running period, τ_{FR}, were assumed to be 24.0 hours, then $(\text{CT } 27.3)_{\text{modulo}\,\tau}$ = CT 27.3 − 24.0 = CT 3.3.] This mode of calculation avoids the possibility that the breakpoint—in some ways a graphical artifact—might be abruptly passed over by an instantaneous phase shift of the A-oscillator demanded by the first assumption.

With these assumptions out of the way, the essence of the entrainment model itself can be considered. If an ESSO has a free-running period in constant darkness of τ hours, then synchronization by an external *Zeitgeber* (e.g., a light cycle) whose period is T (where $T \neq \tau$) must necessarily involve a correction in each cycle so that $\tau \cong T$ (cf. Figures 1-8, 1-9). For example, entrainment of a circadian rhythm having a τ_{FR} of 24.2 hours by a 24-hour light cycle (e.g., LD 12:12) could occur only if the biological oscillation were "shortened" or advanced by 12 minutes (= 0.2 hours) every 24 hours of elapsed real, or *Zeitgeber* time (ZT). These periodic corrections may be regarded as a series of continuing discrete phase shifts satisfying the relation

$$\tau - T = \Delta\phi_{ss} \tag{8-1}$$

where $\Delta\phi_{ss}$ is the phase shift necessary in each cycle to maintain the interval (T hours) between successive eclosion peaks (i.e., to keep the period of the entrained rhythm constant and, thus, at the same phase angle with the imposed light cycle). In a sense, this expression is a tautology—it is merely a more formal and abstract redefinition of entrainment, which can thus be seen to involve the control of both phase and period. Yet, it must be remembered that this daily resetting of an imperfect endogenous biological clock is precisely what might be expected to occur in nature's diurnal light cycles.

Now if this treatment is valid, one ought to be able to formally reduce a "complete" light cycle, such as LD 13:13 (T = 26 hours) to a single 15-minute light pulse falling every 26 hours (a so-called one-point skeleton photoperiod). According to the model, then, Equation (8-1) would demand that this single signal elicit a phase delay ($-\Delta\phi$) of −1.8 hour, assuming a τ_{FR} of 24.2 hours, since the endogenous rhythm must be "stretched out" or lengthened by 1.8 hours every 26 hours of elapsed real time (*ZT*) for entrainment to occur. The crucial point is that the phase shift required to satisfy the equation can be accomplished only if the light signal strikes the A-oscillation (i.e., the underlying oscillation in light sensitivity, depicted by the phase-response curve) at a unique phase point. The *Drosophila's* phase-response curve (Figure 8-2) can be used to predict that a phase delay of −1.8 hour can be obtained only at CT 13.3. One can thus imagine that

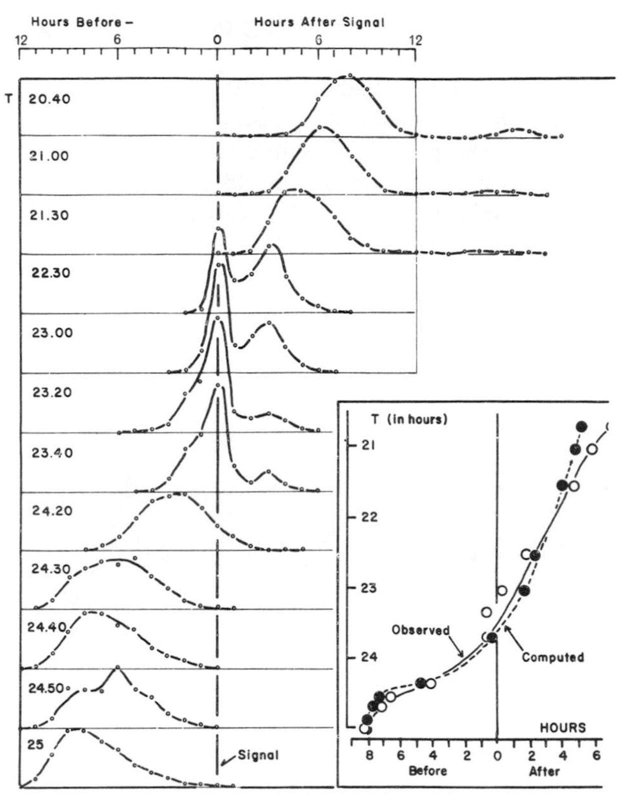

Figure 8-4 The phases of the *Drosophila* eclosion rhythm entrained to various light periods that were either greater or less than the innate free-running period of the rhythm. Inset: medians of eclosion peaks computed and observed. [C. S. Pittendrigh, *in* "Circadian Clocks" (J. Aschoff, ed.), pp. 277–297. North-Holland Publ., Amsterdam, 1965.]

the "fast" biological oscillation (relative to the longer light cycle) would instantaneously be lengthened (i.e., the phase-response curve would move to the right of an observer) by 1.8 hour when the light pulse struck, so that the net result would be a matching of τ with T. The steady-state phase of the entrained rhythm would be independent of the time at which the signal impinged relative to the oscillation: the appropriate $\Delta\phi$ would always be generated.

So much for prediction; the question remains as to whether entrainment can indeed occur by one 15-minute pulse per cycle, and if so, whether the empirically observed phase angle between the light pulse and the eclosion median will be equal to the phase angle difference computed by the model. The answer was resoundingly affirmative: Figure 8-4 shows the remarkably close agreement between the empirical results for the *Drosophila* system and the values predicted by the theory for one-point light cycles having periods ranging from 20.4 to 25.0 hours.

Not content with single pulses of light, Pittendrigh also examined the effects of two-point "skeleton" photoperiods, wherein a LD 12:12 complete cycle, for example, would have the light chopped out and replaced by darkness except for a single 15-minute pulse placed at the beginning and end of the normal light period (to yield, on an absolute time scale, a regime consisting of 0.25 hours light, 11.75 hours darkness, 0.25 hours light, 11.75 hours darkness). If the model were correct, then these two-point skeletons should be sufficient to entrain a

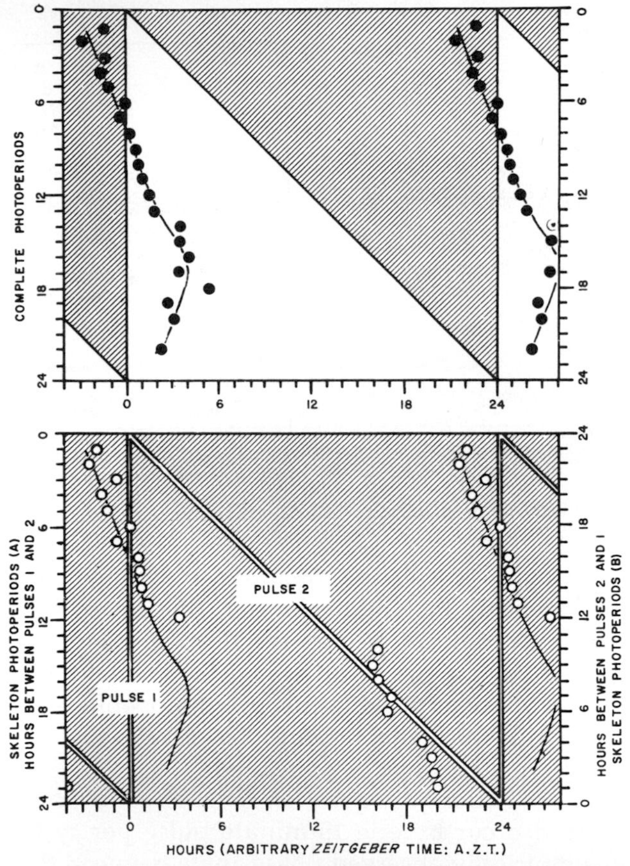

Figure 8-5 The phase of the *Drosophila* eclosion rhythm as a function of complete (upper panel) and skeleton (lower panel) photoperiods. The plotted points are medians of the steady-state distributions of eclosion. In the lower panel, the solid curve is fitted to the medians for complete photoperiods observed in the upper panel. [C. S. Pittendrigh, *in* "Circadian Clocks" (J. Aschoff, ed.), pp. 277–297. North-Holland Publ., Amsterdam, 1965.]

circadian rhythm by satisfying the relation (expanded from Equation 8-1)

$$\tau - T = (\Delta\phi_1) + (\Delta\phi_2) \qquad (8\text{-}2)$$

where $\Delta\phi_1$ and $\Delta\phi_2$ represent the phase shifts engendered by the two successive signals. In the case of a skeleton of a 24-hour light cycle imposed on a rhythm having a τ_{FR} of 24.0 hours, the two pulses must necessarily cause phase shifts of equal magnitude but opposite in sign in order to produce a net $\Delta\phi$ of zero. The phase-response curve could theoretically oblige with a unique set of phase points: pulses falling at about CT 14.0 and CT 23.0 generate phase shifts of +3.05 hours and −3.05 hours respectively (as calculated by computer simulation). A test of all possible skeleton photoperiods of complete light cycles (ranging from LD 1:23 to LD 23:1) revealed that they precisely simulated the action of complete photoperiods and that the predictions of phase by the model were correct (Figure 8-5). The apparent failure of the model to accurately predict the effect of skeletons of photoperiods greater than 12 or 13 hours turned out to be the exception that further proved the validity of the rule: it was due simply to the fact that the organism cannot distinguish between the skeleton of LD 10:14, for example, and LD 14:10, with the result that it always reverts to the phase of the shorter of the two complementary alternatives.

It should now be evident that the validity of at least the basic assumptions of the coupled-oscillator model and the theory of entrainment of a circadian rhythm by light cycles are supported by the experimental results obtained from light-perturbation studies. This model has also proved quite useful in predicting the limits of entrainment, the rate of approach to equilibrium during entrainment, and the maximum skeleton photoperiod to which an organism can be entrained, not only in *Drosophila* but also in other systems. Although it is possible that the same formulation could provide grist for the mill of the supporters of the extrinsic timing hypothesis—for they, too, are willing to accept an endogenous rhythm in light sensitivity (but which would be ultimately driven by an external geophysical time cue)—the entire set of data seems to fit more elegantly into the conceptual framework positing an endogenous, self-sustaining oscillation(s).

Topological Models for the Effects of Light on Circadian Rhythms

The dual oscillator model provided the impetus for the development of a family of more sophisticated mathematical models of the formal, black-box variety. In particular, T. Pavlidis of the Department of Electrical Engineering at Princeton University, A. T. Winfree of the Department of Biological Sciences at Purdue University, and A. Johnsson and H. G. Karlsson of the Department of Electrical Measurements at the Lund Institute of Technology in Sweden have made important new contributions to the field.

Pavlidis, for example, has reformulated the analytical model of Pittendrigh and co-workers in terms of state-space topology. The dynamic behavior of a circadian oscillator can be described by a set of two first-order differential equations involving two variables (as, for example, cats and mice or ion concentration and rate of ion transport) which if known at any given time allow the prediction of the behavior of the system at some later time. These two variables, termed the *state* of the system, can be represented as coordinates in a plane, and the behavior of the systems can be described in terms of plane curves, or *trajectories.* The oscillatory motion of a biological clock, in turn, can be represented as a stable periodic trajectory that closes in on itself and is termed a *stable limit cycle.* Such a system, if disturbed (as, for example, by a light pulse), will always tend to return to an equilibrium configuration.

Given the system with a limit cycle, the problem now was to define a mapping that would map the limit cycle into itself. This would simulate the effect of short light pulses on the *Drosophila* eclosion rhythm. Pavlidis accomplished the mapping by utilizing the experimentally determined phase-response curve (Figure 8-2) and the relation

$$CT_2 = CT_1 + \Delta\phi(CT_2)|_{\text{modulo } \tau} \tag{8-3}$$

which states simply that if CT_1 is the circadian time at which a brief light pulse is supplied to the system, then the new circadian time (CT_2) that will be achieved instantaneously after the pulse can be calculated merely by adding algebraically the phase shift engendered by the pulse to the initial time (CT_1). The phase plane portrait of the limit cycle of a model for circadian clocks determined in this fashion is shown in Figure 8-6. Essestial to the model is the assumption that light drives the system toward a sequence of states (i.e., CT times) that is also traversed during the absence of light between CT 4.0 and 12.0: light pulses applied during this interval cause little or no phase shift (hence, the term dead portion on the phase-response curve). This topological model has provided an important mathematical approach to further predictions.

But having gone this far, why stop here? The building of formal models has become even more refined, always starting out with a known body of experimental data and then (hopefully) making new

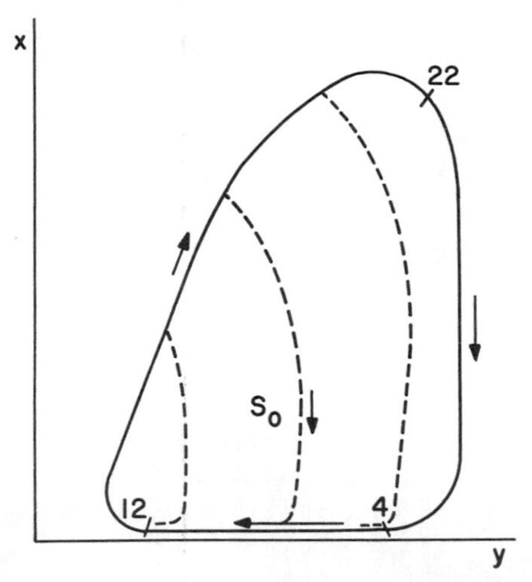

Figure 8-6 Phase plane portrait of the limit cycle of a model for circadian clocks (solid line). The trajectories under light exposure at different circadian times are shown by broken lines; *x* and *y* can be any two variables. S_0 is a state of equilibrium. Numbers indicate approximate circadian time. [T. Pavlidis, *in* "Biochronometry" (M. Menaker, ed.), pp. 110–116. Natl. Acad. Sci., Washington, D.C., 1971.]

and often unexpected predictions which, in turn, lead to a further round of experimentation. A case in point is an extensive series of experiments by Winfree which asked how and why the resetting response in *Drosophila* depends on the duration of a standard light perturbation, as well as on the time at which it is given. He took advantage of the fact that in constant illumination the clock underlying the eclosion rhythm is held at CT 12: if the culture is released into constant darkness (LL/DD transfer), the oscillation always starts up from this phase point. Young pupae previously reared in constant illumination were transferred to darkness, thereby inducing circadian rhythmicity. Then, at some time (*T*) after the LL/DD transfer, they were exposed to a standard dim blue light for a duration (*S*) and the daily peaks in emerging adult flies were monitored 4 to 8 days later. The value of *S* was systematically varied from 15 to 120 seconds and the value of *T* from 0 to 24 hours. The interval from the end of the light pulse to the mean emergence time of any given eclosion peak was termed the centroid time (θ). In actuality, these experiments simply measured the phase-resetting response elicited by the light pulse, but the results could now be plotted in three dimensions since three variables were involved: the stimulus coordinates, *T* and *S*, comprised the independent variables, while the measurement of emergence time, θ, represented the dependent variable.

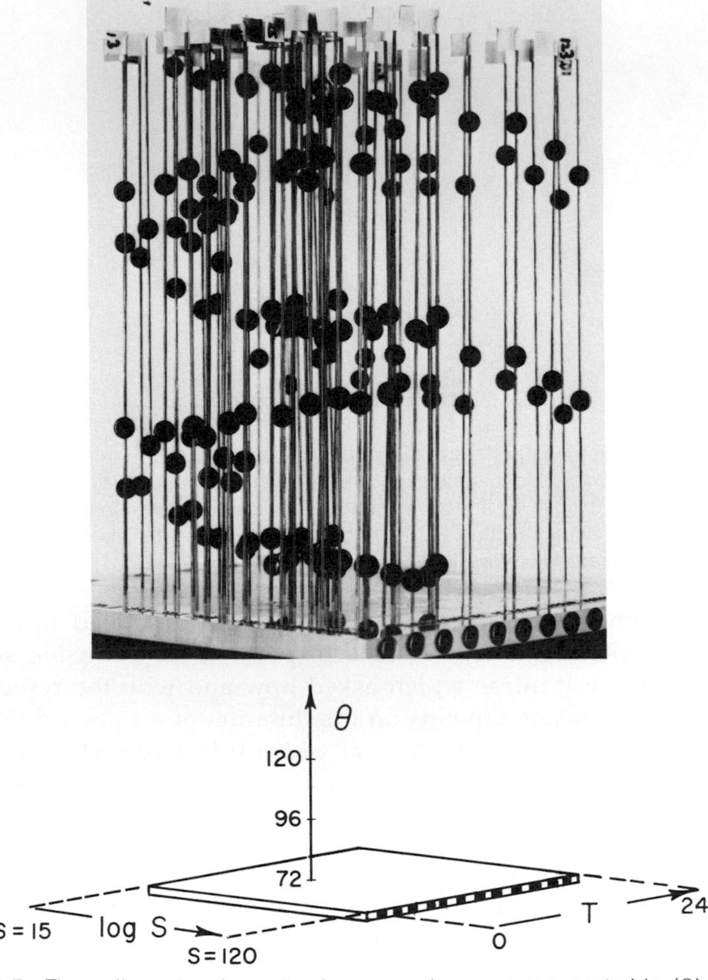

Figure 8-7 Three-dimensional graph of measured emergence centroids (Θ) in *Drosophila* versus the stimulus variables, *T* (time interval between LL/DD transition and light stimulus), and *S* (duration of dim blue-light stimulus). The graph is oriented in the diagram below to emphasize the spiral, corkscrew feature. [A. T. Winfree, *in* "Biochronometry" (M. Menaker, ed.), pp. 81–109. Natl. Acad. Sci., Washington, D.C., 1971.]

The photograph shown in Figure 8-7 illustrates such a three-dimensional graph with T, log S, and θ as the axes. Each vertical wire represents a single experiment in which a population of pupae received a specific stimulus (T,S) and the little buttons on the wires depict the centroid times (θ) of emergence peaks after transients have subsided. The θ axis covers only about $2\frac{1}{2}$ days of emergence data in this case. From this type of plot, the helicoidal nature of the graph is clearly evident: the centroids spiral up around a vertical axis of rotation. A digital computer was then used in an attempt to find a smooth surface, $\theta(T,S)$ called the *resetting surface,* which would fit the complete cloud of centroid points. The resulting surface, although difficult to visualize, is perhaps best described as a vertical corkscrew linking together tilted planes (the latter arising from sets of stimuli that have little effect on the phase of the rhythm).

Unfortunately, there is space for only a brief glimpse into the complexities and implications of this formal "corkscrew" model of a resetting surface. Suffice it to say that as a result of this type of analysis an unusual and heretofore unexpected singularity was discovered. The central axis of the corkscrew (i.e., the center of the helix shown in Figure 8-7) appears to represent a critical stimulus time having the coordinates $T^* = 6.8$ hours, $S^* = 50$ seconds. This feature of the resetting surface allows one to predict that if a light perturbation were given exactly at this singularity point either no circadian rhythmicity in emergence would be observed, or one of unpredictable phase. In other words, the *Drosophila* clock would be unstable at the phase point of the oscillation 6.8 hours after its initiation by the LL/DD transfer (i.e., about CT 18.8), since the oscillation always commences at CT 12.0)—which is precisely the point where "delay" phase shifts switch to "advances" (the breakpoint in the phase-response curve, Figure 8-2). Perturbations on either side of the singularity point would radically change the phase of the clock. This prediction of a tendency toward arrhythmicity following a stimulus at (T^*,S^*) was proved valid empirically: the clock was found to be stopped in the same state that it is in pupae that have been reared from the egg stage in total darkness (cf. Figure 2-24, Panel A).

These formal topological models, then, represent attempts at a geometric description of the dynamics of the putative endogenous clock in both a fixed environment and under imposed light perturbations. It seems reasonable to assume that the light-sensitive mechanism underlying the *Drosophila* eclosion rhythm (and presumably other circadian rhythms, by extension), or the A-oscillator of Pittendrigh and Bruce, is always in some state that, in principle, can be determined by some small number of measurements (e.g., of several flux rates in a biochemical pathway). These measured variables, in turn, can be used as coordinates to locate the state of the clock at any instant as a point in space. If the circadian oscillator is indeed autonomous in constant darkness and temperature, then for each successive state of the clock there would be a correlated spontaneous rate of change represented by a vector. These vectors would define the motion of the clock along smooth trajectories through its state space. This approach, therefore, allows us to reduce simple hypotheses to abstract geometrical propositions whose implications for light-, or temperature-, induced phase-resetting behavior can be intuited without having to resort to exact, and more restrictive, differential equations for an unknown clock mechanism. The heuristic value of such methodology is beautifully borne out by the prediction, and subsequent empirical confirmation, of the singularity point.

Aschoff's Rules and the Wever Model for Circadian Rhythms

The preceding formal models have given particular emphasis to the effects of light pulses and perturbations—so-called differential *Zeitgeber* since the on-off signals are what seem to be important to the clock system. A quite different approach has been taken by J. Aschoff and R. Wever of the Max-Planck-Institut für Verhaltensphysiologie in Erling-Andechs, West Germany. These workers have stressed the continuous effect of light and the role of light intensity on circadian rhythms—termed proportional *Zeitgeber*— and have developed another family of mathematical models to "explain" their findings. Especially germane to this school of thought are a series of empirical generalizations, popularly dubbed Aschoff's Circadian Rules, which suggest a relationship between the nocturnal or diurnal habits of animals and the length of their free-running periods (τ_{FR}) under different intensities of continuous illumination.

The first rule states that light-active (diurnal) animals exhibit a shorter τ in continuous light than in constant darkness, while in night-active (nocturnal) animals the reverse holds true. Rule II is like unto it: τ decreases with increasing light intensity in light-active animals, but is directly proportional to intensity in night-active animals. Published experimental results in a wide variety of mammals, birds, and insects bear out these first two rules remarkably well (see Figure 1-10). The success of these generalizations encouraged their extension to two other properties of the hypothesized underlying oscillation which appear to be correlated with light intensity: (a) the total activity or "level of excitement" during each circadian cycle; and (b) the activity (α); rest (ρ) ratio, between the time interval when the animal is active and the timespan when it is at rest during any cycle. Aschoff's Rule III states that the α/ρ ratio increases with increasing light intensity in light-active animals, but decreases in dark-active animals. For a day-active bird, such as the chaffinch *Fringilla*, the ratio increases as intensity increases for two reasons: first, α increases, leading to the expected decrease in ρ (since $\rho = \tau - \alpha$); and secondly, since τ itself

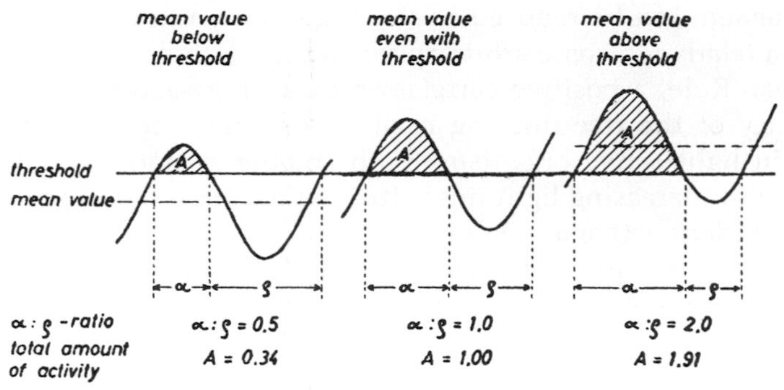

Figure 8-8 The Aschoff-Wever special model for circadian activity rhythms. Schematic diagram of an oscillation whose mean value (dashed line) increases (in three steps) relative to a fixed threshold (heavy line). The section of the oscillation above the threshold (*A*) is designated "activity," the section below threshold, "rest." Note that the ratio of activity time to rest time (α:ρ) and the total amount of activity (A) increase with the mean value of the oscillation. [R. Wever, *in* "Circadian Clocks" (J. Aschoff, ed.), pp. 47–63. North-Holland Publ., Amsterdam, 1965.]

decreases under these conditions according to Rule II, ρ is still further diminished. Finally, Rule IV states that the level of excitement (e.g., the number of perchhops per hour) increases with increasing light intensity for day-active animals, with the reverse holding true for nocturnal animals. Support for these last two rules is also amply documented.

It is clear, then, that any formal theory for a biological clock must ultimately account for these phenomena; Wever was delighted to oblige. The model that he developed assumes that a continuous circadian oscillation underlies the observed discontinuous rhythm of activity and rest (or sleep). Furthermore, it hypothesizes that activity occurs only when the oscillation is above a certain *threshold,* which commences (onset of activity) when the basic oscillation passes upward through this threshold and ending (onset of rest) when it crosses downward through the threshold level about one-half a period (i.e., 180° later, as schematized in Figure 8-8. One can see that the α/ρ ratio is determined by the relative positions of the mean value, or level of the oscillation, and the threshold value: if either the mean value increases relative to the threshold, or the threshold decreases relative to the mean value, then α increases, ρ decreases, and hence the α/ρ ratio increases. Furthermore, as the mean value of the oscillation increases relative to the threshold value, the integrated area under the curve describing the oscillation also increases; this total area under the curve can be used as a rough measure of the total activity, or level of excitement.

According to this model, then, changes in the α/ρ ratio and the level of excitement can be reduced to a change in the mean value of the oscillation relative to some arbitrary threshold. In this so-called modified Circadian Rule, a positive correlation exists between mean value and frequency of the free-running oscillation under constant conditions. Thus, in light-active organisms both frequency and mean value increase with increasing light intensity, but decrease with increasing intensity in dark-active animals.

Once the basic qualitative model had been established, it was only a matter of time before a mathematical formulation was attempted. The basic purpose for deriving a model equation for a circadian clock, of course, is to facilitate the recognition of relationships: given the behavior of an oscillation under a given condition, it is easier to derive the behavior under other conditions by using a simplified model than by using a complicated biological system. (The danger, of course, is the possibility of losing sight of the forest for the trees.) In this case, it was hoped that mathematical analysis might aid in finding a rationale for the empirical generalizations encompassed by Aschoff's Rules, which, in turn, might help one predict the behavior of circadian rhythms under light cycles and other *Zeitgeber.* To this end, a number of parameters had to be investigated. It was assumed, for example, that the circadian system could be self-sustained, as well as capable of being entrained by external *Zeitgeber* (i.e., susceptible to both exogenous and endogenous excitation). Similarly, a value for the coefficient of energy exchange between the system and the environment had to be judiciously selected: the value finally chosen typifies an oscillation approximately midway between the two extreme classes (smooth, sinusoïdal, dedamped, pendulum-type oscillations, on the one hand, and jerky, saw-toothed, relaxation oscillations on the other). Yet another assumption had to be made concerning the relatively effectiveness toward entrainment by proportional and differential *Zeitgeber.* All of these different parameters were then plugged into a special type (van der Pol) of differential equation that is capable of describing a wide variety of self-sustaining oscillations, which looks something like Equation (8-3):

$$\ddot{y} + \epsilon(y^2 + y^{-2} - 3)\dot{y} + (1 + ky)y = \ddot{x} + \dot{x} + x \qquad (8\text{-}4)$$

where x represents the controlling environmental factors (e.g., light intensity), $\dot{x}$ *and* $\ddot{x}$ the first and second time derivatives of x, y the oscillating biological variable (e.g., activity), $\dot{y}$ and $\ddot{y}$ the first and second time derivatives of y, and ϵ the variable coefficient of energy exchange. The latter parameter, as well as the constant, k, must be empirically determined by selecting values which cause the equation, when solved, to agree with known biological observations.

Now that an equation had been derived, it could be solved for various intensities of constant illumination. The results demonstrated effectively that Aschoff's Rules were all predictable consequences of Wever's model. Similarly, the effects of *Zeitgeber* conditions could be stimulated with a computer by allowing the value of x in Equation (8-3) to periodically vary, as it does under natural conditions. Although some workers feel that the selection of a single differential equation as a model for a circadian clock is overly restrictive, one could reply that the strong point of the model is its simplicity: the proposed model equation contains no more specific terms than dictated by the simplest mathematical equivalents of the biological generalizations. But whatever the shortcomings and limitations, the Wever model is instructive in that it focuses on the energy level of the controlling environmental factor, especially under continuous illumination, whereas other formal models have stressed only the differential effect of cycling *Zeitgeber*. A blend of the two approaches is clearly desirable.

CELLULAR AND BIOCHEMICAL CLOCK MECHANISMS

By this time the reader probably has had his fill of formal mathematical models for circadian clocks. (On the other hand, one might ask why biologists continue to experiment on living organisms when the same results could be obtained from computer simulation studies!) But what are the results after expending all this effort? To be sure, abstract models afford all sorts of provocative information concerning the interrelationships among component parts of the black box labeled a biological clock and, indeed, sometimes these models even point the way for an experimental attack, but they can never—by their very nature—establish a direct correlation with known cellular components and biochemical reactions. Although the results obtained with the elegant formal models just briefly examined are certainly consistent with the notion of an endogenous, autonomous, self-sustaining oscillation underlying overt circadian rhythms, they are not compelling. Much more persuasive would be an oscillator model embracing concrete cellular and molecular entities, supported, of course, by appropriate experimental data. To mix a metaphor or two, even a small step in this direction would be a giant stride toward winding up the biological clock problem. Now several different experimental avenues that have been taken to solve this problem will be examined.

Independent Oscillations in Isolated Cells, Organs, and Tissues

One obvious approach toward elucidating the nature of the putative biological clock is to localize it within the organism.

Chapter 2 described the fact that leaves excised from the bean seedling with their petioles intact continue to exhibit sleep movements for as long as 28 days in constant conditions (Figure 2-1). If the leaf blade is excised, leaving only the midrib, pulvinus and petiole, the rhythm is still observed. Even when isolated leaf joint preparations are halved, persisting circadian fluctuations in turgor pressure occur. Similarly, isolated potato plugs exhibit long-term respiratory rhythms (Figure 2-8), the enucleated green alga *Acetabularia* displays a rhythm in O_2 evolution that persists for over a month (but see Figure 2-6), and excised petal tips of the night-blooming jasmine, when floated in a petri dish, continue with their fragrance rhythm. The same approach in animals has revealed, for example, that adrenal glands isolated from the golden hamster, *Mesocricetus auratus,* and cultured in defined medium in a light cycle show 24-hour rhythms of oxygen consumption and of corticosteroid secretion. The period of the rhythms is temperature-compensated with Q_{10} of 0.96 to 1.11 over the range 15° to 37.5°C; furthermore, reversal of the imposed light cycle reverses the phase of the respiratory rhythm. Similarly, the neural discharge (compound action potential) from the optic nerve of an eye isolated from the sea hare, *Aplysia,* fluctuates with a circadian period for over a week in continuous illumination, and the nuclear volume of isolated larval *Drosophila* salivary glands undergoes a circadian variation for up to 10 days in a chemically defined medium. [And it has already been shown that a number of different circadian rhythms can occur simultaneously in the unicellular dinoflagellate *Gonyaulax* (Figure 2-3) or the algal flagellate *Euglena* (Figure 2-16).]

From all these and other facts, one can conclude that cells, tissues, or organs isolated from multicellular organisms are able to oscillate independently. The problem with this approach, of course, is that the carcass eventually becomes a corpse: either the organism is killed or the organ is destroyed in the process so that a loss of rhythmicity does not necessarily imply that one has finally located the site of "the" clock, if such an entity exists. (Consider the difficulty, for example, that a sextuplicately amputated cockroach would have in demonstrating a circadian rhythm of activity, even though the underlying "master clock" almost certainly is confined to the head region.) It would be gratifying, nevertheless, if a circadian rhythm could be shown to occur (or to *not* occur) in an isolated subcellular component, such as an *in vitro* chloroplast or mitochondrial preparation. Indeed, as will be discussed shortly, persisting oscillations in DPNH can occur in cell-free extracts of yeast, but there is a catch: these are high-frequency oscillations with a period length in the neighborhood of 7 minutes or so, and they are neither light-entrainable nor temperature-compensated.

Neuroendocrinological Control of Circadian Rhythms

In the exploration of circadian rhythms, two diverging pathways have come to exist: mathematical analysis leading in one direction and the biochemical approach leading in another. Attempts to bridge the resulting gap have been somewhat successful in the search for specific tissues that might serve as physiological chronometers or even "master" controlling clocks and, at the same time, exhibit all the formal properties of circadian rhythms. Once again, the question is whether a localized, anatomically identifiable mechanism controls the timing of such overt rhythms as locomotory activity or eclosion. The answers to these questions obviously have a direct bearing on the intrinsic versus extrinsic timing hypotheses. Now several of the most intensively studied systems will be briefly examined.

THE OPTIC LOBE CLOCK OF THE COCKROACH

One of the first reports of the localization of a biological clock was the claim by J. Harker at Cambridge University in the late 1950s that the subesophageal ganglion in the head of the lowly, but ubiquitous, cockroach contained the basic clock mechanism underlying its nocturnal, circadian activity rhythm. At the time, she hypothesized that this driving oscillator controlled the effector organs in the thorax (where the legs are attached) through the mediation of hormones produced by neurosecretory cells in the head. As described in Chapter 2, the case for *hormonal coupling* was based on three lines of evidence: (i) that parabiosis indicated factors transported by the blood were involved (in restoring rhythmicity to arhythmic roaches); (ii) that transplantation of subesophageal ganglia (from donor roaches to either headless or out-of-phase recipients) induced and determined the phase of activity rhythms; and (iii) that arhythmicity ensues when the neurosecretory cells of the protocerebrum were removed surgically. Unfortunately, these provocative results have not been confirmed by subsequent work in other laboratories and attention turned to other sites for the localization of the clock and to the possibility that electrical or *neural coupling* might play a significant role.

One of the key problems in this type of approach is that it is most

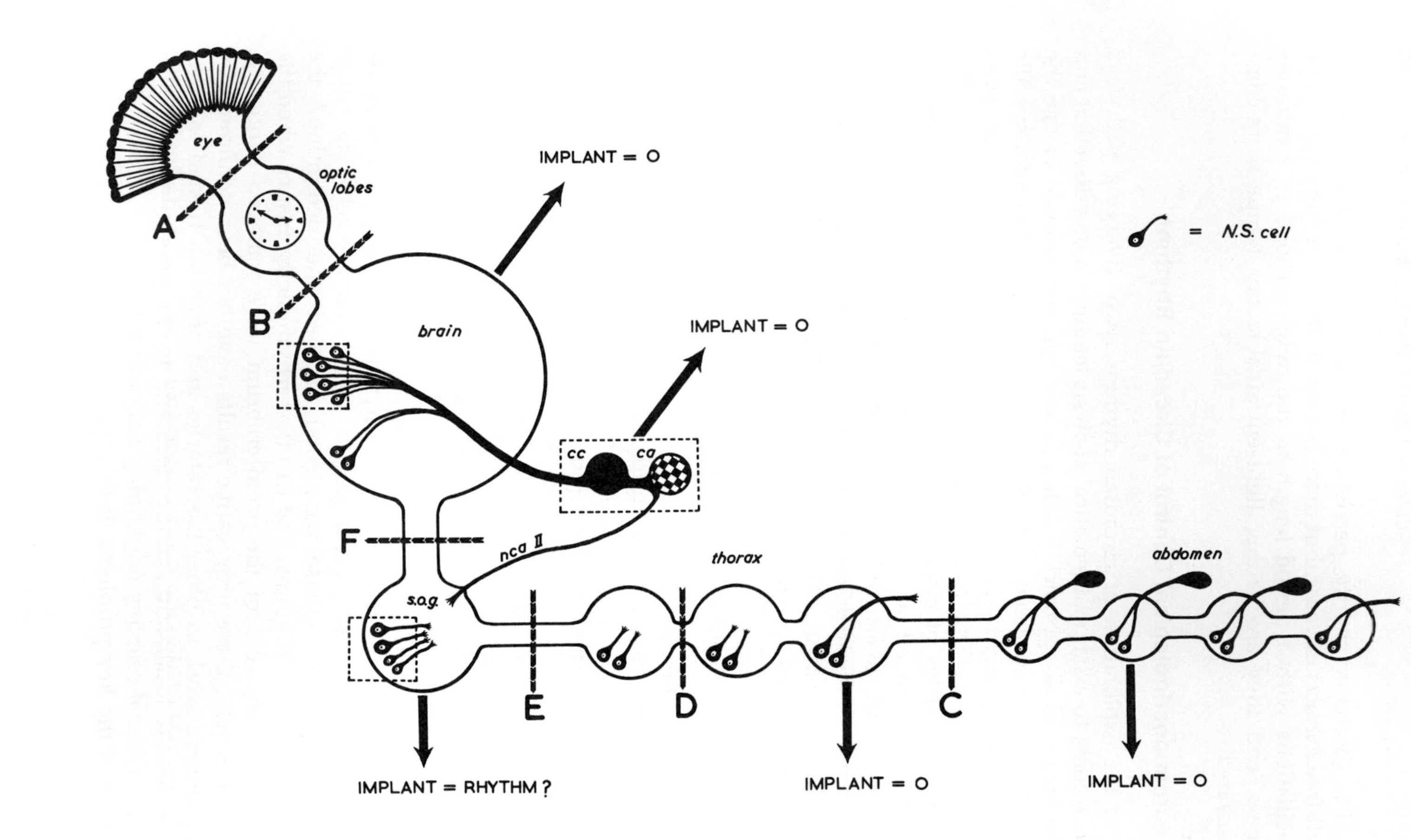

eye
optic lobes
A
B
brain
IMPLANT = O
IMPLANT = O
= N.S. cell
cc
ca
F
nca II
s.o.g.
thorax
abdomen
E
D
C
IMPLANT = RHYTHM ?
IMPLANT = O
IMPLANT = O

difficult to perform surgical operations on the various endocrine organs without also disturbing or disrupting the associated neural connections. The generally accepted facts, however, seem to be the following (illustrated in Figure 8-9). First, no rhythms now seem to be induced or transferred by implantation or transplantation of glands. Second, one can remove virtually all endocrine organs in the head of the roach without disrupting the activity rhythm. These include the corpora cardiaca and corpora allata and much of the neurosecretory tissue of the medial brain and the subesophageal ganglion. Furthermore, a variety of other operations can be performed without stopping the rhythm: severing the compound eyes from the optic lobes, splitting the protocerebrum midsagitally into two halves, removal of an entire protocerebral hemisphere, or severing the ventral nerve cord connectives anywhere posterior to the prothoracic ganglion. Finally, rhythmicity *can* be stopped only by operations that interrupt the neural pathways between the optic lobes and the brain, or between the brain and thorax: beheading (which is not as obvious as it sounds), severing the optic lobes from the protocerebrum, bilaterally splitting the protocerebral hemispheres, removing a large chunk of the pars intercerebralis, or severing the circumesophageal connectives or the connectives between the subesophageal and prothoracic ganglia.

The simplest interpretation of all these observations seems to be, therefore, that a driving circadian oscillator (Pittendrigh's A-oscillator?) is located in each optic lobe and is neurally coupled to the leg muscles via the protocerebrum, nerve cord connectives, and thoracic ganglia. These two pacemakers, in turn, are entrained by the light signals perceived by the compound eyes. (One wonders, in fact, what would happen if the two eyes were subjected to conflicting light cycles when cross-communication between the optic lobes is prevented by medial brain bisection!) It should again be emphasized that none of these experiments tell us anything about the nature of the clock itself; rather, they deal only with the coupling between the clock and the overt

Figure 8-9 Synopsis of experiments on the neuroendocrinological control of the circadian rhythm of locomotory activity in the cockroach. Ganglia of the central nervous system are represented by the linked spheres, with neuroendocrine tissue, including known neurohaemal organs, indicated in black. Dotted boxes represent endocrine tissue that can be removed without altering the rhythm. Arrows show organs transplanted from rhythmic donors to headless arrhythmic recipients: 0 signifies that the host shows no detectable rhythm. Heavy broken lines are cuts made in the nerve trunks: cuts *B, E, F,* or splitting the protocerebral lobes bilaterally apparently stop the rhythm; cuts *A, D, C,* or splitting the pars intercerebralis midsagittally do not. NS, neurosecretory; SOG, subesophageal ganglion; CC, corpora cardiaca; CA, corpora allata. [J. Brady, *Nature* (*London*) **223,** 781–784 (1969).]

rhythm–the gears between the escapement of the clock and its hands. But, at least, some progress toward localizing a possible site of the putative endogenous clock has been made.

THE PROTOCEREBRAL CLOCK OF THE SILKMOTH

In contrast to daily behavioral rhythms of gross motor activity (as in the cockroach), another major class of insect circadian rhythms comprises once-in-a-lifetime gated developmental events such as eclosion in *Drosophila,* where the event—when it finally does occur in the life cycle of the organism—takes place at a particular time interval within the 24-hour time frame, with the necessary consequence that its "rhythmicity" can be observed only in populations of developmentally asynchronous populations. But the fruit fly is small, and its head even smaller, which makes it technically difficult to root around in the brain and to perform the standard extirpation and transplantation experiments so beloved by the endocrinologist. (But in defense of *Drosophila,* one should note that it has been shown in an eyeless mutant strain that the eclosion clock still entrains quite normally and gates eclosion, which indicates the pathway for light information was not via the eyes, in direct opposition to the roach story.)

It was precisely for this reason that J. W. Truman, then at Harvard University, turned to the giant silkworm of the moth family Saturniidae. The larval caterpillar usually overwinters as a diapausing pupa; in the warm and lengthening spring days, diapause terminates, adult development begins, and the large adult moth emerges, or ecloses, a few weeks later. The timing of eclosion is controlled by the imposed photoperiod via a circadian clock and varies from species to species. Thus, in a *LD 17:7* light cycle, *Hyalophora cecropia* characteristically emerges during the first 6 hours of the day, with a peak after dawn, while *Antherea pernyi* emerges during the last 5 hours of the light period (Figure 8-10, Panel A). If a pupal population of either of these species is transferred to constant darkness, subsequent emergence peaks occur at 22-hour intervals, indicative of a free-running circadian rhythm.

In an attempt to localize the underlying circadian clock, the brains were removed from a group of developing moths. This operation resulted (Figure 8-10, Panel B) in random emergence across the 24-hour time span: eclosion no longer was being gated. Fascinatingly, the effect of brain removal could be reversed by simply implanting a brain into the abdomen of the brainless moths (Panel C). The eventual eclosion of these "loose-brain" moths then occurred during the gate typi-

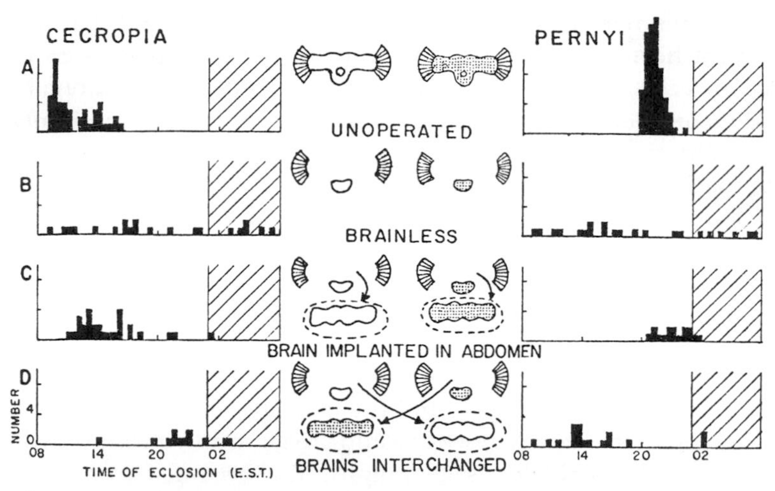

Figure 8-10 The eclosion of *Hyalophora cecropia* and *Antherea pernyi* silk moths in a *LD 17:7* light cycle showing the effects of brain removal, the transplantation of the brain to the abdomen, and the interchange of brains between the two species. [J. W. Truman, *in* "Circadian Rhythmicity" (J. F. Bierhuizen, ed.), pp. 111–135. Centre for Agricultural Publishing and Documentation, Wageningen, The Netherlands, 1972.

cal of the species. Finally, if a cecropia brain was transplanted into a debrained pernyi moth (or vice versa), the recipient adopted the gate characteristic of the donor while retaining its own specific motor patterns (Panel D). Subsequent experiments have demonstrated that the implantation of the cerebral lobes alone is sufficient to gate the eclosion of a debrained moth.

It is also clear that the light information necessary for the entrainment of the silkmoth eclosion clock is perceived directly by the brain—in fact, the protocerebrum—just as was found for the eyeless *Drosophila*. This was demonstrated by an elegantly simple experiment. Pupae (cecropia) were inserted through tight-fitting holes in an opaque partition that separated two small photoperiod chambers. Their heads were exposed to a *LD 12*:12 light cycle; their abdomens were exposed to a similar regime, but with a reversed phase. Further, one group of these pupae had their brains transplanted to their abdomens, while a control group had their brains removed and then immediately reimplanted back to their heads. The only difference between the two groups of moths, therefore, was the phase of the photoperiod to which their brains were exposed. The result: eclosion time was dictated by the photoperiod.

From these experiments it seems reasonable to conclude that the clock-gating silkmoth eclosion resides in the brain, that it is entrained in some unknown manner by direct photoreception, and that it exerts its control hormonally on the sequence of development steps culminating in eclosion. It appears that only the *gating* is hormonally mediated, however, since eclosion still occurs (albeit randomly) in brainless moths. But how does one rationalize the differing clock sites in the roach and in the silk moth? On the basis of these and other findings, Truman was led to propose that circadian clocks exist in two forms: *Type I clocks* would have their photoreceptor and their basic circadian oscillator at the same anatomical location (in the same cell in unicellular organisms), while *Type II clocks* have photoreceptor and oscillator anatomically separate. As a consequence, light would act directly on the oscillator (which might actually comprise a photoreceptive pigment) in Type I clocks, and the clock would stop in continuous light. In contrast, Type II clocks would free run in continuous illumination since the oscillator is geographically separated from the photoreceptor. It is noteworthy that insect clocks tend to fall neatly into one or the other of the two categories. Furthermore, Type I clocks seem to be associated with once in a lifetime events (e.g., silkmoth eclosion), while Type II clocks appear to control daily repetitive behavioral rhythms (e.g., locomotory activity in the roach). Although this framework is useful from a descriptive or taxonomic point of view, its ultimate significance remains to be assessed.

Having located an endogenous site for the silk-moth eclosion clock, Truman was now ready to investigate its response to light perturbations and to develop a semiformal model of how it might operate. In brief, the data can perhaps be most simply interpreted as reflecting a process that has two alternative pathways, conveniently named the *scotonon* and the *photonon*. The former is a dark-dependent process that has a duration of 22 hours, which represents the period of the free-running rhythm. The photonon, on the other hand, is initiated by a light interruption of the scotonon with its duration dependent on the extent to which the scotonon has been completed at the time of the onset of the light period or light signal. The beginning of the scotonon itself is taken as the point of the timing process that coincides with lights-off (i.e., dusk), since if pernyi are transferred from continuous light to darkness, the hormone controlling eclosion is secreted 22 hours after the transition (quickly followed by emergence 1.5 hour later). Thus, in darkness, the clock commences a cycle according to its free-running, scotonon kinetics, much as an hourglass containing 22 hours-worth of sand, and then recycles automatically. With the onset of light in a light cycle, a defined change occurs in the kinetics of the remainder of the cycle, which results in either its lengthening or shortening, depending on the length of the photophase. Indeed, it was found that the completion of some event in the early part of the scotonon was necessary for maximal accuracy of the clock (as reflected in the gate width, or variance in eclosion time). On this basis, the scotonon was divided in turn into two successive periods: the synchronization period that occurs during the beginning hours of the cycle and the dark decay period that encompasses the remainder of the scotonon. Light interruptions during this second period, while effecting a change in kinetics, do not markedly effect clock accuracy.

In summary, then, the silk-moth photochemical hourglass-type circadian clock-gating eclosion appears to comprise three processes: (i) an initial photoreversible process [synchronization period] that takes about 2 hours to complete in darkness and is almost instantaneously reversed by light; (ii) a subsequent dark-decay process [dark-decay period] lasting about 20 hours, during which the substance built up during the photoreversible process spontaneously breaks down; and (iii) a light-decay process [photonon] that requires about 16 hours to destroy all the substance produced during the 2-hour photoreversible period. Although this model undoubtedly will undergo modifications as it is refined, it is a valuable construct since it seems to account for the empirically observed behavior of one of the few circadian clocks that have proved amenable to surgical manipulation.

NEURAL CLOCKS IN THE SEA HARE

Fascinating, indeed, is the sea hare *Aplysia californica* (a molluscan gastropod) to marine biologists and invertebrate physiologists alike, but this creature is even a greater delight to those interested in the neurophysiological control of circadian clocks, in particular F. Strumwasser and A. Eskin of the California Institute of Technology, J. Jacklet of the State University of New York at Albany, and M. Lickey of the University of Oregon at Eugene. The collective results of these workers provide a valuable glimpse into the functioning of neuronal oscillators in the central nervous system and their interaction as a population of cross-coupled clocks.

The first studies of interest to us utilized the parietovisceral (or abdominal) ganglion (PVG) whose relatively large size (several millimeters in diameter) not only allows it to be easily isolated from the sea hare, but also permits one to map some 10 identifiable neurons on its dorsal surface. Thus, the typically largest neuron is termed cell 1, or the giant cell, while cell 3 is referred to as the *parabolic burster cell*. Furthermore, the electrical activity of cell 3 can be automatically and continuously monitored by the careful insertion of a glass micropipette filled with potassium sulfate (as an electrolyte) into the cell body. With this internal electrode, then, one can not only record both the output (action potentials or impulses) and the input (postsynaptic potentials) of the parabolic burster cell, but also depolarize or hyperpolarize the membrane by passing a current through the micropipette, or even perturb the membrane potential by injecting substances directly into the cell. With these techniques, it is possible to isolate a PVG from a sea hare, place it in a thermostatically controlled chamber perfused with seawater, and record the intracellular electrical potentials of cell 3 over the ensuing 48 hours, when it is completely divorced from peripheral receptors, blood-borne hormones, and fluctuating levels of other compounds in the natural circulation.

The results from this experimental setup revealed that the parabolic burster cell of the isolated PVG displays a persisting—albeit rather strongly damped—circadian rhythm of spike output rate whose form and period are "conditioned" by the light regime to which the intact sea hare (which itself exhibits a circadian rhythm of locomotory activity) had been exposed before it was sacrificed for removal of the PVG. Thus, if the animal had been maintained on *LD 12:*12, the peak in spike discharge rate in the isolated PVG occurred at the time that the dark-to-light transition would have occurred with the intact animal. A somewhat modified rhythm in PVG discharge was observed if the donor *Aplysia* had been kept under constant illumination. In addition, a fortnightly lunar rhythm was discovered which expressed itself as a modulation of the usual circadian rhythm. Heat pulses applied to the PVG during a 10-hour period prior to the expected circadian peak caused an earlier expression (i.e., a phase advance) of the peak. Similarly, actinomycin D (an inhibitor of DNA-dependent RNA synthesis that acts by binding to the DNA), applied intracellularly via the micropipette during the heat-sensitive period, also caused a premature release of spike activity; but if the inhibitor was injected into the cell just after the normal spike had occurred, the subsequent peak was delayed.

These results (whose implications will be considered later) can be interpreted as resulting from a premature release of mRNA caused by the binding of the inhibitor to the DNA (or by the heat pulse); this released message would then initiate cytoplasmic production of a polypeptide that depolarizes the neuronal membrane, or alternatively an enzyme that controls the production of the depolarizing substance. If a *pulse* is applied after the peak has occurred, only a small amount of mRNA would be available for premature release, which results in a

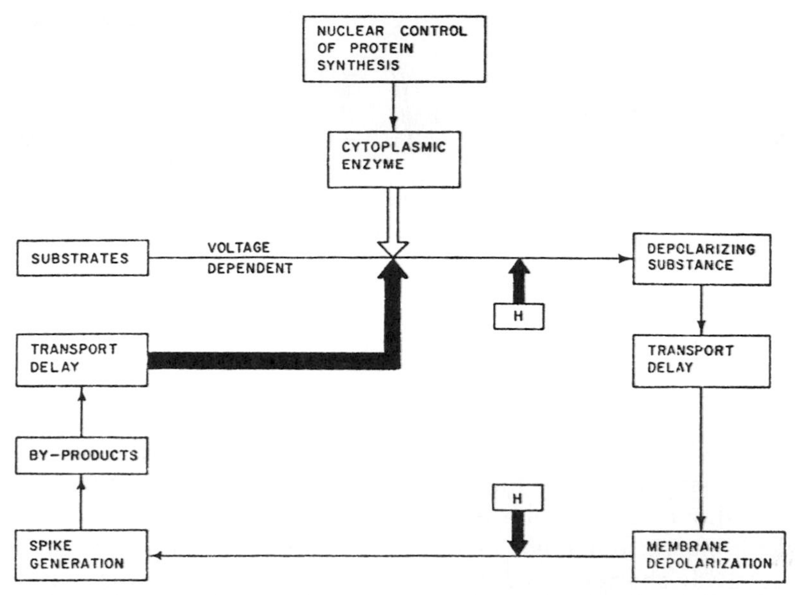

Figure 8-11 A model of the mechanisms giving rise to the circadian rhythm of neural spike discharge in the isolated parabolic burster cell of the abdominal ganglion of the sea hare *Aplysia.* The cell is postulated to endogenously synthesize a depolarizing substance that causes depolarization of the excitable membrane. H, hyperpolarization. [F. Strumwasser, *in* "Circadian Clocks" (J. Aschoff, ed.), pp. 442–462. North-Holland Publ., Amsterdam, 1965.]

phase delay of the next peak. These relationships are shown diagrammatically in Figure 8-11; the circadian oscillator would lie in the top (black!) box.

One might rightly ask how the output of the PVG is normally entrained in the living *Aplysia;* the answer may lie in another clock system found in this animal. It is possible to culture the isolated eye of *Aplysia* and monitor its spontaneous electrical activity via the severed optic nerve. In constant darkness, this preparation shows a circadian rhythm ($\tau = 27.5$ hours) of optic nerve impulses taking the form of trains of compound action potentials (CAP), as depicted for the left and the right eye of a single specimen in Figure 8-12. Peak activity occurs during the projected "subjective dawn" of the light-dark cycle to which the whole animal had previously been entrained. Eyes from sea hares that had been exposed to constant light showed a free-running rhythm. Furthermore, the isolated eye itself is responsive to light cycles; *in vitro* entrainment occurs. Thus, the mechanisms of photoreception, phasing, and basic oscillation all reside in the same organ.

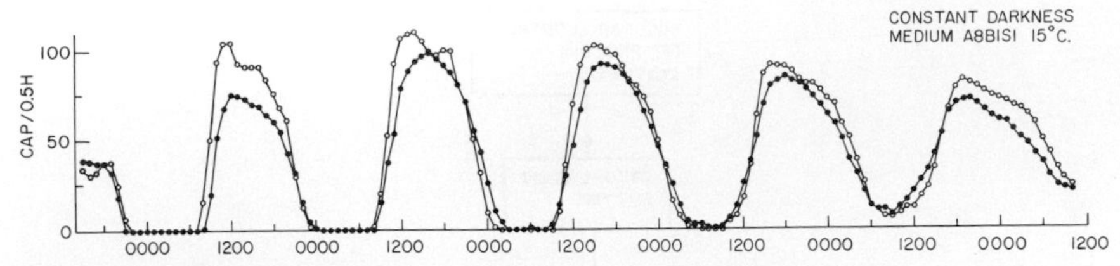

Figure 8-12 Circadian rhythm in compound action potential frequency in each of two eyes isolated from the sea hare *Aplysia* and maintained in organ culture in constant darkness. The first onset of CAP frequency in culture corresponds approximately with the last transition from dark to light at 0800 of the last day seen by the whole animal before the eyes were removed. [J. W. Jacklet, 1974. *J. Comp. Physiol.* **90,** 33–45 (1974).]

Perhaps even more interesting was the finding that the *accuracy* of this isolated clock system *depends upon the number of cells that it contains.* As many as 80% of the retinal cells of the eyecup could be sliced away without affecting the basic circadian period of CAP output by the optic nerve, but then, as the cell population is further reduced, progressively shorter circadian periods and ranges are expressed until finally the population oscillates at ultradian periods (<24 hours) of 1 to 12 hours. There are at least three alternative models for the organization of the endogenously active neuron (oscillator) population of the eye which might explain these provocative results if one assumes the operation itself has not unduly damaged the cells: (i) a population driven by a master circadian oscillator to which the individual cells are enslaved [But then the period should remain constant until suddenly this master clock was cut away]; (ii) a population of circadian oscillators [But then a circadian period should have been displayed right up to the bitter end and not show a change at the critical 20% level]; and (iii) a population of noncircadian, ultradian oscillators which together display a circadian periodicity, perhaps due to inhibitory cross-coupling. This last hypothesis won by default (though it has strong theoretical underpinnings); its important implications to clock timing will be discussed shortly.

THE PINEAL CLOCK

We now turn our attention to mammalian neuroendocrinological clock control mechanisms. In the search for a "master clock" we will bypass the adrenal, pituitary, and thyroid glands and the hypothalamus and proceed directly to where the action is: in the *pineal gland.* This neurochemical transducer rests between the two cerebral hemispheres and weighs about 100 mg in man and about 1 mg in the rat.

Research on this organ constitutes a major field in itself. Indeed, Descartes in 1677 considered the pineal to be the seat of the rational soul; vision passed through the eyes to the pineal by way of the brain "strings," and the pineal then responded by allowing animal humors to pass down hollow tubes to muscles. The philosopher was not far off the mark, as shown by the work of J. Axelrod of the National Institutes of Health, R. J. Wurtman of M.I.T., M. Menaker of the University of Texas at Austin, and many others have more recently demonstrated: the so-called third eye is still with us!

The pineal produces the indolamine melatonin, which is responsible for a host of different physiological functions. It causes contractions of the melanophores in frog and fish skin (and hence makes the skin color blanch); it exerts an inhibitory effect on the gonads of mammals (e.g., delay of vaginal opening and reduction of ovarian weight in young rats); and when injected into birds, it results in a decrease in the weight of the ovaries, testes, and oviduct. It appears to be intimately involved in the photoperiodic control of estrous and menstrual cycles, which may not be surprising in view of the fact that its parenchymal cells are innervated by sympathetic nerves whose cell bodies lie in the superior cervical ganglia that, in turn, eventually connect with the optic nerve and retina. Indeed, amphibian (but not mammalian) pineals have photoreceptive cells that can generate nerve impulses in direct response to environmental light, which thus constitute extraretinal photoreceptors (commonly dubbed ERP or ERR). Finally, in the house sparrow (*Passer domesticus*) and the white-throated sparrow (*Zonotrichia albicollis*), the pineal appears to be essential for the persistence of the circadian rhythm of gross locomotory activity. Strangely enough, though, it does not appear to play a necessary role in the nonvisual entrainment of the activity rhythm by light cycles in blinded birds (as initially thought; see Chapter 2, Figure 2-23), since if a blinded bird was subsequently pinealectomized, entrainment nevertheless occurred. In view of the myriad of activities that the pineal modulates, it is logical to try to determine the underlying control mechanism.

Melatonin is synthesized almost exclusively within the cells of the pineal according to the simplified reaction sequence shown in Scheme 8-1. The enzymes catalyzing the various steps are TROH, tryptophan

$$\text{Tryptophan} \xrightarrow{\text{TROH}} \text{5-hydroxytryptophan} \xrightarrow{\text{AAD}} \text{serotonin} \xrightarrow{\text{NAT}}$$

$$N\text{-acetylserotonin} \xrightarrow{\text{HIOMT}} \text{Melatonin}$$

Scheme 8-1

hydroxylase (hydroxylation); AAD, aromatic amino acid decarboxylase (decarboxylation); NAT, serotonin *N*-acetyltransferase (acetylation); and HIOMT, hydroxyindole *O*-methyltransferase (methoxylation). This last step is particularly critical in the regulatory process. HIOMT is highly localized in the pineal glands of mammals and birds. Further, the activity of HIOMT is reduced in rats kept in constant illumination, paralleled by a concomitant decrease in the weight of the pineal and an increase in the weight of the ovaries and an accelerated estrous cycle. This condition could be prevented by giving the rats injections of pineal extract. Thus, constant light decreases HIOMT activity, thereby decreasing the production of the gonad-inhibiting compound, melatonin, and the mass of the pineal gland. This reduction in melatonin synthesis, in turn, would lead to a removal of the inhibition of estrous. Finally, it has been demonstrated that HIOMT activity is under direct neural control (and is thus responsive to environmental light cycles): continuous light or darkness had no effect on the activity of this enzyme in blinded rats or in rats whose pineals had been denervated (by removal of the superior cervical ganglia) or whose medial forebrain bundles (containing noradrenergic and serotonergic nerves and connected with the optic tract) had been given bilateral lesions.

Even more germane to our search for a clock site and mechanism, however, was the demonstration that not only do the biogenic amines serotonin (Figure 8-13) and melatonin undergo marked circadian variations within the pineal, but also the mediating enzyme serotonin *N*-acetyltransferase (NAT) exhibits a pronounced rhythm which is 180° out of phase with that for serotonin. All of these rhythms (but not that of HIOMT activity) persist in constant darkness (implicating their control by an endogenous circadian clock), but are abruptly abolished by exposure to continuous illumination. Likewise, the rhythmicities in these substances are suppressed by denervating the sympathetic nerves to the pineal, by interruption of the nerve impulses from the central nervous system, or by bilateral lesions in ths suprachiasmatic nucleus of the hypothalmus. These observations indicate that the rhythms are generated most immediately by sympathetic nerve terminals innervating the pineal, mediated by the known diurnal changes in the release of the neurotransmitter noradrenalin, and *ultimately by a biological clock present in or near the suprachiasmatic nucleus in the hypothalmus.* This clock, in turn, would be modulated (entrained?) by inhibition by environmental light.

In sum, Axelrod's detailed model for pineal function holds that an increased discharge of noradrenalin at night stimulates the β-adrenergic receptors of the sympathetic nerve terminals innervating the pineal,

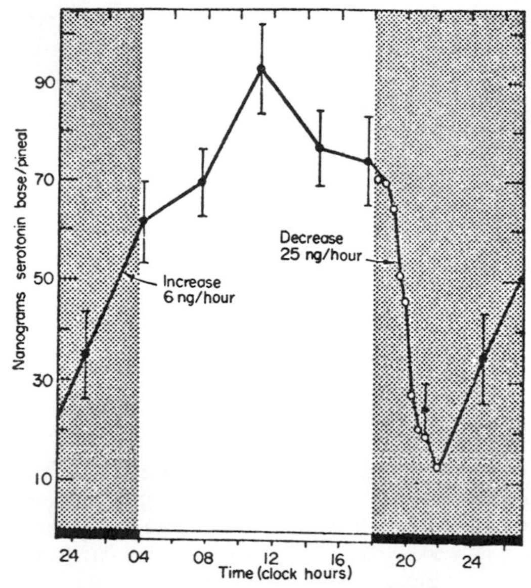

Figure 8-13 Diurnal rhythm in pineal serotonin of adult male rats. Means represented by dots (12 samples per solid dot; 5 per open dot) are plotted according to the time of sampling if the daily periods of darkness (stippled) and light (clear). Vertical lines extend two times the standard error on each side of the means. [W. B. Quay, *Gen. Comp. Endocrinol.* **3,** 473–479 (1963).]

which causes increased synthesis of serotonin *N*-acetyltransferase molecules inside the pineal cells by mediation of an adenylate cyclase system (now known to be at the center of so many hormonal control mechanisms). (Indeed, the responsiveness of the pineal β-adrenergic receptor itself appears to change: the receptor becomes supersensitive after decreased exposure to noradrenalin and other catecholamines and subsensitive after increased exposure to them.) As the activity of NAT rises during the night, the concentration of its substrate serotonin falls and that of the product *N*-acetylserotonin rises. Increased synthesis of the pineal hormone melatonin would then follow (leading to the observed gonadal effects) as a result of methoxylation of *N*-acetylserotonin by HIOMT.

Clocked Cell Cycle "Clocks"

Having been led through the optic lobe, protocerebrum, parabolic burster cell, eyecup retinal layer, pineal gland, and finally the suprachiasmatic nucleus of the hypothalamus in an enjoyable, but largely

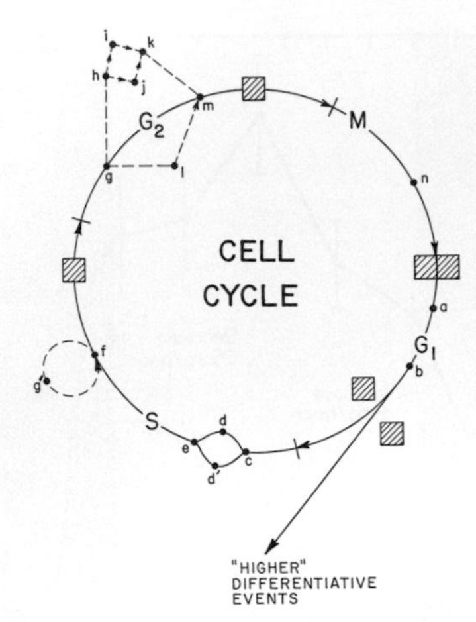

Figure 8-14 Diagram of a generalized cell cycle. The four classical phases are shown: G_1 ("gap 1"), S (synthetic), G_2 ("gap 2"), and M (mitotic). In addition, a number of arbitrarily chosen points are designated by the smaller letters to illustrate possible alternative pathways, branching networks, loops, and blockage points. [L. N. Edmunds, Jr., *in* "Les Cycles Cellulaires et Leur Blocage chez Plusiers Protistes," pp. 53–67. Colloques Internationaux du C.N.R.S., no. 240, Centre National de la Recherche Scientifique, Paris, 1976.]

unsuccessful search for a "master" clock, it is time for us to get back to basics again.

Since it has already been shown in Chapter 2 that a single unicellular organism is capable of displaying a variety of circadian outputs (Figures 2-3, 2-16,), it would seem equally plausible that at least some single cells isolated from the tissues or organs of a multicellular organ such as the pineal gland could manifest similar rhythmicities. (However, it is possible that circadian rhythmicity may be, in certain cases, merely a higher level derivitive or attribute of a population of coupled ultradian oscillators.) And one of the most basic of cellular processes is the cell cycle itself.

The cell cycle of a typical microorganism such as the unicellular algal flagellate *Euglena* comprises a series of relatively discrete morphological and biochemical events, although the specific elements may vary among different systems (Figure 8-14). These developmental sequences are not necessarily linearly ordered, however, since branching networks and feedback loops may provide several alternative pathways, some of which may occur concurrently. Nevertheless, the various processes taken as a whole are ordered temporally.

In this sense, then, the cell cycle itself is a "clock" (though not strictly circadian): the specific events—ranging from mitosis, chloroplast replication, and phototactic response to the timed synthesis of some specific enzyme—correspond to the numerals on the dial and the generation time (g) of the cell reflects the period (τ) of the timing process(es), or oscillation(s). As is well known, this cell cycle clock is quite imprecise and very labile, for it is markedly affected by alterations in temperature, illumination, and nutrients in the medium. For example, in the best of all possible worlds, the generation time of *Euglena* may be as short as 8 hours; more typically, g ranges from 12 to 30 hours in the laboratory; and the upper limit approaches infinity as the stationary phase of population increase is entered. It may be that g is nothing more than a summation of all the individual "reaction times" of the constituent processes comprising the cell cycle and that the latter is simply a statistical averaging machine. Yet, as will be shown, this is an oversimplification.

Now, in exponentially increasing cultures of microorganisms, the phase points of the individual cell cycles of the cell population are distributed randomly; they are *developmentally asynchronous.* A striking contrast is afforded by *developmentally synchronous* populations: cell division (as well as at least some preceding events) in cultures of numerous protists, algal unicells, and cells dissociated from plant and animal tissues can be synchronized by a variety of inductive treatments so that there is a one-to-one mapping of similar phase points of the cell cycles throughout the culture with respect to time. Thus, the imposition of appropriately chosen light cycles is the method of choice for most photosynthetic unicellular algae such as *Euglena* (Figure 8-15). Our question, as usual, concerns the mechanism underlying this synchronization or entrainment process.

A general model for such synchronization by shifts in environmental conditions assumes that under a given set of extrinsic conditions the cell progresses through a sequence of stages and that under a different set of conditions the cell cycle consists of the same sequence, but with the relative time spent between such stages being different. If a series of shifts is performed between these two sets of conditions (to which specific stages of the cell cycle are *differentially sensitive*), such that the period of the imposed regime corresponds to a doubling of cell number, the model predicts a gradual attainment of complete synchronization. Therefore, in the well-synchronized culture, a majority of the cells pass through the same developmental stage at the same time; thus, what is determined for the entire culture can be assumed to obtain as a first approximation for the individual cells. It is for this reason that synchronously dividing cultures of microorganisms have proved such useful tools for elucidating numerous biochemical and physiological problems associated with the cell developmental cycle.

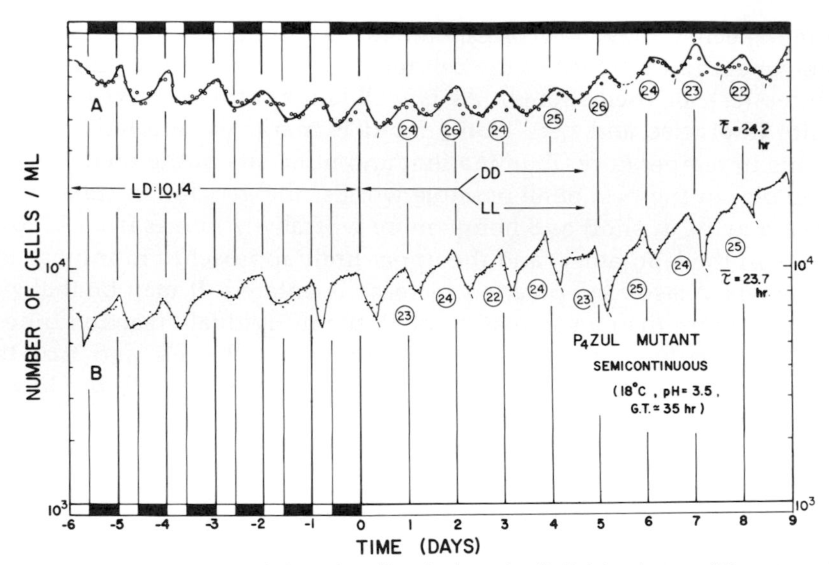

Figure 8-15 Long-term, persisting circadian rhythm of cell division in two different semicontinuous cultures of the P_4ZUL photosynthetic mutant of *Euglena* grown organotrophically at 18°C. The cultures were first entrained by a *LD 10:*14 light cycle (6 days shown) and then placed either in DD (top, curve A) or in LL (bottom, curve B); the first 9 cycles under constant conditions are indicated. Overall generation time (G.T.) of both cultures was calculated from the known dilution rate to be about 35 hours. Successive period lengths are encircled just below each free-running cycle; the average period is given to the right for each curve. [L. N. Edmunds, Jr., *in* "Chronobiology" (L. E. Scheving, F. Halberg, and J. E. Pauly, eds.), pp. 61–66. Igaku Shoin Ltd., Tokyo, 1974.]

Although this extrinsic model does apply in certain situations, it is not sufficient to explain the observed facts and, indeed, may not even be relevant at all under certain experimental conditions. These empirical findings, ascertained in my laboratory over the past 10 years or so, may be summarized as follows: (i) synchronous division in autotrophically batch-cultured wild-type *Euglena* can be precisely entrained to a 24-hour period by repetitive light-dark cycles having a driving period (T) of 24 hours (e.g., *LD 10:*14), although not all cells necessarily divide during any given cycle (as in *LD 6:*18 where the photoperiod does not afford enough light energy for photosynthesis to permit a doubling in cell number); (ii) entrainment by light cycles having $T \neq 24$ hours (e.g., *LD 10:*10 or *LD 14:*14) may also occur within certain limits; (iii) "skeleton" photoperiods comprising the framework of normal "full-photoperiod" cycles (e.g., *LD 3:6:3:*12) will also entrain the rhythm to a precise 24-hour period; (iv) high-frequency light cycles (e.g. *LD 1:*3) and even "randomly" chosen light regimes—which surely can provide no 24-hour informational input into the system—induce *circadian* periodicities that for all intents and purposes are free-running; (v) rhythmic cell division will persist for a number of days (even weeks) with a circadian period length in the autotrophically grown wild-type batch-cultured under continuous dim illumination wherein there are no repetitive shifts in the environmental conditions that can be invoked as a cause of the persisting rhythmicity; (vi) the ultraviolet light-induced, semichlorophyllous P_4ZUL mutant and the totally heat-bleached W_6ZHL mutant of *Euglena,* both incapable of carrying out photosynthesis and hence requiring the supplementation of their medium with an organic carbon source such as glutamic acid, can likewise be synchronized by the identical light cycles utilized for the wild type (Figure 8-15); and finally, (vii) the entrained division rhythmicities observed in the mutants will persist with a circadian periodicity for weeks in continuous illumination or even constant darkness (Figure 8-15). These last results constitute the strongest sort of challenge to the simple extrinsic model for synchronization, since the use of photosynthetic mutants permits one to disentangle the role of light as an energy source for metabolism (as in the wild-type) from the utilization of light and dark as signals furnishing timing and phasing information to the cells. Other unicellular systems for which persisting circadian rhythms of cell division (or cell "hatching") have been documented include the green algae *Chlamydomonas* and *Chlorella,* the dinoflagellates *Gonyaulax* (Figure 2-3) and *Gymnodinium,* the yeast *Candida,* and the ciliates *Paramecium* and *Tetrahymena*. There is even some evidence that there is a circadian rhythm in the growth rates of the bacterial prokaryotes *Escherichia coli* and *Klebsiella aerogenes.*

As a working hypothesis, the author has assumed that an endogenous, light-entrainable, circadian clock—having all the usual properties outlined in Chapter 1 that characterize the oscillatory mechanism(s) underlying persisting 24-hour rhythms—underlies division rhythmicity in light-synchronized cultures of *Euglena* and "gates" cell division to restricted intervals of time during successive 24-hour time spans [in much the same way as with the eclosion rhythm in *Drosophila* (Figure 2-24)]. Although division bursts (increases in cell number) occur approximately every 24-hours in a cell population, the lengths of the cell cycles in individual cells (i.e., g values) can be deduced to be integer-multiples of τ under conditions where a doubling of cell number does not occur each cycle. Thus, the developmental sequence culminating in the act of cell division of the mature *Euglena* cell is conceptually and operationally separable from the circadian oscillation that gates division (when it does occur) in the individual cell to a specific time of day and in the population to intervals of 24 hours. The implication, of course, is that in some, as yet unknown, manner the clock actually inhibits cell division from occurring in a cell that just missed a temporal gate until the next gate opens 24 hours later.

Overt Circadian Rhythms in Unicellular Organisms

Although it has been clearly established that the unicellular cell cycle "clock" can be coupled to a circadian oscillator system in *Euglena* and other microorganisms, it is equally certain that this circadian clock mechanism is not dependent on the driving force of the cell cycle for its functioning. These two phenomena can be effectively divorced from one another by utilizing cultures that have reached the stationary phase of growth in which there is little or no net change in cell number. For example, the author has discovered a 24-hour rhythm of cell settling in *Euglena* which may occur concurrently with, or in the absence of, cell division. In stationary cultures in continuous illumination and constant temperature, the cells actually tend to settle out of the liquid phase and adhere to the vessel walls and then subsequently detach themselves and reenter the medium. This rhythm will persist for at least 9 days with a temperature-compensated, free-running period, which strongly suggests that it is autonomous and self-sustaining. Additionally, we have evidence that the settling rhythm may also occur during both the growth and stationary phases of the P_4ZUL and W_6ZHL photosynthetic mutants of *Euglena*. Persistent motility rhythms have also been reported by other workers, as well as a circadian, temperature-compensated, rhythm of phototactic response (Figure 2-16).

A circadian rhythm in photosynthetic capacity has also been documented for *Euglena*. Cultures were grown photoautotrophically in 8-liter batches and synchronized by a *LD 10:*14 cycle at 25°C. Aliquots were then taken at various time points during the cell cycle for determination of the ability of the cells to incorporate radioactively labeled [^{14}C]sodium bicarbonate when exposed to a light source for 15 minutes and then assayed with a liquid scintillation counter. Capacity peaked at 4 to 8 hours after the onset of light and then decreased back to basal levels (on a per aliquot basis) as darkness and cell division ensued. A similar diurnal pattern was found when oxygen evolution was monitored. This rhythm in photosynthetic capacity occurs also in stationary cultures maintained in the light cycle (Figure 8-16) and will persist for a cycle or so in continuous dim illumination and constant temperature before damping out.

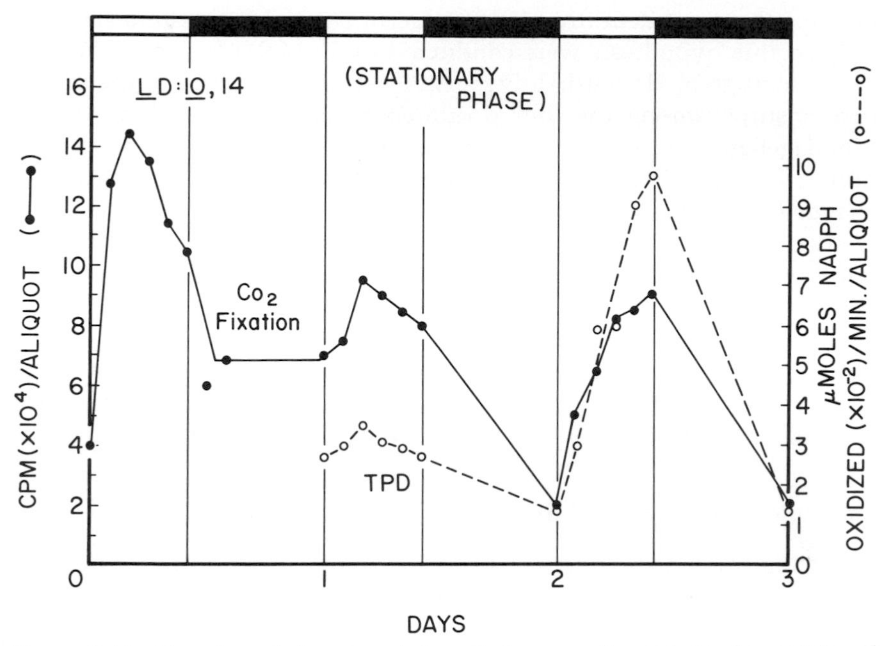

Figure 8-16 Circadian rhythm in photosynthetic capacity and in the activity of glyceraldehyde-3-phosphate dehydrogenase (TPD) in nondividing (stationary phase) cultures of *Euglena* maintained in a *LD 10:14* light cycle. The culture was kept in the stationary phase for 1 week before assays were performed. [W. G. Walther and L. N. Edmunds, Jr., *Plant Physiol.* **51,** 250–258 (1973).]

Finally, a daily rhythm of amino acid ([^{14}C]phenylalanine) incorporation exists in nondividing cultures of *Euglena* maintained autotrophically in a light cycle. This rhythmicity persists for at least two cycles in cultures transferred to constant darkness and is therefore probably truly circadian.

These examples of circadian rhythms discovered in *Euglena* underscore the important fact that circardian temporal organization is not restricted to multicellular organisms and that it does not depend on the cell cycle per se. Indeed, overt persisting circadian rhythms have been documented for *Acetabularia* (Figure 2-5), *Chlamydomonas, Chlorella, Gonyaulax* (Figure 2-3), *Paramecium,* and *Tetrahymena,* among others, and range from bioluminescence to pattern formation, to mating type reversal. In each of these different microorganisms (as for *Euglena*), furthermore, *several different rhythms have been observed concurrently* [and in some cases in individual cells (Figure 2-4) as well as in cell populations], with the attendant implication that all of the overt rhythms are outputs or "hands" of a single "master" clock oscillator.

Perhaps the most definitive set of experiments whose results strongly support this hypothesis were conducted by L. McMurry and J. Woodland Hastings of Harvard University. First, they found that the phase relationships among the four documented circadian rhythms in the dinoflagellate *Gonyaulax polyedra* [glow, luminescence capacity, cell division, and photosynthetic capacity (Figure 2-3)] remained unchanged during several weeks under continuous illumination—unlikely if the different rhythms were each caused by a different underlying oscillator whose periods, if slightly different, would result in a drifting apart of the phases. Second, they discovered that a 6-hour exposure to darkness (a dark "pulse") shifted the phase of all rhythms by an equal amount, a result not expected by the null hypothesis for similar reasons. And finally, the Q_{10} for the period of three of these rhythms is approximately the same (and uniquely, at that: 0.80 to 0.90).

Oscillatory Biochemical Systems

Thus far, only gross physiological and behavioral circadian rhythms at either the multicellular or cellular level of organization have been considered. To discover the basic control mechanism of the putative clock underlying these rhythms, however, it seems eminently reasonable to ask whether endogenous, self-sustaining oscillations can exist at the molecular level, which then might drive the more overt rhythms.

Perhaps one of the best understood examples of a persisting rhythmicity in a biochemical parameter is the yeast system (*Saccharomyces carlsbergensis*) studied by B. Chance, E. Kendall Pye, and co-workers at the Johnson Research Foundation at the University of Pennsylvania, in which glycolytic oscillations have been discovered in both cell cultures and in *cell-free extracts.* Although they are *ultradian,* i.e., have a very short period, they will serve as an instructive model system.

When intact cells grown aerobically in pure culture on an artificial liquid medium containing 2% glucose as the principal carbon source are transferred to anaerobic conditions, the amount of reduced diphosphopyridine nucleotide (DPNH) oscillates with a relatively high frequency. Long trains of sinusoidal oscillations, having a period of about 33 seconds at 25°C, are typically observed. These rapid oscillations were followed by continuously monitoring either the fluorescence or the absorbance attributed to DPNH with a specially designed, temperature-compensated fluorometer or a double-beam spectrophotometric recording system. It is significant that these oscillations were not continuously damped, but instead, often showed a region of

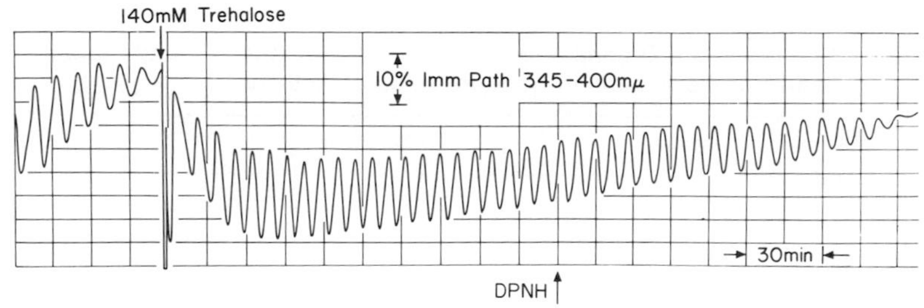

Figure 8-17 The generation of sustained glycolytic oscillations in DPNH in a cell-free extract of yeast by an addition of trehalose. The level of DPNH was continuously monitored with a dual-beam spectrophotometer and percentage transmission recorded directly. Time proceeds from left to right (one unit on the abscissa corresponds to 15 minutes). The train of 42 oscillations lasted over 4.5 hours and had a stable frequency of 0.15 minutes^{-1} and a period of about 7 minutes. [E. K. Pye, *Can. J. Bot.* **47,** 271–285 (1969).]

increasing amplitude after the initial period of damping; this observation rules out certain simple feedback mechanisms.

Even more important perhaps was the discovery that sustained oscillations in DPNH could occur also in *cell-free extracts* from yeast, as illustrated in Figure 8-17. Indeed, the supernatant fraction obtained by centrifuging sonicated cells at 105,000 *g* for 90 minutes retained its oscillatory capacity for as long as 6 months when stored at −20°C. By adding the disaccharide trehalose to the medium (which is accumulated by the yeast as a reserve carbohydrate and which provides a relatively slow, continuous input of glucose into the glycolytic scheme as a result of its hydrolysis by the enzyme trehalase), a prolonged train consisting of over 43 oscillations was obtained. These oscillations lasted over 4 hours, displayed little damping, and exhibited a remarkably stable period of about 7.1 minutes at 25°C; a further addition of trehalose, after these oscillations had subsided, produced another train of oscillations. Indeed, in one experiment, a continuous train of over 90 cycles lasting over 9 hours was obtained with a single trehalose addition; again, the free-running period was about 7 minutes. Finally, the addition of small quantities of adensosine diphosphate (ADP) to the oscillating cell-free extract at different times during the cycle caused phase shifts whose signs and magnitude were clearly phase-dependent in a manner reminiscent of the effects of light signals on the free-running rhythms of higher circadian systems (cf. Figure 8-2).

These glycolytic oscillations at the intracellular level of organization, then, possess many of the characteristics of other endogenous biorhythms: stable frequency, a self-sustained nature, a susceptibility to phase shifting (at least by appropriate effector molecules), fade out,

Table 8-1

Some Constituents in the Organs and Tissues of Higher Animals and Plants Exhibiting a 24-Hour Fluctuation in Concentration or Activity

Organ or tissue	Constituent
Adrenal gland	Biotin (free and bound), coenzyme A, pantothenate, succinic dehydrogenase
Blood plasma or serum	Sugar, cholesterol, bilirubin, tyrosine, copper, iron, Ca^{2+}, Na^{+}, PO_4^{2-}, 5-hydroxytryptamine, hydrocortisone, ACTH, STH, 17-ketosteroids, glutamic oxalacetic transaminase, 11-oxycorticosterone, hexosamine, protein, testosterone, sialic acid
Brain	5-Hydroxytryptamine
Heart	Glycogen
Kidney	Transamidinase
Leaves	Aldolase, amylase, phosphatase, K^{+}, Na^{+}
Liver	Bile, glycogen, adrenalin, phospholipids, RNA, DNA, ATPase, cycloleucine, citrate cleavage enzyme, glucose-6-P-dehydrogenase, tyrosine transaminase, serine dehydratase, alkaline phosphatase
Pancreas	Glycogen, exocrine secretion
Pineal gland	Melatonin, serotonin, norepinephrine, tyrosine hydroxylase, hydroxyindole-*O*-methyl transferase, serotonin *N*-acetyltransferase
Skeletal muscle	Glycogen
Spleen	Glycogen
Urine	Water, Na^{+}, K^{+}, Mg^{2+}, Cl^{-}, PO_4^{2-}, urea, acetone, creatine, creatinine, urobilin, coproporphyrin, amino acids, citrate, pyruvate, uropepsin, urokathepsin, 5-hydroxyindoleacetic acid, suprarenal cholesterol, 17-hydroxycorticosteroids, β-hydroxybutyric acid, epinephrine, norepinephrine, aldosterone

damping, and reinitiation. They differ from the circadian rhythms surveyed thus far, in having a very short period that is highly dependent on temperature and apparently in not being light-entrainable. As will be seen, however, a biochemical driving oscillation for a circadian rhythm could be logically derived from a set of reactions similar to the yeast system by merely choosing different kinetics for the reaction sequence; by coupling a reaction inhibited by increasing temperature to one enhanced by increasing temperature, one could easily provide the additional feature of temperature compensation of the free-running period.

For the present, however, the question is whether actual 24-hour rhythms of biochemical parameters can occur in living systems. As evidenced by the data in Table 8-1, one might conclude our question should perhaps be rephrased: Is there any constituent in the organs and tissues of higher animals that does not exhibit a diurnal fluctuation? [The word diurnal is pointedly used here since only a few of the parameters have been measured in continuous illumination (or darkness) and constant temperature due to practical and technical limitations; therefore, it is not clear whether all of the observed fluctuations are truly endogenous, self-sustaining circadian oscillations.] But perhaps this is not so surprising after all, in view of the large number of overt circadian rhythms documented at the physiological level: gross motor activity must have biochemical correlates. Nevertheless, not all aspects of cellular chemistry oscillate.

Similarly, circadian rhythmicity in the activity and the inducibility of the enzyme tyrosine aminotransferase and in the incorporation of radioactive leucine has been observed in cultured liver cells dissociated from the livers of baby rats with tetraphenylboron and prevented from reaggregating by continual stirring. Although the daily rhythms of this enzyme in the intact liver of adult rats has been shown to be dependent on the cyclic intake of exogenous tryptophan supplied in the diet (recall that most animals have a 24-hour feeding rhythm), individual cells (at least in cell culture) also appear to possess an autonomous oscillation that can persist with a circadian period for up to 2 weeks in continuous illumination. Indeed, after 3 weeks under these conditions, the rhythm of hepatic tyrosine aminotransferase could be reinitiated by exposing the cell culture to a single 1-hour dark "pulse." Thus, this biochemical oscillation is not dependent on tissue- or organ-level organization.

Even "cleaner" data for 24-hour rhythms in enzyme activity are available from work with *unicellular systems.* Already discussed is the fact that the activities of the extractable enzyme (luciferase) and of extractable substrate (luciferin) for the luminescence rhythm of the dinoflagellate *Gonyaulax* show circadian variations in both light cycles and under "constant" conditions (Figure 2-10). Likewise, a 24-hour rhythm in the activity of the enzyme glyceraldehyde-3-phosphate dehydrogenase (GPD) has been documented in nondividing cultures of *Euglena* that have attained the stationary growth phase (Figure 8-16). The similarity between both the phase and the pattern of the activity of this key enzyme of the Calvin cycle photosynthetic scheme and the circadian rhythm in photosynthetic capacity previously described cannot help but suggest a possible control of the latter by GPD; but, of course, the relationship may be merely correlative rather than causal. Finally, 24-hour oscillations in the activities of alanine, lactic, and glucose-6-P dehydrogenases, L-serine and L-threonine deaminases, and acid and alkaline phosphatases have also been demonstrated in stationary cultures of *Euglena* entrained by an imposed *LD 10:*14 light cycle. The use of these nondividing populations effectively divorces

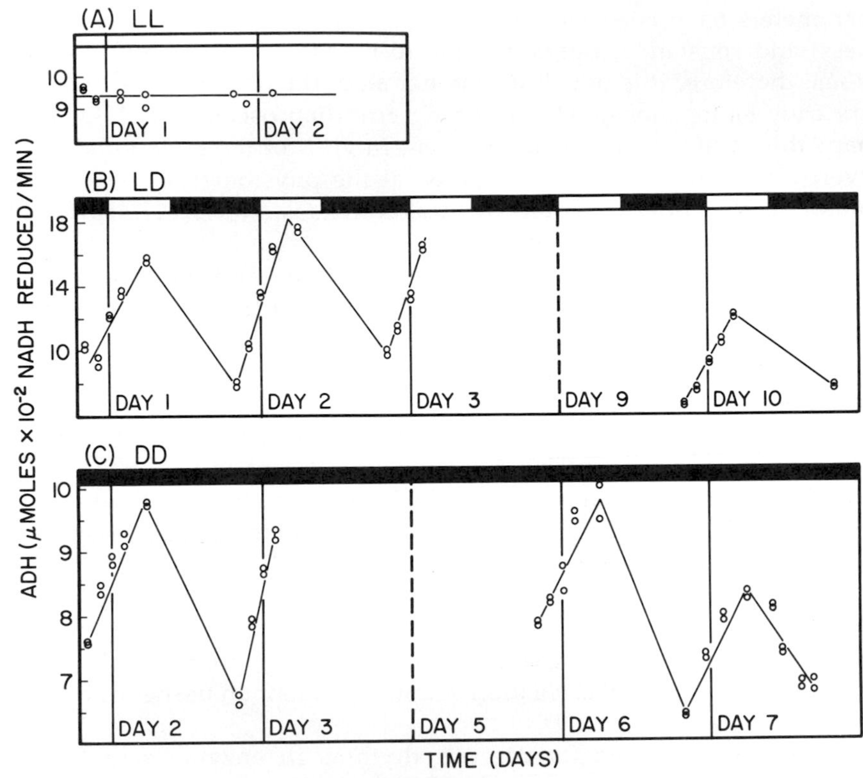

Figure 8-18 Circadian oscillations in the activity of alanine dehydrogenase (ADH) in nondividing, organotrophically batch-cultured *Euglena*. (A) ADH activity in culture maintained under constant bright illumination [LL]. (B) ADH activity in culture in *LD 10:*14 light cycle. Data from day 1, 2, 3, and 10 are shown. (C) ADH activity in constant darkness [DD] after many cycles of *LD 10:*14. Data from days 2, 3, 6, and 7 following the transition from LD to DD are given. Vertical lines are spaced 24 hours apart. Double points represent duplicate determinations. [F. M. Sulzman and L. N. Edmunds, Jr., *Biochem. Biophys. Res. Commun.* **47,** 1338–1344 (1972).]

autogenous enzymatic oscillations from those directly generated by the driving force of the cell cycle itself and the accompanying periodic replication of the genome. Even more interesting, however, was the finding that the activity of alanine dehydrogenase (ADH)—an everyday, kitchen variety, "housekeeping enzyme"—continues to oscillate for at least 14 days in continuous darkness (but *not* in continuous illumination) and constant temperature, and thus constitutes a *bona fide* circadian rhythm in itself (Figure 8-18).

After discovering this circadian biochemical oscillation, the author and his co-workers initiated experiments to determine its genesis on the hypothesis that it was another hand of the ubiquitous and pervasive biological clock. The possibility that fluctuations in pools of substrates of products could change the stability of ADH during its extraction and thus "trivially" generate the observed rhythm was ruled out. Results from mixing experiments likewise do not suggest the presence of fluctuating pools of effector molecules that also might produce the rhythm by altering the activity of the enzyme, nor were there differences in pH optimum, K_m value, or electrophoretic mobility of enzymes extracted at different phases of the oscillation. On the other hand, activity determinations of ADH extracted from the maximum and minimum points of the rhythm and partially purified by ammonium sulfate fractionation and polyacrylamide gel electrophoresis suggest that periodic *de novo* synthesis and degradation of ADH may generate the observed variations in its activity.

Thus, circadian oscillations in enzymatic activity and in numerous other biochemical parameters can and do occur in organs, tissues, isolated cells, and in unicellular organisms and at least some of these constituents most probably underlie (i.e., drive) the overt circadian rhythms surveyed in earlier chapters. The focus next will be on attempts to experimentally dissect the mechanism(s) that might generate the oscillations. And this will finally lead to models for the ultimate clockwork itself.

Dissection of the Clock: Use of Chemical Inhibitors

Because of the difficulties in interpretation of results, it has been said that "only the uninhibited use inhibitors." Nevertheless, this approach has been of considerable value in gaining further information about the nature of the elusive clock. Earlier work on *Phaseolus*, the green alga *Oedogonium*, *Euglena*, *Gonyaulax*, and other experimental material were relatively nonspecific. Virtually the entire chemical shelf has been thrown at biorhythms: EDTA (a chelator), $AgNO_3$, $CaCl_2$, KCl, and $FeCl_3$ (all metabolic "poisons"); the growth factors, giberellin and kinetin; the mitotic inhibitors, urethane and 5′-fluoro-2′-deoxyuridine; respiratory inhibitors, such as arsenate, cyanide, and PCMB (*p*-chloromercuribenzoate); arsenate and DNP (2,4-dinitrophenol), both uncouplers of oxidative phosphorylation; and the herbicides CMU and DCMU (monochloro- and dichlorophenyl dimethylurea), which specifically inhibit photosynthesis. Likewise, at the clinical level, there are not many pharmaceuticals that have not been tried on mammalian rhythms. In many cases, particularly in the case of algal systems, these chemicals could be applied and then washed out (by centrifuging the cells and resuspending them in fresh medium or buffer), thus constituting a chemical analogue of light pulses utilized for phase-shifting overt physiological rhythms.

Unfortunately, however, very little could be concluded from these studies except that the biological clock seems to be remarkably insensitive to perturbations by ordinary chemical agents. In some cases, phase shifts of sorts occurred; while in others, the entire rhythm damped out. But how does one differentiate between merely affecting the expression of the rhythm (e.g., by uncoupling the "hands" from the clock), on the one hand, and actually affecting the underlying oscillation itself? In one case, CMU- and DCMU-blocked photosynthesis (and hence the rhythm of photosynthetic capacity) in *Gonyaulax*, as expected, but the phase of the bioluminescent glow rhythm was unchanged. Thus, photosynthesis is clearly not necessary for clock function in a photoautotroph, and stoppage of this clock-controlled process has no feedback on the clock itself (on the assumption that the several rhythms in this dinoflagellate are controlled by the same basic circadian oscillation).

Workers then turned to specific inhibitors of macromolecular synthesis (conveniently being discovered at the time) on the hypothesis that the sequence whereby mRNA is synthesized on DNA template and subsequently directs protein synthesis might be responsible for the generation of circadian oscillations. J. Woodland Hastings and M. W. Karakashian, for example, discovered that very low concentrations of actinomycin D, an inhibitor of DNA-dependent RNA synthesis, virtually abolished rhythmic glow luminescence (although a low level of glow still occurred), as well as cell division (Figure 8-19). One subsequent peak did occur, indicating that determination of the peaks must occur about 24 hours prior to their expression. This discovery suggested that newly synthesized RNA acts as a messenger for the synthesis of a protein which, in turn, controls the rhythm of luminescence. Further, if clock function does relate to newly synthesized RNA, then the inhibition of new DNA synthesis should not disturb the rhythm. This conclusion was supported by the effects of both amethopterin and novobiocin on *Gonyaulax* (Figure 8-19); the delayed inhibition of the glow rhythm by mitomycin C was thought to be due to the slow breakdown of DNA in its presence, so that it could no longer serve as primer to RNA synthesis.

Not all the evidence, however, was as clear-cut. In the first place, other rhythms in *Gonyaulax* do not react to the presence of actinomycin D as does the glow rhythm. Thus, the rhythms of photosynthetic capacity and of stimulated luminescence are not at all changed

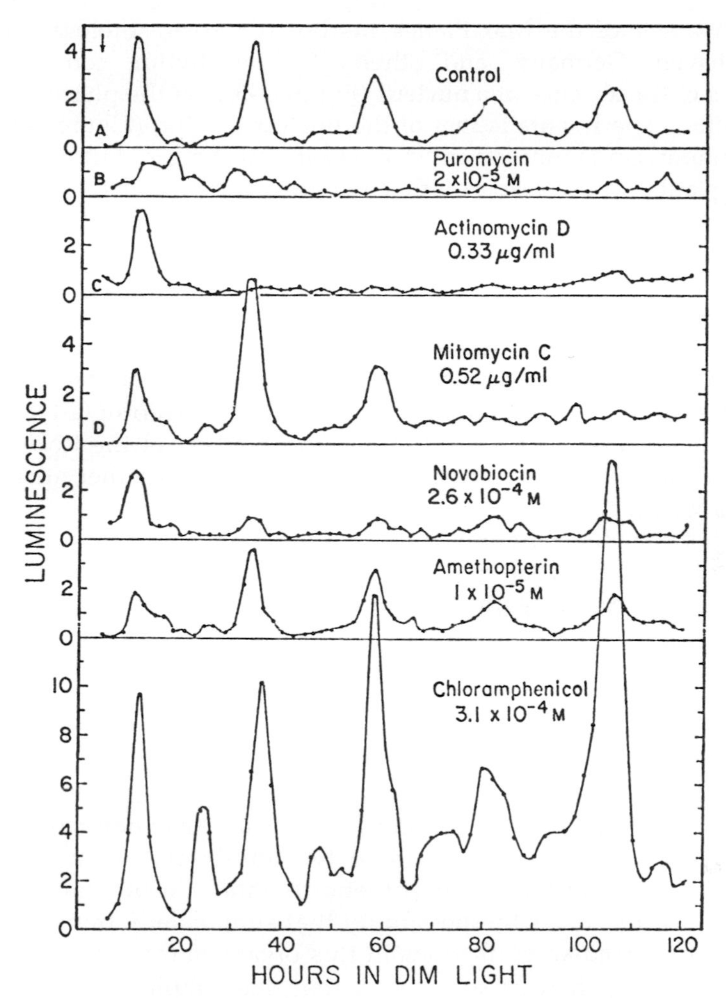

Figure 8-19 Recordings of the glow emitted from the dinoflagellate *Gonyaulax polyedra* during exposure to substances that affect macromolecular synthesis. The cell suspensions were kept in continuous illumination except while recording was in progress. The inhibitors were added at the time indicated by the arrow. Luminescence in arbitrary units. [M. W. Karakashian and J. W. Hastings, 1963. *J. Gen. Physiol.* **47,** 1–12 (1963).]

at concentrations of inhibitor which completely eliminate the glow rhythm. (It penetrates the cells since cell division is inhibited in its presence.) Likewise, a most unusual set of paradoxes exist as a result of the work on the photosynthetic rhythm in the marine alga *Acetabularia* by T. Vanden Driessche of the Université Libre de Bruxelles, B. M. Sweeney of the University of California at Santa Barbara, E. and H.

G. Schweiger of the Max-Planck Institut für Meeresbiologie at Wilhelmshaven, Germany, and others: (i) the rhythm continues for months in the absence of a nucleus (Figure 2-5), yet the phase in transplantation experiments is that of the nuclear portion (Figure 2-6); and (ii) actinomycin D inhibits rhythmicity in intact algae, but not in anucleate plants. In the case of either paradox, the presence of a nucleus clearly alters the state of circadian rhythmicity in the rest of the cell. These curious results perhaps could be accounted for by a much higher turnover rate in whole cells of the specific mRNA postulated to be involved in rhythmicity. In any case, it would appear that daily transcription from the nuclear DNA is not necessary for the oscillation to continue in anucleate cells. Furthermore, it has been demonstrated that rifampicin, an inhibitor of transcription of chloroplast DNA in *Acetabularia* (it competitively inhibits the binding of the RNA to the RNA polymerase), also does not alter rhythmicity in either nucleate of enucleated cells.

On a somewhat different but, nevertheless, related tack, the question occurs as to what role protein plays in the functioning of circadian clocks. Once again, the story quickly becomes a bit muddled. On the one hand, puromycin, a potent inhibitor of protein synthesis, completely blocks the glow rhythm in *Gonyaulax;* yet, in the presence of another inhibitor, chloramphenicol, not only are the period and phase of the rhythm unaffected, but the amplitude of the rhythmic output is greatly increased (Figure 8-19), contrary to what might be expected if the oscillation involves protein synthesized on an RNA template. These seemingly contradictory results have never been satisfactorily explained. Similarly, these two inhibitors do not alter the period or phase of the photosynthetic rhythm in a nucleate *Acetabularia,* which precludes the possibility that simple daily translation can account for circadian oscillations in this organism (but see below).

Nevertheless, there is some work that does implicate protein synthesis in circadian clock function. In his doctoral research at Princeton University, J. F. Feldman discovered that the period of the circadian rhythm of phototactic response in *Euglena* is increased (reversibly) by the addition of cycloheximide (actidone), another strong inhibitor of protein synthesis (Figure 2-16). Furthermore, he elegantly demonstrated that the effects of this drug are on the clock itself rather than on some parameter controlled by the clock or on some "uncoupling" mechanism between the clock and the parameter. This was confirmed by experiments that assayed the position of the light-sensitive (A-) oscillation by 4-hour resetting light signals after cycloheximide addition in continuous darkness (Figure 8-20), as predicted from the

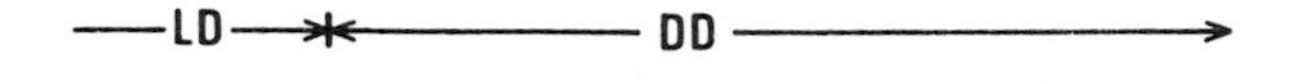

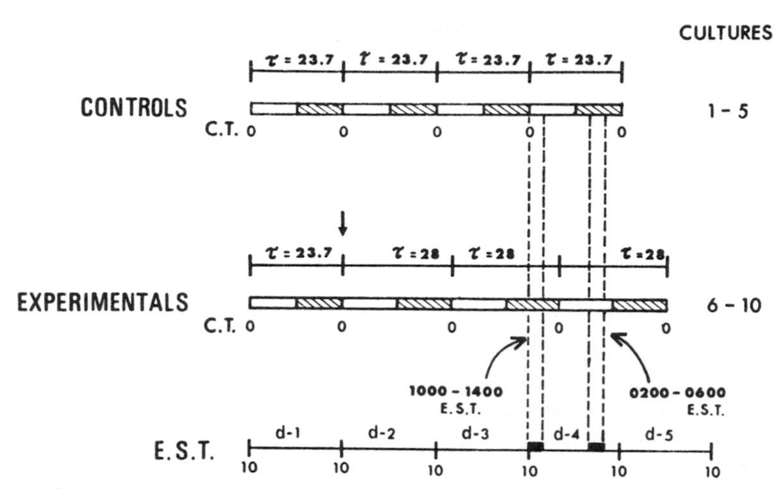

Figure 8-20 Testing for the lengthening of the free-running period of the rhythm of phototactic response in *Euglena* after the addition of cycloheximide in constant darkness. The phase of the underlying clock was assayed by measuring the phase-resetting effects of 4-hour light pulses in reference to those predicted by the phase-response curve determined previously for this rhythm. The bottom line indicates Eastern Standard Time (i.e., absolute elapsed time); in the prior light cycle, lights were on from 1000 to 2000 EST; *d-1*, *d-2*, etc., indicate successive days in constant darkness. The bar labeled "controls" indicates (open bars) and subjective night (hatched bars) of the control cultures; similar notation for the "experimentals" in which cycloheximide was added at subjective dawn of the second day of DD (arrow). Solid blocks on the EST scale at day 4 of DD give times at which each of the two, 4-hour light pulses were administered to separate cultures. Note that the first pulse strikes the control cultures in their early subjective day whereas the second pulse hits during the subjective night; but the same pulses strike the experimentals with the reverse phase relationship, which indicates that the phase has been changed due to the lengthening of the period of each cycle by the inhibitor. [J. F. Feldman, *Proc. Natl. Acad. Sci. U.S.A.* **57,** 1080–1087 (1967).]

phase-response curve that had been previously determined for this unicell (Figure 2-17). Yet, even this "clean" set of results must be interpreted cautiously: cycloheximide not only may have other primary effects in addition to the inhibition of protein synthesis, but may also produce secondary metabolic effects such as alteration of pools of intermediates or the levels of energy pools.

Finally, a very recent and thorough study of the effects of different inhibitors of transcription and translation on the expression and control of circadian rhythm of oxygen evolution in *individual* cells of *Acetabularia* has been reported by D. Mergenhagen and H. Schweiger at Wilhelmshaven. As was found previously, actinomycin D inhibited the rhythm slowly in nucleate cells but not in anucleate ones, while rifampicin had no effect. Since the latter inhibitor specifically inhibits RNA polymerase in chloroplasts (and mitochondria and prokaryotic cells) while actinomycin D affects both nuclear and organellar DNA, it seems clear that transcription from organellar DNA is not involved for the function of the circadian oscillator; and since actinomycin D has no effect in enucleated cells, transcription from nuclear DNA also appears unnecessary. Turning to translation then, these workers discovered that cycloheximide (which specifically inhibits protein synthesis on the 80 S ribosomes of the cytosol) and puromycin (which attacks both 80 S ribosomes and the 70 S ribosomes in the cell organelles) inhibited the photosynthesis rhythm in both nucleate and anucleate cells. (Note the discrepancy with respect to the much earlier results obtained with puromycin on groups of cells.) These experiments thus show that the observed inhibition is associated with the 80 S ribosomes but do not exclude the involvement of the 70 S ribosomes. This question was resolved by using chloramphenicol (which specifically inhibits 70 S ribosomes): The inhibitor was ineffective in both nucleate and anucleate cells. The simplest conclusion that can be drawn from this entire study, then, is that translation on the cytosol 80 S ribosomes is required for the operation of the circadian oscillator. The necessity for continual translation of mRNA suggests that either protein synthesis is part of the clock mechanism itself or that s short-lived component of the clock requires continual resynthesis.

Thus, inhibitor studies have been inconclusive about the mechanism responsible for circadian oscillations. Yet, perhaps there is some cause for optimism: several chemical species have only recently been discovered to markedly and similarly affect phase and period—the only measurable features that can be unequivocally assigned to the circadian oscillator(s) at present—in three seemingly quite disparate systems. First, E. Bünning and I. Moser of the University of Tübingen in Germany have investigated the influence of the antibiotic valinomycin (an ionophore known to cause permeability changes, particularly in the turnover of K^+) on the circadian leaf movements of the bean plant *Phaseolus* (cf. Figure 2-1). Plants were exposed to valinomycin via the transpiration stream for 5-hour time spans at various phases of the free-running rhythm in continuous illumination and constant temperature. Steady-state phase shifts resulted whose sign and magnitude were dependent on the subjective circadian time at which the antibiotic "pulses" were administered. Most significantly, the *phase-response curve thus obtained* (Figure 8-21) *was similar to that*

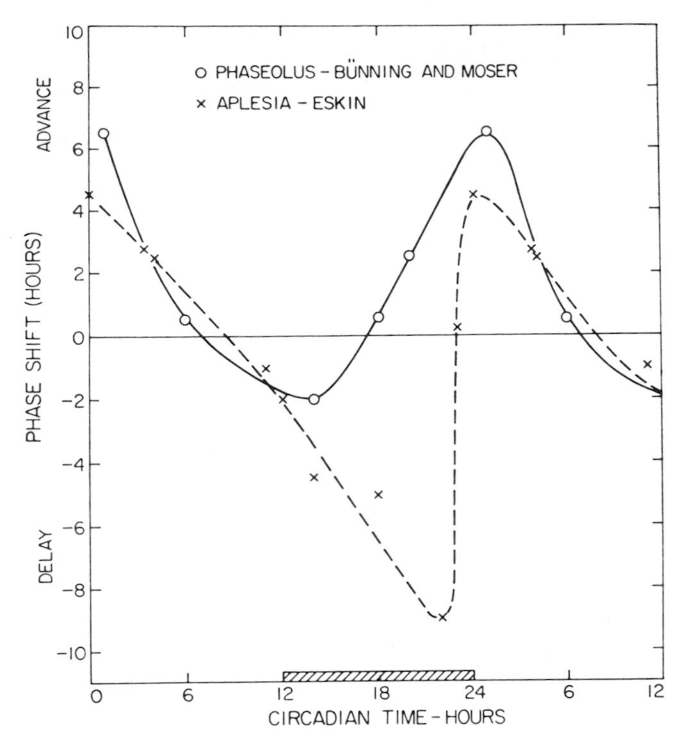

Figure 8-21 Phase-response curve for the effect of valinomycin (in 0.1% ethanol) on the phase of the leaf movement rhythm in *Phaseolus* calculated with reference to the ethanol control (open circles). [From E. Bünning and I. Moser, *Proc. Natl. Acad. Sci. U.S.A.* **69,** 2732–2733 (1972).] Phase-response curve for the effects of high potassium ion pulses on the phase of the rhythm in firing rate of the optic nerve of isolated *Aplysia* eyes (crosses). [From A. Eskin, *J. Comp. Physiol.* **80,** 353–376 (1972).] The data have been replotted so that circadian time 0 corresponds to dawn in each case and so that phase advances are plotted upward. Subjective night (the time of the dark period in the previous entraining light cycle) is shown as a shaded bar on the abscissa. [B. M. Sweeney, *Plant Physiol.* **53,** 337–342 (1974).]

found for light pulses in Phaseolus (Figure 8-21). Further, the effects of 4- to 5-hour pulses of high concentrations of KCl and of 2-hour pulses of ethyl alcohol given at different circadian times were examined. The phases of the leaf-movement rhythm yielding advances (shortening of the period) corresponded to those that responded to light signals with advances (even though K^+ pulses induced only advance phase shifts and no delays).

Eskin has also demonstrated that 4-hour pulses of high K^+ medium (constituting depolarizing stimuli) advance or delay the circadian rhythm of spontaneous optic nerve impulses in the isolated eye of the

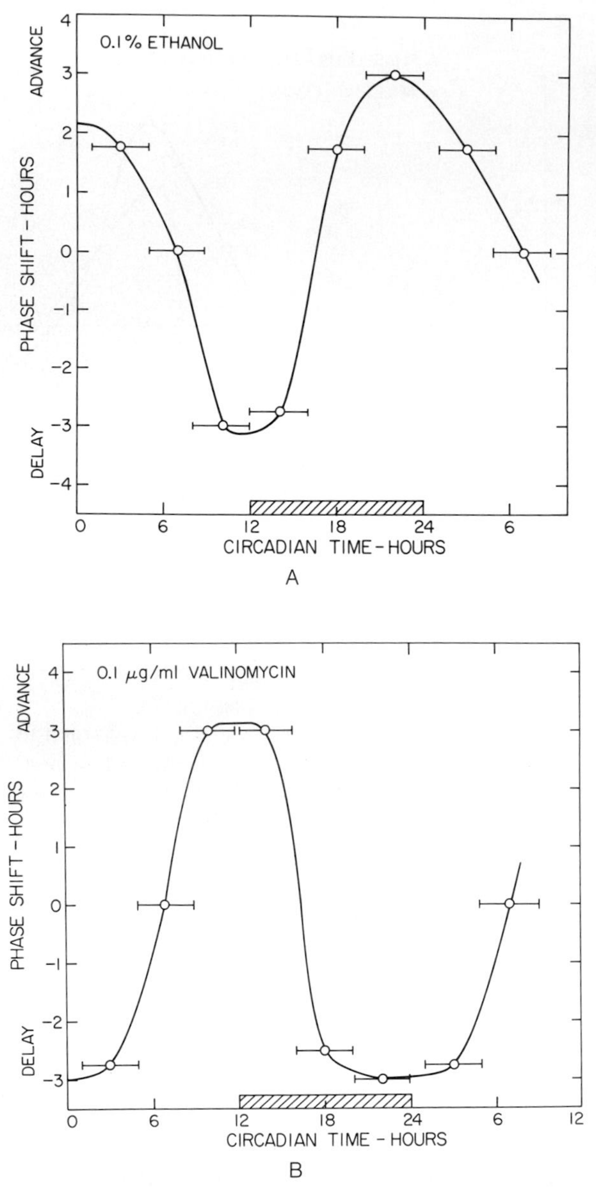

Figure 8-22 Phase-response curves for the effect of 4-hour exposures to 0.1% ethanol (A) and to 0.1 μg/ml valinomycin in 0.1% ethanol (calculated with reference to ethanol only) (B) on the phase of the circadian rhythms in stimulated bioluminescence in *Gonyaulax.* Note that phase advances (+) are plotted upward. Subjective dawn corresponds to circadian time 0; the subjective dark period is represented by the hatched bar on the abscissa. [B. M. Sweeney, *Plant Physiol.* **53,** 337–342 (1974).]

sea hare *Aplysia* (cf. Figure 8-12), depending on the phase of the circadian cycle at which the pulses were applied (Figure 8-21). These results supported his general hypothesis that light cycles, which can entrain *in vitro* the rhythm of the eye preparations, couple to the intracellular clock mechanism through membrane depolarization (see also Figure 8-11).

Finally, Sweeney reports essentially similar findings for the rhythm of stimiulated bioluminescence in *Gonyaulax* (cf. Figure 2-9). Not only is there a circadian rhythm in the intracellular level of K^+ in this dinoflagellate, but also the phase-response curve for ethanol pulses (Figure 8-22A) closely resembles that for light signals (Fig. 8-2). Furthermore, these ethanol treatments lowered the concentration of K^+ in the cell. Interestingly, when valinomycin was administered in addition to ethanol, the phase of the bioluminescence rhythm is returned to that of the untreated cell suspension (Figure 8-22B), and K^+ concentration was only slightly lowered. Valinomycin thus negates the effect of ethanol alone. Since both ethanol (a lipid solvent known to stabilize red blood cells against hemolysis) and valinomycin might be expected to exert effects on the structure and permeability of biological membranes, it seems quite possible that the phase-shifting effects of light—mimicked to a surprising degree by these substances—may also be membrane mediated. And indeed, a model will be considered in which circadian oscillations are generated by changes in membrane properties.

Genetic Dissection of the Clock

If one recalls the tremendous impetus given to the field of genetics and, later, molecular genetics during this century, by such model systems as *Drosophila, Neurospora,* bacteria, and bacteriophage, it would seem almost axiomatic that a similar approach would facilitate the elucidation of the complex biological clock mechanism. Indeed, a search actually is underway in several laboratories for circadian clock single-gene mutants whose isolation would allow us to focus on one element of the timekeeping process at a time.

In the introductory chapter (Figure 1-17), it has been noted that the phase of a typical circadian rhythm can be genetically selected: early- and late-eclosing strains of *Drosophila* have been derived in the laboratory. Similarly, the period length and even capacity for rhythmicity itself are clearly under genetic control in the fruit fly. Thus, S. Benzer and R. J. Knopka of the California Institute of Technology at Pasadena, as part of a larger effort to genetically dissect behavior, have isolated three mutants in which the normal 24-hour rhythm is drastically altered. One mutant is arrhythmic; another has a period of 19 hours; and a third has a period of 28 hours (Figure 1-18). Both the eclosion rhythm of a population and the locomotory activity of individual flies are affected. Quite provocatively, all of these mutations appear to involve the same functional gene of the X chromosome.

Circadian clock mutants have also recently been isolated in the common bread mold, *Neurospora crassa*, a system that holds much promise for subsequent biochemical analysis. Three mutants, designated "frequency" (*frq*), were obtained after mutagenesis of the band (*bd*) strain with *N*-methyl-*N'*-nitro-*N*-nitrosoguanidine which exhibited altered periods in their free-running circadian rhythm of conidiation (cf. Figure 2-14). In constant darkness at 25°, *bd* has a period of 21.6 hours; under the same conditions, *frq-1* displays a period of 16.5 hours; *frq-2*, 19.3 hours; and *frq-3*, 24.0 hours. Each of the mutants segregates as a single nuclear gene. Thus, in the cross between *bd frq-1* (i.e., band) and *bd frq-1*$^+$, half of the progeny showed normal period length and half showed the mutant short period of 16.5 hours (Figure 8-23); there were no intermediate period lengths. All three mutants appear very tightly linked to each other, although it is not yet clear whether they are allelic. No significant changes in the light and temperature responsiveness in any of the strains were found. It seems likely, therefore, that these mutants represent alterations in the basic timing mechanism itself, although it is not necessarily the case that the *Neurospora* clock is determined by only one gene.

Finally, V. Bruce of Princeton University has just isolated a number of mutants affecting the period of the circadian clock underlying the phototactic rhythm in the green alga *Chlamydomonas* (cf. Figures 2-15, 2-16). Four strains (designated *per-1, per-2, per-3,* and *per-4*), obtained by nitrosoguanidine mutagenesis, exhibited periods ranging from 27 to 29 hours at 25°C. (All were found to have a temperature coefficient of <1.0), running slightly *faster* at 16°C that at 22° or 25°C.) Apparently unlike *Drosophila* or *Neurospora*, the long period characteristic of the mutants seems to be controlled by several single genes at separate loci. Thus, crosses between single mutants, as well as crosses involving three or four mutant genes, yielded progeny with both parental and recombinant period lengths, including not only normal (wild-type) periods, but also extra long periods (double, triple, and quadruple mutants). Perhaps the most interesting finding was that the period lengthening effect is additive: if one gene lenghtens the period by m hours and a second by n hours, then the period of the double mutant is lengthened by $m + n$ hours. This additive effect would be a logical consequence of "tape-reading"-type models for a circadian clock (dis-

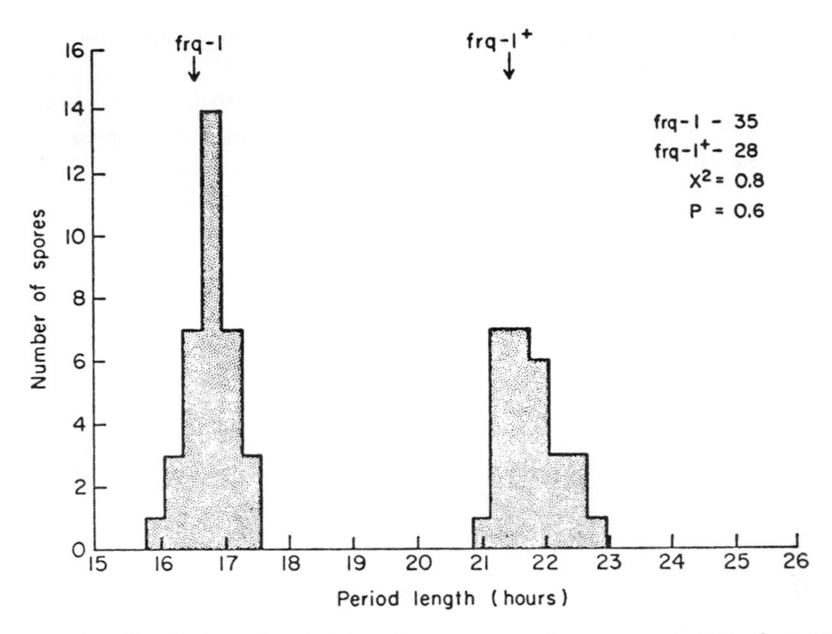

Figure 8-23 Distribution of period lengths among random spore progeny of a cross between the *Neurospora* clock mutant *bd frq-1*$^+$ ("band") and the double-mutant *bd frq-1* (*frq* strain). Arrows indicate parental period lengths. The mean period length for the *frq-1* progeny was 16.8 ± 0.4 hour; for the *frq-1*$^+$ progeny, 21.8 ± 0.5 hour. [J. F. Feldman, *Genetics* **75,** 605–613 (1973).]

cussed in the next section) in which mutations affecting the period would involve the addition or deletion of tape segments.

Biochemical and Molecular Models for the Circadian Clock

There are several different *categories* of models for an endogenous, self-sustaining circadian clock, which are neither mutually exclusive nor jointly exhaustive. Their basic features will now be considered.

FEEDBACK "LOOP" MODELS FOR OSCILLATIONS IN INTERMEDIARY METABOLISM

The short-period glycolytic oscillations (Figure 8-17) that B. Chance and co-workers discovered in both intact yeast suspensions in stationary phase and in cell-free extracts have been documented. In addition, the mechanism underlying these oscillations is also relatively well understood and will serve as a model for the "escapement" of one type of biological clock.

For a chemical mechanism to exhibit oscillatory behavior, certain general types of reaction pathways must exist. If X and Y represent any two substances whose net rate of production are V_x and V_y, respectively, then oscillations can occur if the following three conditions are met: (i) one of the chemicals (X) must activate its own production (assuming that X remains fixed); (ii) the other substance (Y) must tend to inactivate its own net production, normally true of most chemical reactions since increasing concentration increases the rate of removal of that chemical; and (iii) there must be a *cross-coupling* of opposite character so that increasing the concentration of X activates the

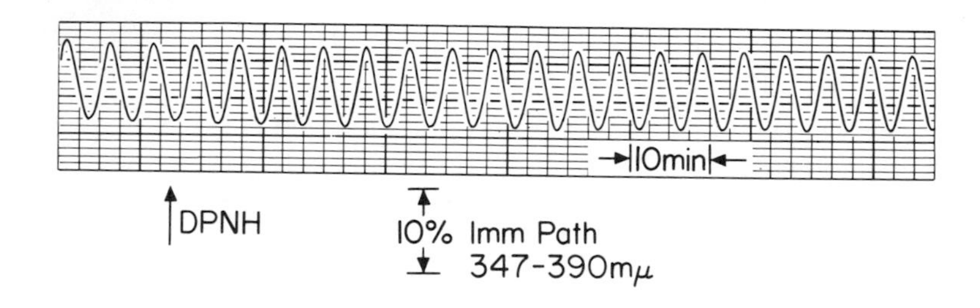

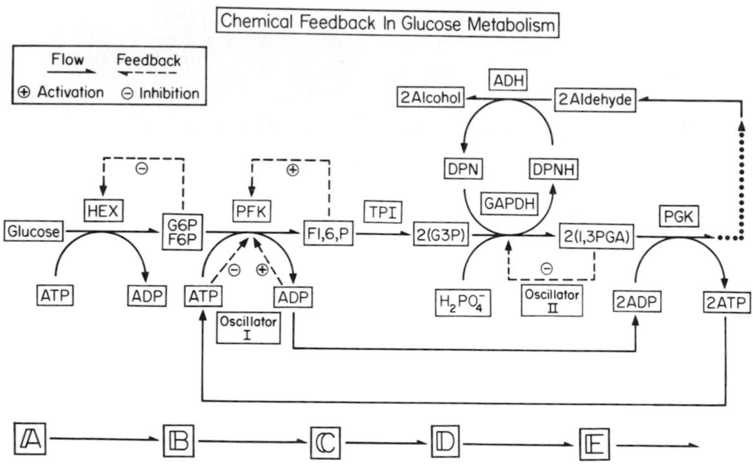

Figure 8-24 Feedback model for the sinusoidal oscillations of DPNH in a cell-free extract of the yeast enzyme system. The period is about 5 minutes. The amplitude of the curve represents about a 10% change. A simplified diagram of the pathway from glucose to ethyl alcohol with the consequent reduction and oxidation of DPN is given in the lower panel. Feedback activation and inhibition are designated by the broken arrows (+, or −). Hypothesized oscillator sites are represented by boxes. Letters A, B, C, D, and E at the bottom of the lower graph refer to intermediates in the pathway. See detailed explanation in text. [B. Chance, K. Pye, and J. Higgins, *IEEE Spectrum* **4,** 79–86 (1967).]

net production of Y, and increasing Y inhibits the net production of X (or vice versa). The period of a resulting oscillation will be dependent at the very least on the kinetic rate constants associated with these reactions.

So far so good; now let us consider a more specialized case of this general mechanism to account for the oscillatory behavior of the glycolytic system in yeast. Figure 8-24 indicates some of the key enzymes that operate in the series of reactions that convert glucose to alcohol in this microorganism; only a few of the 30-odd enzymes and five intermediates (A–E) in this pathway are included. The basic reactions of concern here can be tabulated sequentially in both a general (left column) and a specific (right column) form as shown in Table 8-2. Reaction (a) represents the conversion of glucose (continuously provided to the system directly or via hydrolysis of trehalose and designated as intermediate A in Figure 8-24) to fructose 6-phosphate (F6P) (intermediate B). In the second step, [Reaction (b)], E_1^* represents an activated form of the enzyme phosphofructokinase (PFK) which combines with the substrate F6P to form an enzyme-substrate complex; this complex then breaks down by Reaction (c) to release the free activated enzyme and the product of the reaction, fructose 1,6-diphosphate (F-1,6-P) (intermediate C), with the concomitant production of adenosine diphosphate (ADP) as shown in Figure 8-24. Finally, in Reactions (d) and (e), F-1,6-P serves as the substrate for a second enzyme E_2, triosephosphate isomerase (TPI), to eventually produce glyceraldehyde 3-phosphate (G3P) (designated as intermediate D); this product then continues through the glycolytic pathway leading to the formation of alcohol.

Thus far the enzymatic reactions constitute simply a linear sequence and contain no step that would cause oscillation. The crucial step is given by Reaction (d) in which F-1,6-P reacts with an inactive form

Table 8-2

Sequential Enzyme Reactions Characterizing the Oscillatory Behavior of the Yeast Glycolytic System[a,b]

General	Specific
(a) $A \rightarrow B$	(a′) $GLU \rightarrow F6P$
(b) $B + E_1^* \rightarrow E_1^*{\cdot}B$	(b′) $F6P + PFK^* \rightarrow PFK^*{\cdot}F6P$
(c) $E_1^*{\cdot}B \rightarrow E_1 + C$	(c′) $PFK^*{\cdot}F6P \rightarrow PFK + F\text{-}1{,}6\text{-}P$
(d) $C + E_1 \leftrightarrow E_1^*$	(d′) $F\text{-}1{,}6\text{-}P + PFK \leftrightarrow PFK^*$
(e) $C + E_2 \rightarrow E_2{\cdot}C$	(e′) $F\text{-}1{,}6\text{-}P + TPI \rightarrow TPI{\cdot}F\text{-}1{,}6\text{-}P$
(f) $E_2{\cdot}C \rightarrow E_2 + D$	(f′) $TPI{\cdot}F\text{-}1{,}6\text{-}P \rightarrow TPI + 2(G3P)$

[a] Refer to Figure 8-24 and discussion in text.
[b] Based on J. Higgins, *Proc. Natl. Acad. Sci. U.S.A.* **51,** 989–994 (1964).

(E_1) of the enzyme phosphofructokinase to produce the active form (E_1^*). [In point of fact, it appears that ADP, the other by-product of Reaction (c) in addition to F-1,6-P, is actually the critical activator.] This ability of F-1,6-P (or ADP) to activate PFK results not only in a net increase in its own production by Reactions (b) and (c), but also a decrease (inhibition) in the net rate of production of F6P by Reaction (a). Thus, the coupled reaction sequences now satisfy the general requirements for oscillatory behavior, with F-1,6-P (or ADP) corresponding to hypothetical substance X, and F6P corresponding to substance Y.

Operationally, then, if the concentration of ADP in the cell extract or suspension happened to be low (under constant glucose infusion), then the activity of PFK would also be low due to the lack of ADP-activation [by Reaction (d)] and F6P would start to pile up [via Reaction (a)]. As a result, the rate of Reactions (b) and (c) would increase since F6P is the substrate for the catalyzing enzyme PFK and, in turn, the concentrations of the products F-1,6-P and ADP would also rise. The increase in ADP concentration would then activate PFK [Reaction (d)] and cause the former to increase even more and thus provide further activation of PFK. Eventually, the velocity of Reactions (b) and (c) will exceed the rate of glucose production [Reaction (a)] and cause the concentration of F6P to fall and then limit the PFK-mediated reaction. Finally, the concentration of ADP would be depleted, and the cycle is complete.

If one ascribes appropriate rate constants and glucose concentrations to the set of reactions given in Table 8-2, represents the basic feedback property inherent in the system by two differential equations [$d(E_1^* \cdot B)/dt$ and dC/dt], and then simulates the entire process on an analog computer, sustained oscillations occur in the stationary state similar to those empirically observed. (Outside of this limited range of chosen constants, the system simply proceeds to a nonoscillatory state, i.e., the rhythm damps out.) The phase-shifting effects of pulsing in intermediary metabolites also can be studied in this manner, and the kinetic equations describing the model even afford some degree of temperature compensation. Unfortunately, however, periods of only a few minutes are usually obtained in this enzymatic system. Although one can demonstrate that periods of up to an hour or so might be achieved by decreasing the enzyme activity, it is probable that the circadian period is generated by an entirely different set of rate constants than those found for the high-frequency glycolytic oscillations, or by a frequency reduction of an entirely different (and unknown) character. The important point that emerges from these elegant studies is that

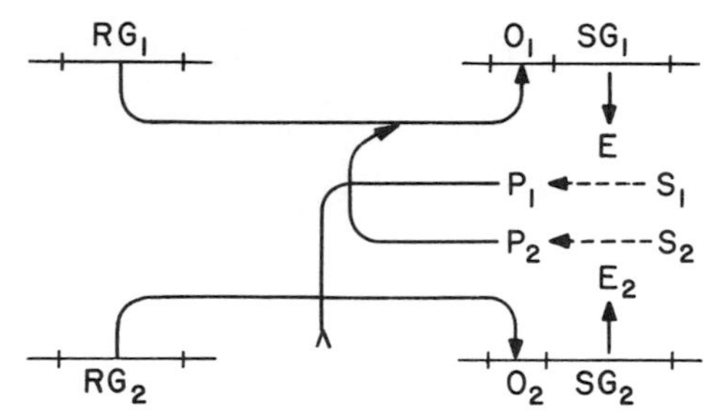

Figure 8-25 Model of regulatory control circuitry for cyclical biochemical phenomena. Synthesis of enzyme E_1, genetically determined by the structural gene SG_1, is blocked by the repressor synthesized by the regulator gene RG_1. Synthesis of another enzyme E_2, controlled by structural gene SG_2, is blocked by another repressor synthesized by regulator gene RG_2. The product P_1 of the reaction catalyzed by enzyme E_1 acts as an inducer for the synthesis of enzyme E_2, while the product P_2 of the reaction catalyzed by enzyme E_2 acts as corepressor for the synthesis of enzyme E_1. [J. Monod and F. Jacob, *Cold Spring Harbor Symp. Quant. Biol.* **26,** 389–401 (1961).]

the mechanism of an endogenous, self-sustaining, fully autonomous (ultradian) biological clock can be accounted for without having to resort in any way to alternative, extrinsic timing hypotheses. Can a circadian clock be far behind?

But aye, there is the rub: What is the determinative cause for a biological rhythm having a period of approximately 24 hours? To start, an attempt can be made to be a little more general in formulating our feedback control circuitry than was done for the specific mechanism for glycolytic oscillations. Thus, J. Monod and F. Jacob of the Institut Pasteur in Paris propose an idealized model for cyclic phenomena which utilizes known cellular components and genomic regulatory elements (Figure 8-25); it shares basic similarities with the general model for chemical oscillations described earlier. In this circuit, the product of one enzyme is an inducer of the other system, while the product of the second enzyme is a corepressor. Provided that adequate time constants are chosen for the decay of each enzyme and of its product, the system will oscillate from one state to the other.

By using this basic control loop, more complex models can be built. Thus, in his masterful theoretical consideration of temporal organization in cells, B. Goodwin (followed by many others) modifies the simple scheme by permitting repression to occur between two different components as well as within single components, so that strong interactions are obtained as shown in Figure 8-26 (top panel). Here a metabolite controlled by Y_1 (denoting ribosomal protein synthesis) interacts by repression with another genetic locus L_2, while a reciprocal

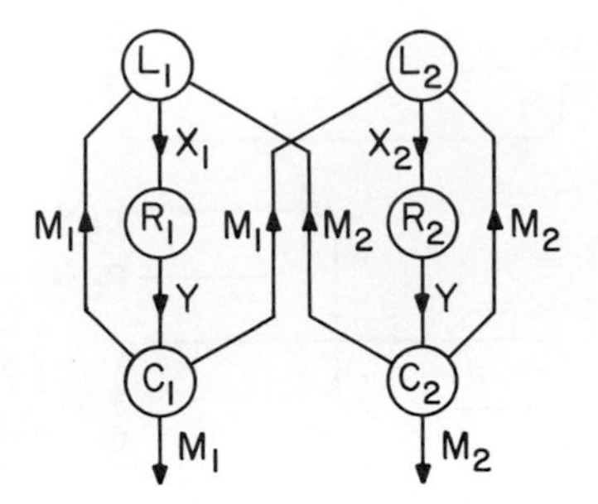

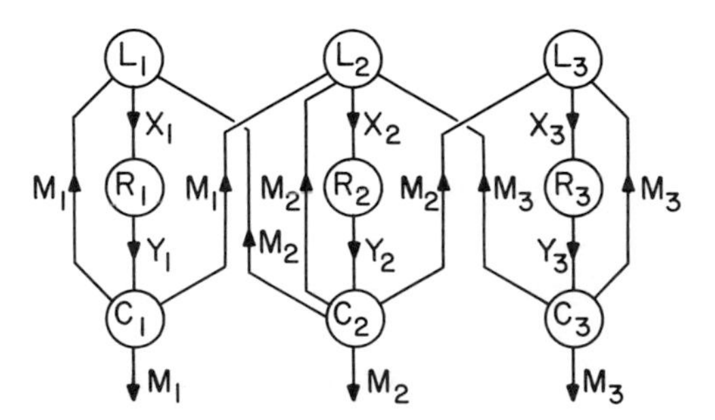

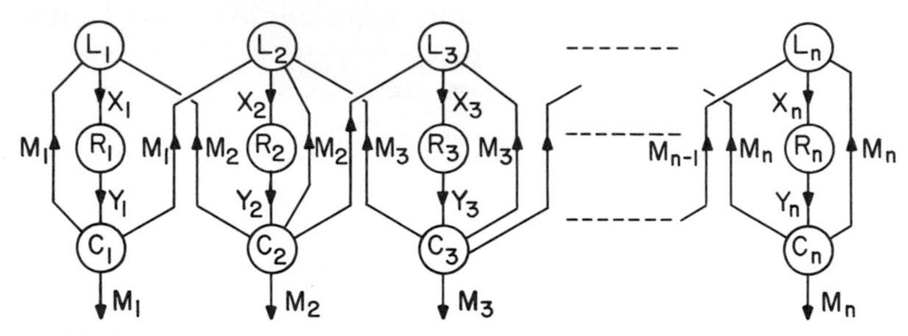

Figure 8-26 Metabolic feedback control circuitry for the generation of biochemical oscillations. L_i represents a genetic locus that synthesizes mRNA in quantities represented by the variable X_i. This specific "signal" encounters a cellular structure R (a ribosome), where its activity results in the synthesis of a particular species of protein in quantities denoted by the variable Y_i. The protein then travels to some cellular locus, *C,* where it exerts an influence upon the metabolic state either by enzyme action or by some other means. This activity, in turn, generates a metabolic species in quantity M_i, a fraction of which closes the control loop by returning to the genetic locus L_i, where it may act as a repressor. The three panels depict increasingly more complex cross-coupled control loops (which may have either an intracellular or an intercellular basis). (B. C. Goodwin, "Temporal Organization in Cells." Academic Press, New York, 1963.)

interaction occurs from L_2 to L_1. Even more complex parallel repression networks involving many interacting, coupled components can easily be generated (Figure 8-26, lower panels). Significantly, these "components" could be within one cell or could involve *"cross talk"* (i.e., intercellular chemical communication) among individual cells comprising a tissue [such as in the isolated *Aplysia* eye (Figure 8-12)]. In such situations, inhibitory coupling between a large number of oscillators could result in a much lower frequency (i.e., longer period) of the overall oscillation observed in the population than that of the constituent oscillators. The addition of time-consuming diffusion steps for key substrates could further lengthen the period of the overt rhythm and perhaps impart a measure of temperature compensation. Finally, if one postulates that one or more key reaction sequences are photoactivable, the fundamental light-sensitive properties of circadian rhythms can be simulated.

So much for "simple" feedback loop models, then. Although they are feasible (and quite convincing at the ultradian level—indeed, a 4-hour period has been estimated for the feedback-controlled oscillation on protein synthesis in higher organisms, disregarding possible inhibitory coupling), they become a bit vaguer and fuzzier around the edges as one tries to force them into a Procrustian fit of the circadian bed. A more direct (though not necessarily any more likely) approach is afforded by another category of models.

"TAPE-READING" TRANSCRIPTION MODELS

Another provocative approach—and a more stringent one than the metabolic feedback loop mechanism just discussed—to the problem of transducing the higher-frequency "ticks" of metabolic events into the 24-hour "tocks" of the circadian escapement has as its basis the notion that the distance between genes could be utilized for timing. In other words, *transcription along the DNA tape could serve as a measure for biological time.*

Thus, at 37°C the rate at which RNA polymerase molecules move along the DNA templates of the chromosome is approximately 30 to 40 nucleotides per second. J. D. Watson of the Cold Spring Harbor Laboratory in New York has calculated that at this rate the total possible transcription time for all the chromosomes of *E. coli* would be 33 hours (assuming that several operons are *not* simultaneously transcribed). Consequently, only 1% of the total genome would be needed to separate periodic events occurring once every 20-minute cell cycle and even the 3 to 4 hours necessary for spore germination could easily be directly timed on the DNA tape. Furthermore, these longer intervals do not necessarily have to be measured by the transcription of a single contiguous piece of DNA: *several operons, located on one or more chromosomes,* could *each* code for a specific σ factor (a subunit of RNA polymerase that recognizes specific sites on DNA for initiation of RNA synthesis) necessary for the reading of the subsequently transcribed operon (Figure 8-27). Further regulatory control and delay could be built into the model by the presence of *anti-σ factors* (also coded for on the genome) which specifically prevent certain σ factors from functioning; these are already known to exist in T4 phages. Finally, the cell cycles of higher organisms (which can undergo circadian modulation), though much longer, have correspondingly greater amounts of DNA; the haploid content of human genes would require an estimated 1000

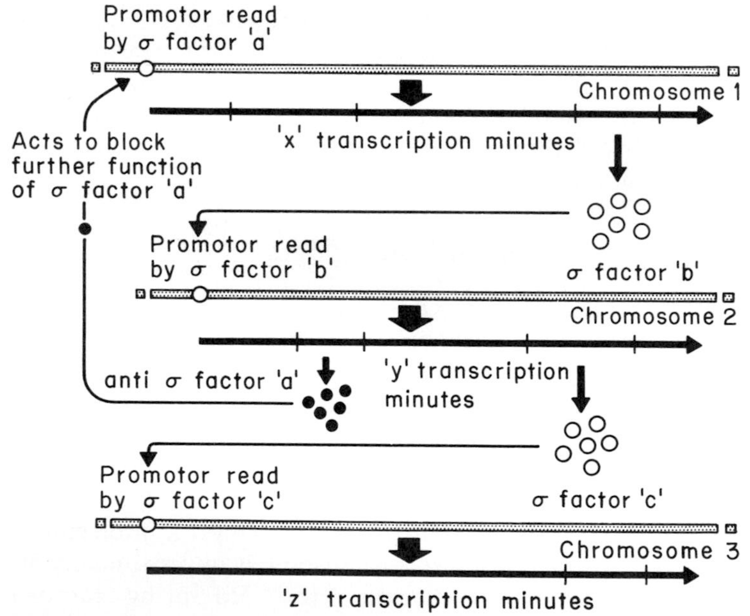

Figure 8-27 A mechanism by which sequential synthesis of different σ factors might be used to count time. Biological clocks might depend on the regular appearance of anti-σ factors, which are gene products that specifically prevent certain σ factors from functioning in the transcription of an operon. Such factors provide a direct way of curtailing the synthesis of specific proteins to restricted time spans in the life cycle of some phages and, by extension, might operate similarly during a circadian cycle. (J. D. Watson, "Molecular Biology of the Gene," 2nd ed. Benjamin, New York, 1970.)

days at 37°C to be completely transcribed by a single molecule of RNA polymerase. Perhaps 24 hours is not so long a time span after all!

The most explicit, and consequently most speculative, transcriptional model that has appeared on the market, however, is the *chronon model* proposed by C. F. Ehret and the late E. Trucco of Argonne National Laboratory in Illinois. In essence, this theory for circadian timekeeping postulates that within every cell there exist hundreds of replicons of a special type, called *chronons,* each of which comprises a polycistronic strand of nuclear (or organellar) DNA some 200 to 2000 cistrons in length (Figure 8-28). These chronon replicons are the longest and therefore are the rate-limiting components of the basic

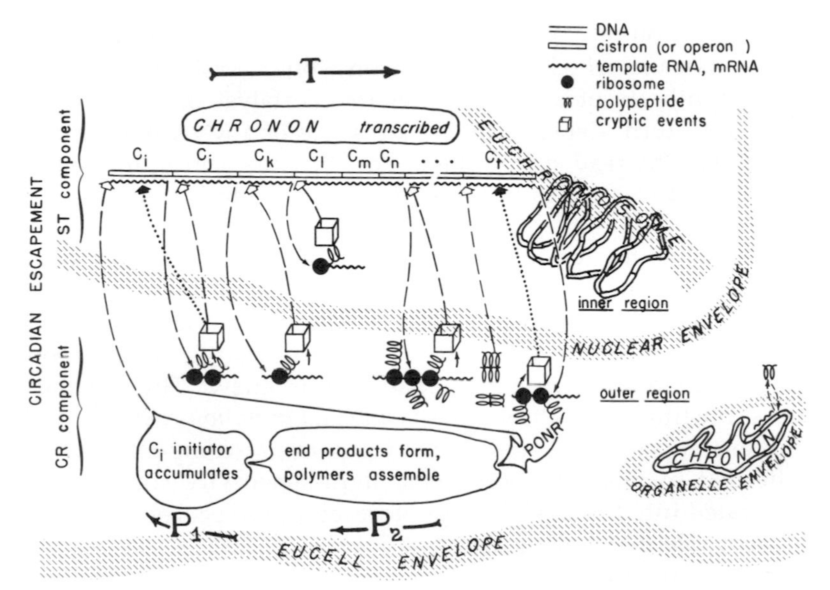

Figure 8-28 The chronon model for a circadian clock. The circadian cycle consists of pretranscriptional (P_1), transcriptional (T), and posttranscriptional (P_2) phases, which are analogues of G_1, S, and G_2 phases of the cell mitotic cycle. The sequential transcription (ST) component is the chronon, one of the hundreds of very long DNA polycistron complexes on a single eukaryotic chromosome (or in some cases in a cell organelle). Transcription of template RNA proceeds from left to right, starting at the initiator cistron (C_i). The overall rate for this process is slow and relatively temperature-independent because mRNA formed by each cistron must diffuse out to cytoplasmic ribosomes and then direct protein synthesis, whereupon some of the ribosomal products diffuse back to the chronon and initiate transcription of the next cistron in the sequence. This cycling process continues until the terminator cistron (C_t) is reached. An initiator substance then accumulates and restarts the transcription cycle again some 24 hours later. [C. F. Ehret and E. Trucco, *J. Theor. Biol.* **15,** 240–262 (1967).]

transcriptional cycle, represented by the sequence

$$P_1 \rightarrow T \rightarrow P_2$$

where P_1, T, and P_2 designate the pretranscriptional, transcriptional, and posttranscriptional phases, respectively [note the analogy to the G_1, S, and G_2 phases of the typical cell cycle (Figure 8-14)]. These chronons, then, comprise the *sequential transcription component* of the circadian escapement and have achieved circadian length through natural selection over the aeons:

$$\frac{\text{Chronon DNA template distance (nm)}}{\text{RNA transcription rate (nm/hour)}} \simeq 24 \text{ hours}$$

The transcription rate would be limited by time-consuming, diffusion feedback loops in which a given cistron would not be transcribed before the synthesis of a specific precursor catalyst or enzyme whose synthesis, in turn, depends on the presence of a RNA transcript from the most recently read cistron (Figure 8-28). Thus, the entire cycle is itself composed of a series of "nested do-loops," to borrow from computer jargon.

Finally, when the terminator cistron is reached after almost 24 hours of sequential transcription has taken place, an initiator substance is synthesized on the ribosomes, accumulates, and then diffuses back through the nuclear membrane to the initiator cistron, whereupon the entire circadian transcriptional cycle begins anew. This second key process constitutes the chronon *recycling and initiation component* of the circadian escapement. A measure of temperature-compensation would be afforded by the diffusion steps; light-sensitivity could easily be incorporated into the model via photoreceptor molecules (or organs) or photoactivable metabolic reaction sequences.

Because of the richness of detail in the chronon model, a number of predictions—some relatively easily testable, others untestable until a better understanding of the structure and function of eukaryotic DNA is attained—have emerged, which shall be briefly considered: (i) *The longest replicons (chronons) of all plant and animal circadian systems should be of approximately equal length.* Unfortunately, this challenging consequence of the chronon theory is difficult to confirm (or better yet, to disconfirm) since the chronon DNA has not yet been isolated. (ii) *Circadian clocks should be confined to eukaryotic cells.* Although this prediction is not necessarily demanded by the model in one of its more "relaxed" forms, it would seem to be a reasonable assumption because of the importance of the nuclear envelope in providing a compartment for the diffusion circuitry. At the time the model was first proposed (1967), no circadian rhythms had been reported and substantiated in the bacteria or other prokaryotes; but, more recently, evidence has been obtained that there is, at least, a diurnal oscillation in the growth rate of two bacterial strains cultured in a long race tube. (iii) *The approximate temperature independence of the escapement is a consequence primarily of the diffusion steps; impediments to diffusion, therefore, should slow down the sequential transcription component considerably.* In this regard, it is interesting to note that deuterated circadian systems ("intoxicated" with heavy water) display almost universally a lengthening of the free-running period (see Figures 2-16, 2-20, 2-22), but these findings are consistent with several other models for circadian timekeeping in addition to the chronon theory. The diffusion circuits, on the other hand, could easily be replaced by temperature-compensated enzyme systems or by some other appropriate mechanism. (iv) *Different mRNAs should be synthesized at different times of the circadian cycle.* This consequence is one of the most fascinating and had undergone at least a degree of empirical verification. Cultures of the ciliates *Paramecium* or *Tetrahymena* were synchronized by a light cycle while slowly growing in the infradian mode ($g > 24$ hours) and then pulse-labeled with tritiated uridine or ^{32}P at different circadian times to label the RNA molecules being synthesized at those times. The RNA was next extracted from the cells and mixed with previously isolated, single-stranded, native DNA that had been immobilized on filters. Finally, the degree of molecular hybridization resulting from this annealing reaction was determined by measuring the radioactivity on the washed filters with a liquid scintillation spectrometer. In a number of experiments comparing reaction kinetics, saturation, and competition capacities of the different RNAs, Ehret and co-workers found clear evidence for temporally characteristic RNA species (designated circadian transcriptotypes). Indeed, in competition-hybridization experiments in which an unlabeled RNA competitor from a given timepoint is included in the annealing mixture with the DNA filter and labeled RNA preparation, unlabeled circadian t_{24} RNA not only competed best with its labeled homologue stock for the available template sites against unlabeled t_6, t_{12}, t_{18}, and t_{24} RNAs (as might have been anticipated), but also proved to be the best competitor against all labeled RNAs used. (v) *Enzymatic activities and metabolically related functions should also display a temporal phenotype or "chronotype."* This they definitely do, in both single cells and in higher organisms (see Table 8-1), but the fact that they do so is certainly not uniquely predicted by the chronon model. And finally, (vi) *one might expect to find correlations between the events and mechanisms that regulate the circadian cycle, on the one hand, and the cell (mitotic) cycle on the other.* Circadian modulation of (coupling to) the cell cycle has already been discussed in an earlier section and the similarity between the $P_1 \rightarrow T \rightarrow P_2$ and $G_1 \rightarrow S \rightarrow G_2 \rightarrow M$ cycles has been noted. Conceptually, the same chronons that limit *T* in the circadian cycle might also limit *S* in the cell cycle. Indirect support for this hypothesis is afforded by the phenomenon called by its acronym the "G-E-T Effect" [*Gonyaulax, Euglena, Tetrahymena*]: cell cultures in the infradian growth mode, regardless of their generation time, are synchronizable by diurnal light cycles or light-dark transitions and are capable of circadian outputs. The validity of this and the preceding five predictions awaits further empirical confirmation.

MEMBRANE MODELS FOR THE CIRCADIAN CLOCK

The models considered thus far embody two different conceptual approaches: the "feedback network" models conceive of the clock as a biochemical network with self-sustained oscillations arising from feedback control circuitry within the biochemical system; at the other pole, the "transcriptional tape" models envisage the clock as sequential gene expression. It is difficult for either of these two categories of clock models to stand alone; in fact, they do tend to merge. What is needed is a unifying scheme that will provide a structural home for a diverse biochemical species.

A membrane model for the circadian clock recently proposed by D. Njus, F. M. Sulzman, and J. W. Hastings of Harvard University addresses itself precisely to these needs. Their model incorporates the network concept, but identifies ions and membrane-bound ion transport elements with the biochemical clock, and thus with the primary oscillations. Feedback control would arise in the system by the effect of ion concentrations (implying changes in transmembrane ion gradients) on the functioning of the ion-transport structures themselves, which, in turn, would affect the distribution of those ions. As will be seen, temperature-compensation would derive from the lipids within the membrane (known to be capable of adapting to environmental temperature fluctuations) which would determine the rate-limiting kinetics of transport activity. Finally, light would affect the clock by causing an ion "gate" in the membrane to open which may be either photosensitive itself or coupled hormonally or neurally to an anatomically distinct photoreceptor.

That ions and membranes are indeed implicated in circadian timekeeping is supported by the results from a wide variety of experiments, mostly of recent origin. Thus, valinomycin, a highly specific carrier for the K^+ ion, has been shown to cause pronounced phase shifts in *Phaseolus* and *Gonyaulax* (Figures 8-21, 8-22). Similarly, ethanol effects (perhaps caused by modulation of the permeability of the membrane to ions) have been demonstrated in the same two organisms as well as in isopods. Lithium—another chemical known to affect membrane transport—lengthens the free-running period in the succulent plant *Kalanchoe.* And finally, strong K^+ pulses have been noted to phase shift the circadian rhythm of spontaneous neural discharge in the isolated eye of *Aplysia* (Figure 8-12), as well as of the rhythm of spike discharge in the parabolic burster cell of this organism (see Figure 8-11). Likewise, the fact that valinomycin and ethanol treatments in *Gonyaulax* could be correlated with changes in intracellular K^+ concentration and that a similar correlation can be made between leaf movements in *Phaseolus* and other plants (and underlying changes in turgor pressure of the motor cells of the pulvinule) and ion concentration is most suggestive. Can it be merely fortuitous that the phase-response curves for light, ethanol, valinomycin, and K^+ pulses share so many common features (Figures 8-2, 8-21, 8-22)?

If, then, the basic circadian clock is a feedback oscillator comprising ions and membranes, then the consequent transmembrane ion fluxes might be expected to be mediated primarily by either the synthesis and degradation or the activation and inhibition of membrane proteins. Since protein synthesis is probably not required to *drive* the oscillation, the latter control process seems to be more likely. It is likely that the kinetics of the resulting oscillation depend, at least partially, on the physiochemical properties of the lipids that compose the membrane lipid bilayer into which these proteins are embedded and are relatively free to migrate about. A simplistic, structural representation of the way in which this *fluid mosaic membrane* complex might serve as the circadian clock is given in Figure 8-29. This diagram shows changes in both the arrangement of the intercalated particles and their sizes caused by the varying distributions of K^+ ions. In turn, the state of these proteins determines the direction and activity of membrane transport. A photoreceptor which serves as a K^+ gate or ion channel is also depicted.

Now what might be the effect of a perturbing *light pulse* on such a system? For the sake of illustration (Figure 8-30), let us assume that a rhythm involves just one ion and only two membrane states. The transmembrane ionic gradient will oscillate as shown in the top panel; membrane transport will fluctuate as depicted in the bottom panel.

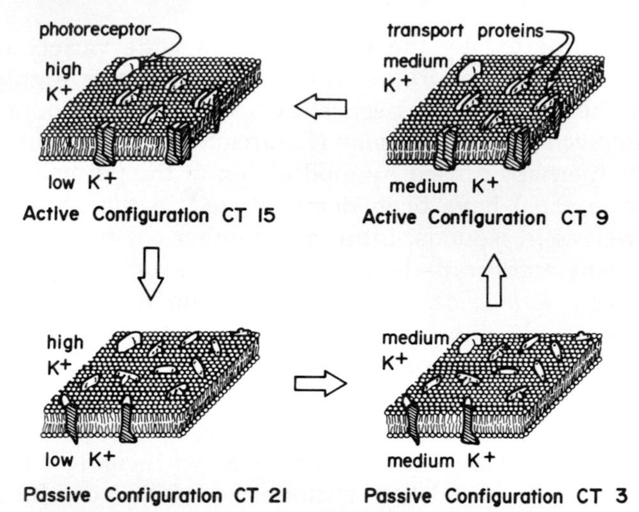

Figure 8-29 A membrane model of the circadian clock. Schematic representation of the way in which a fluid mosaic membrane might keep time. Changes in both particle arrangements and sizes caused by ion distributions are shown. The state of these particles (hypothesized to be membrane proteins intercalated into the lipid bilayer which are capable of migration within the plane of the membrane), in turn, determines the direction and activity of membrane transport. A photoreceptor that serves as a K^+ gate is included. This ionic clock would couple biochemically to the diverse overt circadian rhythms. See text. [D. Njus, F. M. Sulzman, and J. W. Hastings, *Nature* (*London*) **248,** 116–120 (1974).]

The impinging light pulse would open the ion gate and thus lead to the depletion of the ionic gradient (to x_0). The consequence of this drop in the gradient, however, depends on the state of the membrane (y) at the time of the pulse: applied before CT 18, it will generate a phase delay because the membrane is in the "active" mode and the ion gradient (x) will build back up again; but imposed just after CT 18, it will cause a phase advance since the membrane is in its "passive" mode and x would proceed further into its state of discharge. [Note the intended analogy with the "break-point" of the typical phase-response curve for light pulses (Figure 8-2)]. For a step-up in temperature, known to cause phase advances in overt rhythms, the lipid (fatty acid) composition would change over the course of a few hours so as to compensate for the temporarily increased rate of ion transport and the initial free-running period would be restored. The opposite adaptive change in the membrane lipids would soon negate the delaying effects of a temperature step-down. Since a temperature pulse or cycle can be treated as a series of step-changes, a similar analysis would hold for the resultant period in the entrained steady state. Finally, this

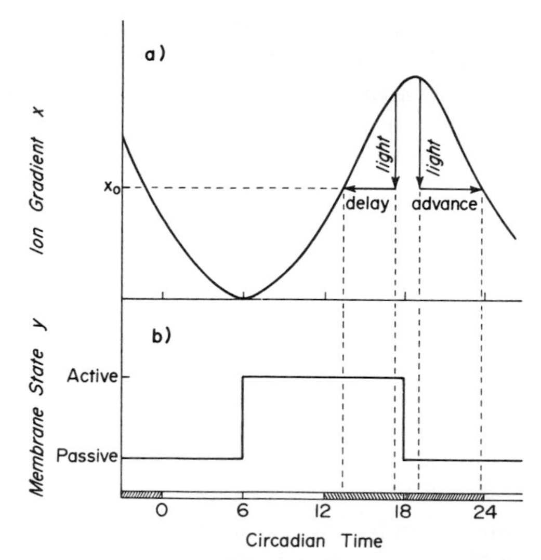

Figure 8-30 Hypothetical oscillations in a simple one-ion system (a) with only two membrane states (b). A light pulse applied just before CT 18 results in a phase delay since it brings the ion-membrane system to a state equivalent to that at about CT 13. Similarly, an identical light pulse impinging just after CT 18 generates a phase advance. Maximum phase shifts are obtained during the subjective night. [D. Njus, F. M. Sulzman, and J. W. Hastings, *Nature* (*London*) **248,** 116–120. (1974).]

membrane clock would be coupled biochemically to the diverse overt rhythms surveyed in earlier chapters: some could be driven directly (e.g., leaf movement or phototaxis), some perhaps by ion-mediated activation and inactivation of enzymes, and some by ion-stimulated inducers or hormones.

This membrane model also provides several avenues for experimental verification. (i) Clock mutants might well phenotypically exhibit altered ion transport capacities or differences in membrane lipid composition. (ii) Changes in lipid composition should be detectable (perhaps by gas chromatography) during adaptation to changing temperatures. (iii) Perturbation of lipid composition (by chemicals or nutritional changes) should be reflected in the output period length of the system. (iv) Phase shifting by light in unicellular organisms should cause a change in the intracellular ionic distributions. (v) Ferritin labeling and freeze-fracture electron microscopy might permit one to directly observe circadian oscillations in the arrangement of the postulated membrane-intercalated particles. And finally, (vi) electrical events in the cell may be correlated (causally?) with the changes in K^+ flux which, in turn, lead to a change in transmembrane potential.

Indeed, the membrane model has withstood the first round of testing. At the recent Dahlem Conference on "The Molecular Basis of Circadian Rhythms" (Berlin, November 1975) several exciting new lines of evidence supporting several of these predictions and thereby further implicating membrane function in biological clocks were the focus of attention. S. Brody and co-workers at the University of California at San Diego have found that the ratio of saturated to unsaturated fatty acids in *Neurospora* varies in a circadian fashion. Furthermore, in perhaps one of the most elegant experiments to date showing the effects of lipid composition on the biological clock, these workers varied the saturated to unsaturated fatty acid content by medium supplementation in a banding mutant of *Neurospora* defective in fatty acid synthesis and discovered that the period between conidiation bands (or rings) varied from 21 hours to 40 hours as a function of increasing levels of unsaturated lipids. In a similar vein, J. Feldman has found that aminophylline, theophylline, and caffeine—all inhibitors of cyclic AMP phosphodiesterase (known to be membrane-bound)—lengthen the period of the conidiation rhythm in *Neurospora* by as much as 4 hours. Lastly, B. Sweeney reported results from freeze-fracture studies of *Gonyaulax* demonstrating a circadian rhythm in the number and size distribution of particles on one of the faces of the peripheral vesicle, thus supporting the postulated participation of membranes in the circadian oscillator.

Interplay among Membrane Transport, Cell Cycles, and Circadian Clocks

A picture is beginning to emerge of the mutual interaction among cell cycle, biological clock, and membrane-transport control systems. These relationships are illustrated in a highly diagrammatic fashion in Figure 8-31. Between any two of the three basic elements a "forward" and a "reverse" formal relation exists; the six resulting interactions have already been discussed in some detail in the previous sections. Thus, periodic changes in membrane transport capacity occur across the cell cycle of various types of cells (Relation A), and conversely, these discrete changes in permeability themselves may initiate and control the various events that comprise the cell cycle (Relation B). Analogously, oscillatory, clock-controlled changes in membrane transport are found in a variety of systems (Relation C), while recent evidence has increasingly implicated the role of membranes in the functioning of circadian clocks (Relation D). Finally, it is clear that not only do cell cycles themselves constitute labile biological clocks of sorts, but also

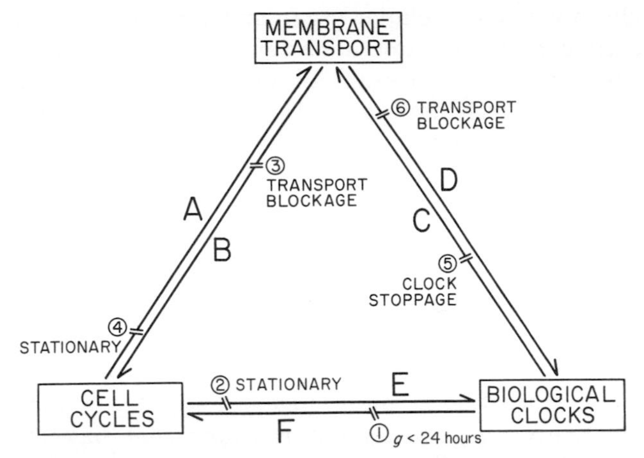

Figure 8-31 Formal relationships existing among membrane transport, cell cycles, and biological clocks. Between any two elements a pair of relations is indicated (A↔B, C↔D, E↔F) as discussed in the text. Further, any two elements can be uncoupled or dissociated formally from each other at the blockage point (1, 2, . . ., 6) pertaining to each of the two coupling relations. Experimental conditions that might generate such an uncoupling are given for the six blockage points. These relations constitute a most unholy trinity: no pretense is made that these elements are jointly exhaustive. [L. N. Edmunds, Jr. and V. P. Cirillo, *Int. J. Chronobiol.* **2,** 233–246 (1974).]

that in a number of unicellular and multicellular systems true circadian clocks gate cell division and other events of the cell cycle (Relations E and F, respectively).

This is not to say, however, that each of these three functional entities is necessarily a prerequisite for the operation of the other two. Thus, one can uncouple (or otherwise prevent the circadian clock from expressing itself) (Blockage 1) in cultures of *Euglena* by forcing the cells into fast, ultradian growth (g = 10 hours) with appropriate intense illumination, elevated temperature, or nutrient supplements, and yet still observe cell division. In a sense, then, the operational cell cycle controls override the regulatory effects of the circadian oscillator, just as seen in *Drosophila,* where the developmental sequence culminating in the act of eclosion of the adult fly from its pupal case is conceptually and empirically separable from the circadian oscillation which gates eclosion in the individual fly to a specific time of the 24-hour day and in populations to intervals of about 24 hours. In this system, the underlying clock can be arrested under constant illumination, yet the developmental cycle still occurs (though flies now emerge at all times of the day). Conversely, one can block (Blockage 2) the cell developmental cycle in many unicellular organisms and still find that a circadian clock still continues to operate as evidenced by a variety of circadian outputs. In *Euglena,* for example, persisting circadian rhythms of phototaxis, motility, cell settling, and photosynthetic capacity, as well as oscillations in enzymatic activity, have all been observed in stationary cultures in which virtually no cell division is occurring.

It is not clear, however, to what extent either cell cycles or circadian rhythms can be separated from periodic changes in membrane permeability. The meager evidence available suggests that if transport is blocked by appropriate inhibitors (Blockage 3), the cell cycle will also be arrested. But if the cell cycle is arrested, as in stationary cultures of microorganisms (Blockage 4), it appears likely that oscillatory changes in transport of at least some molecules will still occur, particularly in view of the many overt persisting rhythms that have been documented in nondividing cells.

Even more difficult to attack is the transport—clock axis. Thus, an oscillator might underlie fluctuations in transport capacity (Relation C), perhaps mediating its control via periodic activation or synthesis of appropriate carriers (permeases). Alternatively, the regulatory mechanisms involved in cyclical changes in membrane permeability might themselves constitute a biological clock (Relation D), as discussed earlier in the membrane model for circadian timekeeping. Conclusive (i.e., uniquely verifying) inhibitor experiments are sorely needed: the problem, of course, in attempting to dissociate these parameters is to devise a foolproof method of "stopping" the operation of any and all biological clocks (Blockage 5) or of inhibiting all biological transport (Blockage 6) *without* killing or otherwise injuring the organism or impairing its function.

THE INTRINSIC VERSUS EXTRINSIC TIMING PROBLEM

As was emphasized in the introduction to this chapter on models and mechanisms for endogenous timekeeping, the relative merits of the "endogenous" and "exogenous" timing hypotheses have not been tackled head-on for two reasons: (i) the proper experimental questions have not been developed to definitely resolve the issue, given the present state of the art—indeed, it is difficult to even contrast the theories, given the different ways in which they have been formulated and the different languages in which they have been expressed. (ii) The author does not believe that such an approach will be fruitful. Naturally, these two reasons are related!

Rather, the author has attempted to be as persuasive as possible with existing data and models that endogenous, *self-sustaining* (*autonomous*) circadian clocks can and do exist; that, indeed, they are an inevitable consequence of temporal adaptation in living systems; and that they are fully explicable in terms of known, concrete biochemical and molecular parameters. To the extent that this approach has succeeded, alternative timing hypotheses are unnecessary. This is not to say that in some instances extrinsic timing does not occur; it does (even during entrainment of an autonomous clock by an external *Zeitgeber*!). Also there is evidence to suggest that subtle geophysical variables affect organisms in a variety of ways. But the author strongly asserts (on the basis of the evidence presented) that these pervasive cues play at best only a subsidiary role in the timing of the overt rhythms and biochemical oscillations documented in this chapter [bird migration, for example, may be a different matter all together, with magnetic fields playing an important role]. If it transpires that all of my own work and that of my colleagues involved in a similar approach to rhythms merely describes the "autophasing machinery," we will, nevertheless, have had the satisfaction of making a significant contribution.

Only time will tell.

SUMMARY AND CONCLUSIONS

1. Biological problems pivot upon the complexities of biological organization, which may be temporal as well as spatial. That organisms can and do measure time is evidenced by four categories of phenomena: persisting rhythms, time sense, seasonal photoperiodism, and celestial orientation and navigation. Since it seems likely that the last three types are more sophisticated evolutionary variants on the first theme, primary attention should be given to the basic mechanism underlying overt persisting (especially circadian) rhythms. A basic assumption shared by many workers in the field is that this clock has a cellular and biochemical origin.

2. One class of models dealing with circadian clocks is concerned only with the formal relationships between the various components parts of the system—not with the identity of the parts themselves. A typical approach is to perturb the "black box" with a brief light pulse and observe what then happens to its output of overt rhythms. The discovery of transients and the differing responses of the fruit-fly emergence rhythm, for example, led to the dual oscillator model, in which one light-sensitive, autonomous oscillator is considered to be coupled to a second light-insensitive, but temperature-sensitive oscil-

lator that more or less directly underlies the overt rhythm. This model, in turn, led to a detailed formal model for the entrainment of a circadian rhythm by light cycles in which a daily correction occurs of the slightly imperfect, circadian timekeeper. Still more recent and sophisticated formal models have attempted to provide a topological description of the basic rhythm in light sensitivity reflected in the phase-response curves of rhythms to light pulses. Other models have addressed themselves to Aschoff's Rules for the effects of differing light intensities on the period of rhythms.

3. The other basic approach to circadian rhythms has been an attempted elucidation of the biochemical and molecular nature of the circadian escapement itself. This is the area in which the most important recent advances have been made.

4. Independent circadian oscillations abound in isolated cells, organs, and tissues. Thus, timekeeping does not require the entire organism.

5. Much effort has been expended on the neuroendocrinological control of circadian rhythms. The question: Does a localized, anatomically distinct mechanism control the timing of overt rhythms such as gross locomotory activity or eclosion? The search for such a site has led to the optic lobe clock of the cockroach, the protocerebral clock of the silk moth, and the neural clocks of the sea hare; and in mammals and birds, to the pineal gland clock. In the latter case, the quest led back out to the suprachiasmatic nucleus in the hypothalamus. In sum, there appears to be a hierarchy of clocks, embracing "master" clocks and their driven "slaves"; yet, a single cell from the lowliest slave is still capable of independent circadian outputs.

6. The cell cycle (comprising the classic G_1, S, G_2, and M phases along with a host of other marker events) is a labile clock of sorts representing a developmental sequence (in the same way as the life cycle of a higher organism constitutes a clock). Yet, this "clock" can be modulated by a true circadian oscillator so that cell divisions are "gated" at 24-hour intervals. Many other overt circadian rhythms also occur in nondividing unicellular populations.

7. Oscillations in biochemical parameters (both ultradian and circadian) occur at every level of biological organization. The high-frequency glycolytic fluctuations that occur in both intact yeast cells and in cell-free extracts are particularly instructive with regard to the potential generating mechanism.

8. Circadian rhythms are particularly resistent to chemical perturbation. Inhibitors of macromolecular biosynthesis have been utilized without too much success to dissect the clock; only cyclohexi-

mide has proved to have a clear-cut effect (lengthening) on the free-running period. On the other hand, inhibitors of membrane function such as lithium and valinomycin have a marked effect on the period; a phase-response curve for pulses of these substances is quite similar to that for light signals. Potassium ion pulses also generate phase shifts.

9. Attempts to genetically dissect the clock have centered on the isolation of clock mutants (having free-running periods significantly different from 24 hours) in *Drosophila, Neurospora,* and *Chlamydomonas.* In some cases, arhythmic mutants have been obtained. Most of these mutations appear to involve only one or two functional genes.

10. Three classes of biochemical and molecular models have emerged for the circadian clock: (i) *Feedback loop networks* (the best understood of which is the glycolytic system of yeast, constituting a truly endogenous, self-sustaining oscillation, but with a very short period). Inhibitory cross-coupling among individual intracellular or cellular oscillators would permit the overall oscillation observed in the population (cell culture, tissue) to display a much longer (e.g., circadian) period than that of the constituent oscillators. (ii) *"tape-reading," transcriptional models,* in which the fixed distance between the genes on the DNA tape could serve as a measure of time. The most explicit construct in this class is the chronon model, in which long, polycistronic pieces of DNA require about 24 hours to sequentially transcribe due to the presence of many diffusion loops. (iii) *Membrane models,* the most detailed of which incorporates the network concept, but identifies ions and membrane-bound ion transport channels with the biochemical clock. In this model, light pulses would effect phase shifts by opening up K^+ gates embedded within the lipid bilayer of the fluid mosaic membrane, and temperature-compensation would be derived from adaptive changes of the membrane lipid composition to the environmental temperature changes. All three of these types of models make empirically testable predictions; the clock(s) mechanism will probably comprise elements of each.

Selected Readings

Axelrod, J. (1974). The pineal gland: A neurochemical transducer. *Science* **184,** 1341–1348.

Brady, J. (1974). The physiology of insect circadian rhythms. *Adv. Insect Physiol.* **10,** 1–115.

Brown, F. A., Jr., Hastings, J. W., and Palmer, J. D. (1970). "The Biological Clock. Two Views." Academic Press, New York.

Bünning, E. (1973). "The Physiological Clock," 3rd rev. ed. Springer-Verlag, Berlin and New York.

Burton, A. C. (1971). Cellular communication, contact inhibition, cell clocks and cancer. *Perspect. Biol. Med.* **14,** 301–318.

Edmunds, L. N., Jr. (1971). Persisting circadian rhythm of cell division in *Euglena:* Some theoretical considerations and the problem of intercellular communication. *In* "Biochronometry" (M. Menaker, ed.), pp. 594–611. Natl. Acad. Sci., Washington, D. C.

Edmunds, L. N., Jr. (1976). Temporal differentiation in *Euglena:* Circadian phenomena in non-dividing populations and in synchronously dividing cells. *In* "Les Cycles Cellulaires et Leur Blocage Chez Plusieurs Protistes," Colloq. Int. CNRS No. 240, pp. 53–67. CNRS, Paris.

Edmunds, L. N., Jr., and Cirillo, V. P. (1974). On the interplay among cell cycle, biological clock and membrane transport control systems. *Int. J. Chronobiol.* **2,** 233–246.

Ehret, C. F. (1974). The sense of time: Evidence for its molecular basis in the eukaryotic gene-action system. *Adv. Biol. Med. Phys.* **15,** 47–77.

Ehret, C. F., and Trucco, E. (1967). Molecular models for the circadian clock. *J. Theor. Biol.* **15,** 240–262.

Goodwin, B. C. (1963). "Temporal Organization in Cells." Academic Press, New York.

Halberg, F. (1959). Physiologic 24-hour periodicity; general and procedural considerations with reference to the adrenal cycle. *Z. Vitam.-, Horm.- Fermentforsch.* **10,** 225–296.

Hastings, J. W., and Schweiger, H.-G., eds. (1976). "Dahlem Workshop on the Molecular Basis of Circadian Rhythms." Dahlem Konferenzen, Berlin (in press).

Hess, B., and Boiteux, A. (1971). Oscillatory phenomena in biochemistry. *Annu. Rev. Biochem.* **40,** 237–258.

Menaker, M., ed. (1971). "Biochronometry." Natl. Acad. Sci., Washington, D.C.

Menaker, M. (1972). Nonvisual light reception. *Sci. Am.* **226,** 22–29.

Njus, D., Sulzman, F. M., and Hastings, J. W. (1974). Membrane model for the circadian clock. *Nature (London)* **248,** 116–120.

Pavlidis, T. (1963). "Biological Oscillators: Their Mathematical Analysis." Academic Press, New York.

Pavlidis, T., and Kauzmann, W. (1969). Toward a quantitative biochemical model for circadian oscillators. *Arch. Biochem. Biophys.* **132,** 338–348.

Pittendrigh, C. S. (1961). On temporal organization in living systems. *Harvey Lect.* **56,** 93–125.

Pittendrigh, C. S. (1965). On the mechanism of the entrainment of a circadian rhythm by light cycles. *In* "Circadian Clocks" (J. Aschoff, ed.), pp. 277–300. North-Holland Publ., Amsterdam.

Pittendrigh, C. S., and Bruce, V. G. (1959). Daily rhythms as coupled oscillator systems and their relation to thermoperiodism and photoperiodism. *In* "Photoperiodism and Related Phenomena in Plants and Animals (R. B. Withrow, ed.), Publ. No. 55, pp. 475–505. Am. Assoc. Adv. Sci., Washington, D.C.

Satter, R. L., and Galston, A. W. (1973). Leaf movements: Rosetta stone of plant behavior? *Bioscience* **23,** 407–416.

Sweeney, B. M. (1969). "Rhythmic Phenomena in Plants." Academic Press, New York.

Sweeney, B. M. (1972). Circadian rhythms in unicellular organisms. *In* "Circadian Rhythmicity" (J. F. Bierhuizen, ed.), pp. 137–156. Centre for Agricultural Publishing and Documents, Wageningen, The Netherlands.

Truman, J. W. (1972). Circadian rhythms and physiology with special reference to neuroendocrine processes in insects. *In* "Circadian Rhythmicity" (J. F. Bierhuizen, ed.), pp. 111–135. Centre for Agricultural Publishing and Documents, Wageningen, The Netherlands.

Wever, R. (1965). A mathematical model for circadian rhythms. *In* "Circadian Clocks" (J. Aschoff, ed.), pp. 47–63. North-Holland Publ., Amsterdam.

Winfree, A. (1971). Corkscrews and singularities in fruitflies: resetting behavior of the eclosion rhythm. *In* "Biochronometry" (M. Menaker, ed.), pp. 81–209. Natl. Acad. Sci., Washington, D.C.

Glossary

Amplitude A measurement of the height of the peaks relative to the troughs of a cycle (see Cycle)

Autophasing Self-engendered phase shifting in an individual, carried out in constant conditions through the actions of phase-response mechanisms responding to a variety of *Zeitgeber*

Biological clock The mechanism that is thought to time those organismic rhythms that persist in constant conditions. The clock has not yet been identified, but the properties thus far elucidated lead to two hypothetical clockwork schemes: the escapement- and nonescapement-type living clocks (see Escapement clock and Nonescapement clock)

Biological horologe A living clock (see Biological clock)

Black box A container with walls still opaque to the inquiring eyes of science, in which all thus far unexplained processes take place

CC An abbreviation for constant conditions (see Constant conditions)

Chronomutagenic agent (*Chrono*, time; *mutatio*, change; *genic*, producing) Any substance that produces an alteration in the phase or period of a biological rhythm when administered to an organism

Chronon A polycistronic strand of DNA, the transcription of which measures off an interval of about 24 hours

Circadian rhythm (*Circa*, about; *diem*, day; + -an) Rhythms about a day in length. Used formally, it stands for solar-day rhythms that persist in constant conditions with a slight deviation from the 24-hour period displayed in nature

Circalunadian A basic lunar-day rhythm—almost always bimodal—that persists in constant conditions with a period either slightly longer or shorter than 24.8 hours

Circamonthly rhythm A persistent synodic monthly (see Symodic month) rhythm whose period differs slightly from 29.5 days

Circannual rhythm A persistent annual rhythm whose period differs somewhat from 365 days

Circatidal rhythm A basic tidal rhythm that persists in constant conditions with a period deviating slightly from 12.4 hours. The same as *circalunadian* (see Circalunadian)

Constant conditions A laboratory setting in which at least the levels of illumination and temperature do not vary

CR Abbreviation for *circadian rhythm* (see Circadian rhythm)

CT Abbreviation for *circadian time;* a time scale covering one full circadian period. The zero point is defined arbitrarily and, in practice, is usually the instant of change over from dark to light (equivalent to "dawn")

Cycle A sequence of events that repeats itself through time in the same order and at the same interval

Day-neutral plant Those plants whose flowering response is not dependent on photoperiodism

DD Abbreviation for *constant darkness*

Desynchronization A change, usually temporary, in the phase relationships of one or more rhythms in an individual

Diurnal solar-day rhythm One in which the major peak(s) comes during the hours of light

Endogenous clock See Escapement clock

Entraining agent Same as *Zeitgeber* (see *Zeitgeber*)

Entrainment The coupling of an organismic rhythm to an external oscillation (*Zeitgeber*) which causes the rhythm to display the frequency of the *Zeitgeber*

Escapement clock Any clock, such as a wristwatch, grandfather, cesium, or living one, that generates its own interval of time autonomously

Exogenous clock See Nonescapement clock

Form The shape of the curve describing a cycle

Free-running period The same as *natural period* (see Natural period)

Frequency The number of cycles per unit time; the inverse of the period

Infradian rhythm One with a period somewhat longer than 24 hours

Jet lag A melange of symptoms, dominated by a disrupted sleep pattern, occurring when one's physiological rhythms are out of phase with the ambient light-dark cycle (and each other) after rapid transmeridional flights

LD Abbreviation for light-dark cycle. The notation is often followed by a numerical ratio indicating the duration in hours of illumination and darkness (in that order) and a parenthetic statement as to the intensity of the light used. For example, LD 8:16 (15 foot candles) signifies a light-dark cycle of 8 hours of light whose intensity is 15 foot candles, alternating with 16 hours of darkness

LL Abbreviation for *constant light;* i.e., a laboratory situation where the light is never turned off and the intensity of the illumination is held constant

Long-day plant Those plants that flower in photoperiods essentially longer than 12 hours

Lunar day The 24.8-hour interval between consecutive moonrises; one rotation of the earth on its axis in relation to the moon

Natural period The fundamental period of a clock when it is not entrained to some forcing oscillation (*Zeitgeber*)

Nocturnal solar-day rhythm One in which the major peak(s) come during the hours of darkness

Nonescapement clock Any clock, such as a sundial or an electric clock (or a living clock), that does not generate its own interval of time, but instead simply signals an interval relayed to it from some outside source (e.g., keeping with the above examples, the passage of the sun overhead, or the 60-cycle alternating line current)

Oscillation Used here to mean the same as *cycle* (see Cycle)

Period The time interval of one complete cycle

Periodic Used here to mean the same as cyclic

Persistent rhythm Any organismic rhythm that will continue to be displayed in constant conditions

Phase Some arbitrarily chosen fraction of a cycle, e.g., the peak. It is also a relative term used to describe where a particular phase of one cycle is in relation to another; for example, "the active phase of the mouse's rhythm is in phase with the hours of darkness," or "the crabs rhythms are about 6 hours out of phase with one another"

Phase angle A relative term measuring in degrees (or sometimes in units of time) the distance between a particular point in a cycle and some arbitrarily chosen constant reference point. For example, if a natural diurnal rhythm is subjected to reversed light-dark cycles in the laboratory, the rhythm inverts and is then said to be 180° out of phase with the old natural light-dark cycle. Or, if the moment of "light on" in a light-dark cycle is arbitrarily given a value of 0, and the 24-hour period of a rhythm is divided into 360° (a permissible practice, since oscillations are linear projections of circular motion), and if the major peak of the rhythm comes 6 hours after the dark-to-light transition, the phase angle is 90°

Phase-response curve (rhythm) A waveform plot describing the direction and amount of phase change produced in a rhythm subjected to an appropriate short stimulus given at all points in a circadian cycle

Photoperiod The illuminated portion of a light-dark cycle

Photoperiodism A response, such as reproduction or migration, of and organism to the relative length of day and night

Q_{10} Temperature coefficient = rate at temperature X + 10°C/rate at temperature X

Short-day plants Those plants that flower in photoperiods essentially shorter than 12 hours. Same as long-night plant

Siderial day The interval produced by the rotation of the earth on its axis in relation to an arbitrary point on the star sphere. The interval is 23 hours and 56 minutes

Solar day The 24-hour interval between consecutive sunrises. One rotation of the earth on its axis in relation to the sun

Subjective day That span of hours in constant conditions that corresponds to what had been daytime for the experimental organism before being placed in constant conditions

Subjective night That span of hours in constant conditions that corresponds to what had been daytime for the experimental organism before being placed in constant conditions

Sun-compass orientation The ability of an organism to orient in a desired direction using the position of the sun in the sky as a reference point. Commonly used by arthropods, vertebrates, and Boy Scouts

Synodic month The 29.5-day interval between consecutive new moons

Temperature coefficient See Q_{10}

Transients In some organisms, after exposure to a single pulse of light or temperature in otherwise constant conditions, one or more intermediate phase angles are displayed before a new steady-state phase is adopted. The intermediate angles are termed transients

TT Abbreviation for *constant temperature*

Ultradian rhythm One with a period somewhat shorter than 24 hours

Zeitgeber Time giver; any external stimulus that will entrain or rephase a biological rhythm

ZT *Zeitgeber* time. A time scale covering one full *Zeitgeber* period. The zero point is defined arbitrarily

Greek-letter symbolism as used in the biorhythm literature. τ, (Tau) period; τ_{FR}, free-running period; ϕ, (Phi) phase; Δ, (Delta) means change in (can be used with any of the above, e.g., $\Delta\phi$ = phase shift).

Index

Italic numbers indicate pages containing a figure.

Acetabularia
photosynthetic capacity rhythm, 38–*41*
effect of chronomutagens, 49, 331, 332, 334
in enucleated cell, *40*
Actinomycin D, 48, 49, 312, 330, *331*
Activity rhythms
African waxbill, *68*
bluefish, 3, *4*
chaffinch, 300
fiddler crab, 1, *2*, 3, 94, 108, *109*, *213*, *214*, *215*, 216, 245
flying squirrel, *63*, *240*, 241, 288
golden hamster, 86, *262*, *266*, 270, *271*, *288*
green shore crab, 97, 98, *99*, *100*, *112*, 113
kangaroo rat, *251*
mouse, 260
effect of D_2O, *64*, 65, 66
effect of gamma radiation, 234, *235*, 257
method of study, 61, *62*
nickle coated glass slide, 234
penultimate hour crab, 95, *96*, *97*, 108
rat, *251*, *256*, *257*, *258*, 264, *265*
roach, 81, 82, 305, *306*, 307
sandhopper, 98, *99*, 100, 101, *102–104*, 108, 115, *116*, *117*, 189
white-crowned sparrow, *67*, 68, *70*
Adrenal glands, isolated, oxygen consumption rhythm, 304
African waxbill, activity rhythm, *68*
Alcohol, 20, 34, 49, 101, *335*, *336*, 337
Alcohol metabolism rhythm, 120, *121*, 162
Algae, nitrate reduction, annual rhythm, 270, *272*
American eel, sensitivity to electric fields, 223
Aminophylline, 354
Amphetamine, rhythmic sensitivity, *87*, 88
Annual rhythms
algae, 270, *272*
bean, 219, *221*, 270, *272*

golden-mantled ground squirrel, 4, *5*
innate nature of, *5*
palolo worm, 3
planaria, 266, *267*, 270, *272*
Quelea, *206*
woodchuck, 270, *272*
Antarctica, rhythm studies in, 86
Aphid, photoperiodic response, 203
Aplysia, *see* Sea hare
Arabinosyl cytosine, rhythmic sensitivity, 89, 90
Arctic tern, circumpolar navigation, 180
Aschoff's rules, 250, 300
Autophasing, *13*, 14, 16, 17, 25, 244, *247*, 248, *249*, 260
Avena, *see* Oat
Bean, oxidative metabolism rhythm, 41–*44*, 45, 219, *221*, 270, 272
sleep movement rhythm, *32*–36, 86, *199*, 288
effect of chronomutagens, 34, 334, *335*
effect of light on form, 238, *239*
role of magnetism on entrainment, 231, 232
water uptake, annual rhythm, 223, 270, *272*
monthly rhythm, 266, *268*, 269
role of magnetic field, 223
Beach flea, *see* Sand hopper
Bees
direction finding, 175–179, 222
time sense, 75–*77*, 238, 239
Biological clock
accuracy during replication, 25, 78, 79
adaptive significance, 24
basic frequency, 5–7, 14, 110, *111*
chronomutagenicity, 19, *20*, 25–27, 29, 34, 48, 49, *57*, 59, *64*, 65, 67, *68*, 101, *201*, 245, 312, 329–337, 349, 351, 354
circa nature, 8, *13*, 14, 16, 94, 241
evolution of, 23–25
genetic basis, *22*, *23*, 54, 55, 78, 79, 252, 321, 337–*339*
hypothesis, 19, 24–26, 29, 34, 45, 273–278, *347*, 348–354
innate nature of, *5*, 20, *21*, 76, 98, *100*, 321
location in organism, 25, 27, 81, 82
multiple or master, 27, 33, 34, 50–52, 110, *111*, 304, 323, 324
properties of, *17*, 19, 26
Bioluminescence rhythm, 27, *38*, 45–49, *288*
chemistry of, *47*
effect of chronomutagens, 48, 49, 330, 331, *336*, 337
effect of temperature, *46*, 48
Bird migration, stimulus for, 180
Birthtime, clock control, 153, 270
Bluefish, activity rhythm, 3, *4*
Body temperature, *see* Temperature rhythms
Bread mold, membrane composition rhythm, 354
zonation rhythm, 53, *54*, 55, 86
in Antarctica, 86
chronomutagenicity, 354
effect of temperature, 54
genetic basis, 54, 55, 252, 338, *339*
Caffeine, 354
Cancer
effect on cell division rhythms, *80*, 81
role of bioclock, *80*, 81, 82, 89, 90
Candida, cell division rhythm, 321
Cannabis, photoperiodic response, 194
Carcinus, *see* Green shore crab
Carrot, oxidative metabolism rhythm, *220*, 255
Carbon tetrachloride, rhythmic sensitivity, 88
Carpodacus, *see* House finch
Cave studies, 35, 76, 125, 126, 135
Cell cycle, *318*, 319
Cell division rhythms
Candida, 321
Chlamydomonas, 321
Chlorella, 321
Escherichia, 321, 349
Euglena, *320*, 321, 322
Gonyaulax, *38*, *46*, 48, 49, *50*
Gymnodinium, 321
Klebsiella, 321, 349
Paramecium, 79, 321
rodents, 79, *80*, 81
vascular plants, 49, 51, 143
Chaffinch, activity rhythm, 300
Chemicals, effects on rhythms, 19, *20*
aminophylline, 354

actinomycin D, 48, 49, 312, 330, *331*
alcohol, 20, 34, 49, 101, *335*, *336*, 337
caffeine, 354
chloramphenicol, 48, 49, *331*, 334
cyanide, 19, *20*
cycloheximide, *57*, 59, 245, 332, *333*, 334
deuterium oxide, *20*, 34, 49, 59, *64*, 65, 67, *68*, 101, 134, *201*, 245, 349
lithium, 34, 351
puromycin, *26*, 27, 48, *331*, 334
theophylline, 354
valinomycin, 34, 49, *335*, *336*, 337
Chenopodium, photoperiodic response, *201*
effect of D_2O, *201*
Chicken, oxidative metabolism, 257, 262, *263*, 264
Chlamydomonas
cell division rhythm, 321
phototactic rhythm, genetic basis of, 338
Chloramphenicol, 48, 49, *331*, 334
Chlorella, cell division rhythm, 321
Chloroform, rhythmic sensitivity, 88
Chromatophores, 83, *84*
Chronomutagenetic altering of period, 19, *20*, 25–27, 29, 34, 48, 49, *57*, 59, *64*, 65, 67, *68*, 101, *201*, 245, 312, 329–337, 349, 351, 354
Chronon model, *347*, 348–350
Circalunadian definition, 94
Clock, *see* Biological clock
Cockroach, *see* Roach
Coleus
photoperiodic response, *200*
sleep-movement rhythm, *200*
Color-change rhythm
effect of cosmic radiation, 83, 85
effect of translocation, 85, 216
fiddler crab, 83, *84*, 85, *212*
green shore crab, 83
temperature independence of, 83, 213
Color sensitivity rhythm, 172, *173*
Commuter diatom, 98, 101, *105*, *106*
Copulation frequency, *171*
Cosmic radiation, effect on rhythms, *25*, 83, 85, 219, *220*
Coupler, clock, 26, 27, 50, 66, 79, 81, 110, *111*, 353
Cricket, rhythmic sensitivity, 88
Cyanide, 19, *20*
Cycloheximide, effect on period, *57*, 59, 245, 332, *333*, 334
Daldinia, reproduction rhythm, 52, *53*, *54*
Daylength, annual change, *193*
DDT, rhythmic sensitivity, 88
DDVP, rhythmic sensitivity, 88
Death, clock control
annual modulation, 154, *156*
daily modulation, 154, *155*
Desynchronization of rhythms, 33, 34, 150, 151, 241, 242
Deuterium oxide, 20, 34, 49, 59, *64–68*, 101, 134, *201*, 245, 349
Diapause, photoperiodic control, *202*
Direction finding, role of the clock,
bees, 175–179, 222
birds, 179–187, 222, 223
genetic basis, 190
planaria, 222, 223, *224*, 225, *226*, 227, *228*, 229, 232–*234*, 266, *267*, 270, *272*
sand hoppers, *188*, *189*, 190
snails, 222, 232, 233
DPN reduction rhythm, 305, 324, *325*, 327, *340*
Drosophila, *see* Fruit fly
Early birds, body temperature rhythm of, *140*
Eclosion rhythm
fruit fly, *71*
entrainment by light-dark cycles, 69, 292, *293*, *294*
genetic basis, *22*, *23*, 252, 337, 338
innate nature, *21*
method of study, 69
phase response curve, 69, *72*, 74, *286*, 287, *288*, 289, 293, 294
silkmoth, 308, *309*
Efficiency rhythms, *see* Psychomotor/performance
Electric field, *157*, 232, 233
10 Hz, *158*, *159*
Endotoxin, *E. coli*, rhythmic sensitivity, *87*
Entrainment, after longitudinal travel, 162–169
by inundation cycles, 101, *102–104*, *112*, 113, *147*, 152

by light-dark cycles, 8, 9, *10*, *11*, *12*, *13*, 33, *40*, *41*, *64*, 67–*70*, 108–*110*, 135, 151, 152, *164*, *183*, 245, *247–249*, 292, *293*, *294*, 295
by magnetism, 231, 232
by mechanical agitation cycles, 115, *116*
by 10 Hz field, *158*, *159*
by temperature cycles, 11, *15*, 16, *23*, *73*, *113*
by pressure cycles, *114*
Enzyme rhythms, *47*, 316, *323*, 324–329
ESSO hypothesis, 289, *290*, 291
Ether, rhythmic sensitivity, 88
Euglena
amino acid incorporation rhythm, 323
cell division rhythm, *320*, 321, 322
enzyme rhythm, *323*
motility rhythm, 322
photosynthetic capacity rhythm, 322, *323*
phototactic rhythm, 55–60
effect of cycloheximide, *59*, 331, *332*
method of study, *56*
phase-response rhythm to light, *58*
phase-response rhythm to temperature, *74*
temperature entrainment, *73*
temperature independence, 83, 213
Excirolana, *see* Sand hopper
Excretory rhythms, *128*, 129, *150*, *261*, 326
after kidney transplant, 152
entrainment to artificial "days," 151
Extraoptic entrainment, 67, 68, *70*
Ferret, photoperiodic response, 207
Fiddler crab, 2, 4
activity rhythm, 1, 2, 3, 94, 108, *109*, *213*, *214*, *215*, 216, 245
color-change rhythm, 83, *84*, 85, *212*
effect of cosmic radiation, 83, 85
effect of translocation, 85, 216
temperature independence, 83, 213
oxidative metabolism rhythm, 95, 219, *220*, *246*, *258*
Flowering inhibition rhythm, 198, 201
Flying squirrel, activity rhythm, *63*, *240*, 241, *288*
Free-running rhythm definition of, 3
Fringilla, *see* Chaffinch
Fruit fly
eclosion rhythm, *21*, *22*, *23*, *71*, 289–291
entrainment by light-dark cycles, 69, 292, *293*, *294*
genetic basis, *22*, *23*, 252, 337, 338
in antarctica, 86
method of study, 69
phase response curve, 69, *72*, *74*, *286*, 287, *288*, 289, 293, 294
salivary glands, nuclear volume rhythm, 304
Fucus, oxidative metabolism rhythm, *220*, *258*
Gamma radiation, 233, *234*, *235*, 255, 257, 270, *272*
Genetic basis of clock, *22*, *23*, 54, 55, 78, 79, 252, 321, 337–*339*
Geophysical forces, roles in rhythms, *25*, 83, 85, 86, 157, *158*, *159*, 184, *185*, 219, *220*, 222–239, 255, 257, 259, 270, *272*
Glycogen rhythm, 326
Goat, photoperiodic response, 207
Golden-mantled ground squirrel, annual hibernation rhythm, 4, *5*
Gonyaulax, *37*, 199
cell division rhythm, *38*, 49, *50*
effect of temperature on period length, *46*, 48
effect of temperature on phase, 48
irritability rhythm, 47
luminescent flashing rhythm, *38*, *288*
chemistry of, *47*
effect of chronomutagens, 48, 49, *336*, 337
effect of temperature on period, *46*, 48
effect of temperature on phase, 48
luminescent glow rhythm, 48
effect of chronomutagens, 48, 49, 330, 331
membrane particle rhythm, 354
photosynthetic capacity rhythm, 36–*38*, 330
potassium concentration rhythm, 337
Green shore crab, 96, 97, *98*
activity rhythm, 97, 98, *99*, *100*, *112*, 113

color-change rhythm, 83
Growth rhythms, *Gonyaulax*, *38*, 49, *50*
Escherichia, 321, 349
Klebsiella, 321
oat, 51, *52*
penis, *144*
roots, 51
stems, 51
Gymnodinium, cell division rhythm, 321
Hamster
activity rhythm, 86, *262*, *266*, 270, *271*, *288*
adrenal rhythm, 304
photoperiodic response, 207
Hantzschia, *see* Commuter diatom
Heart-rate rhythm, *145*
development in infants, 148
during space travel, 147, *148*
in transplanted hearts, 146, *147*
role of temperature, 145, 146
Heavy water, *see* Deuterium oxide
Henbane, photoperiodic response, 197, *198*
Hormonal rhythms, 81, 82, 316, *317*, 326
Hourglass hypothesis of photoperiodism, 203
House finch, photoperiodic response, 204, *205*
activity rhythm, *70*
Innate nature of clock, *5*, 20, *21*, 76, 98, *100*, 321
Inundation cycles, entrainment by, 101, *102–104*, *112*, 113, *147*, 152
Iris, photoperiodic response, 192
Isolated tissues, rhythms in, 27, 33, 35, 36, 39, *40*, 43, *44*, 82, 304
Jet lag, *see* Longitudinal travel
Juncos, photoperiodic response, 204
Kalanchoe
petal movement rhythm, *288*
photoperiodic response, 196, *197*
effect of lithium, 351
Kangaroo rat, activity rhythm, *251*
Libido, monthly rhythm, 170, 171
Librium, sensitivity rhythm, 88
Light
effect on form of rhythms, 238, *239*
effect on period length, *14*
effect on phase, 8, 9, *10–13*, 33, 58, 69, *72*, 74, 135, 152, *167*, *240*, *241*, 248, *249*, *251*, *286*, 287, *288*, 289, 290, 292–300
entrainment by, 8, 9, *10–13*, *40*, *41*, *64*, *67–70*, 108–*110*, 151, *164*, *183*, 245, *247*, 248, *249*, 292, *293*, *294*, 295
inhibition of rhythms by, *14*, 38, *39*
initiation of rhythms by, *21*, 51
Light phase setting rhythm, 9–*13*, 69, *72*, 74, 164, *167*, *240*–244, *247*, 248, *249*, *251*, *286*, 287, *288*, 289, *290*, 202–300
Light sensitivity rhythm, *196*, *197*, *198*, 202, 205, 207
Lithium, 34, 351
Locomotor rhythm, see Activity rhythm
Longevity, role of light-dark cycles, 75
Longitudinal travel, and biological rhythms, 160–169
adjustment, time required, 162–168
avoiding jet lag, 168
importance of clock, 162–163
role of light phase-setting rhythm, *167*
Luminescence rhythms, *see* Bioluminescence
Lunar day, definition, 93, *94*
Magnetism
effect on orientation, 184, *185*, 222, 223, *224*, 225
effect on rhythms, 86, *226*, 227, *228*, 229, 231, 232, 259
Maryland mammouth, *see* Tobacco
Mating reversal rhythm, *see* Sex reversal
Mealworm, oxidative metabolism rhythm, 255
Mechanical agitation cycles, entrainment by, 115, *116*
Megoura, *see* Aphid
Melatonin rhythm, 316
Menstrual cycle, length of, 170, *269*, 270
Mice
activity rhythm, 260
effect of D_2O, *64*, 65, 66
effect of gamma radiation, 234, *235*, 257
method of study, 61, *62*
oxidative metabolism rhythm, 255
rhythmic sensitivity, *87*–90
Migratory restlessness, 66, *67*, 181
Mite, rhythmic sensitivity, 88
Monthly rhythms in,
beans, 266, *268*, 269

hamster, 270, *271*
man
birthtime, 270
color sensitivity, 172, *173*
copulation frequency, *171*
libido, 170, *171*
menstruation, 170, *269*, 270
orgasm frequency, *171*
pain, 171, *172*
senses, 170
volunteering, 170
Palolo worm, 3
planaria, *226*, 227, *228*, 229, 266, *267*
snails, 222
Moods, rhythms in, 143
Moon compass orientation, in birds, 187
sand hoppers, *188*–190
Motility rhythm, *Euglena*, 322
Nasonia, see Wasp
Nassarius, see Snail
Nectar-secretion rhythms, 76
Neuroendocrine systems, 81, 82, 316
Neurospora, see Bread mold
Nickle-coated glass slide, 234
Night blooming jessamine, 35
Night owls, body temperature rhythm of, *140*
Nitrate reduction in algae, annual rhythm in, 270, *272*
Noctiluca, 45
Oat, growth rhythms, 51, *52*
Onion seeds, oxidative metabolism rhythm, 45, 264
Optic nerve discharge rhythm, sea hare, 304, 313, *314*, *335*, 337
Orgasmic frequency, monthly rhythm in, *171*
Orientation, *see* Direction finding
Oxidative metabolism rhythms in, adrenals, hamster, 304
bean, 41–45, 219, *221*, 270, *272*
carrot, *220*, 255
chicken, 257, 262, *263*, 264
fiddler crab, 95, 219, *220*, *246*, *258*
Fucus, *220*, *258*
mealworm, 255
mouse, 255
onion, 45, 264
potato, *42–44*, 219, *220*, *221*, 234, *235*, 236, 255, *258*, 270, *272*
Quahog, *220*, *258*
snail, *258*
Oyster, shell-opening rhythm, *206*, 215
Pain tolerance rhythms
daily, *122*
monthly, 171, *172*
Palolo worm, reproductive rhythm, 3, 4
Parabolic burster cell discharge rhythm, 311–*313*
Paramecium
cell division rhythm, 79, 321
sex reversal rhythm, *78*, 79
Penguin, Adelie
navigation, 186
orientation, 184, 185, *186*
entrainment by "constant" light, 187
Penis growth rhythm, *144*
Pentobarbitol, rhythmic sensitivity, *87*, 88
Penultimate hour crab, 95, *96*, *97*, 108
Petal movement rhythm, 35, 36, *288*
Phase response curve, *see* Phase setting rhythm
Phase setting, effect of
inundation cycles, 101, *102–104*, *112*, 113, *147*, 152
light cycles, 8, 9, *10*, *11*, *12*, *13*, 33, *40*, *41*, 64, 67–69, *70*, 108–*110*, 135, 151, 152, *164*, *183*, 245, *247*, 248, *249*, 292, *293*, *294*, 295
magnetism, 231, 232
mechanical agitation cycles, 115, *116*, *117*
pressure cycles, 114
temperature cycles, *11*, 15, 16, *23*, 48, *73*, *74*, *113*, 249, 250, 253, *286*, 287, 291, 312
Phase setting rhythm
general, *72*, *167*
importance of form, *13*, *167*
in genesis of circadian period, *13*, 14, 16
light, 9, 10, *11–13*, 69, *72*, 74, 164, *167*, *240–244*, *247–249*, *251*, *286*, 287, *288–290*, 292–300
temperature, *11*, 15–17, 19, 23, *74*, 249, 250, 253, *286*, 287, 291
Phaseolus, see Bean
Photosynthetic capacity rhythms, 36–41
in *Acetabularia*, 38–41
effect of chronomutagens, 49, 331, 332, 334

role of the nucleus, *40, 41*
in *Euglena,* 322, *323*
in *Gonyaulax,* 36–*38,* 330
in natural populations, 36
Phototactic rhythm
in *Chlamydomonas,* 338
in *Euglena,* 55–60, 331, *332*
Pigeons, homing orientation, 183
use of magnetic field, 184, *185,* 222, 223
Pineal gland, 68, 314–*317*
Planaria
annual rhythm, 266, *267,* 270, *272*
monthly rhythm, *226*–229, 266, *267*
orientation, role of magnetism, 222–225
effect of electric fields, 232, 233
effect of gamma radiation, 233, *234*
Pleasure rhythm, 123
Poinsettia, photoperiodic response, 194
Polymorphism, photoperiodic control of, *203*
Postprandial dip, *142,* 162
Potassium concentration rhythm, *Gonyaulax,* 337
Potato, oxidative metabolism rhythm, *42,* 43, *44,* 219–*221,* 234, *235,* 236, 255, *258,* 270, *272*
effect of gamma radiation, 234, *235*
method of study, *42,* 43
sidereal day rhythm, 235, 236
Pressure cycles, entrainment by, *114*
Prokaryotes, growth "rhythms," 321, 349
Psychomotor/performance rhythms, 138, *139,* 141, *142,* 162
effect of temperature, *139,* 141
Puffinus, see Shearwater
Puromycin, *26,* 27, 48, *331,* 334
Quahog, shell opening rhythm, 215, *216, 220, 258*
Quelea, annual reproductive rhythm, *206*
Ram, photoperiodic response, 207
Rat
activity rhythm, *251, 256, 257, 258,* 264, *265*
pineal rhythms, 316, *317*
rhythmic susceptibility, 87
Recrudescence, testes
hamster, 207
house finch, 204, *205*
juncos, 204
Quelea, 206
Reproduction rhythms
bread mold, 53, *54,* 55, 86, 354
Daldinia, 52–54
hamster, 207
house finch, 204, *205*
junco, 204
palolo worm, 3, 4
Quelea, 206
Shearwater, slender-billed, 206
Roach
activity rhythm, 81, 82
neural basis, 305, *306,* 307
cancer in, 82
rhythmic sensitivity, 88
Rodent, cell-division rhythm, 79, *80,* 81
effect of cancer, *80,* 81
Saccharomyces, see Yeast
Sand hopper, 98–*104,* 108, 115–*117, 188*–190
Sea hare
optic nerve discharge rhythm, 304, 313, *314, 335,* 337
effect of chronomutagens, *335,* 337
parabolic burster cell discharge rhythm, *311–313*
Semimonthly rhythm, activity, 213, *215,* 243, *246*
Senses, human monthly rhythms in, 170
Sensitivity rhythms to
amphetamine, *87,* 88
aspirin, 157
cancer therapy, 88–90
carbon tetrachloride, 88
chloroform, 88
color, 172, 173
DDT, 88
DDVP, 88
endotoxin, 87
ethanol, *87*
ether, 88
house dust, 156
librium, 88
light, *196*–198, 202, 205, 207
pentobarbitol, *87,* 88
strychnine, 88
x-irradiation, 88
Serotonin *n*-acetyl transferase rhythm, 316
Serotonin rhythm, 316, *317*
Sesarma, see Penultimate-hour crab

Sex reversal rhythm
  genetic basis of, *78*
  persistence during cell division, 78, 79
Sexual desire, 170, *171*
Shark, sensitivity to electric fields, 223
Shearwater, slender billed, annual reproductive rhythm, 206
Shell opening rhythm, oyster, 215, *216*, *220*, *258*
Shift work and biorhythms, 169
Sidereal day rhythm, 235, *236*
Silkmoth, eclosion rhythm, 308, *309*
Sleep-movement rhythms, 31–36, 86, *199*, *200*, 238, *239*, 288
  chronomutagenicity of, 34, 334, *335*
  in caves, 35
  magnetism, entrainment by, 231, 232
  mechanism of, 32, 33
  phase setting with light, 33
Sleep wakefulness rhythm, 125, *126*, *128*, *129*, *150*, *158*, 162, 241, 269
  development of, 129, *130*
  entrainment to abnormal "days," 151, 152
  entrainment by light-dark cycles, *164*
  "entrainment" by 10 Hz field, *158*, *159*
Snails, mud
  monthly rhythm, 222
  orientation
    effect of magnetic field, 222
    effect of electric fields, 232, 233
  oxidative metabolism rhythm, *258*
Snowshoe rabbit, photoperiodic response, 207
Solanum, *see* Potato
Soybean, photoperiodic response, 193, 194
Space, persistence of rhythms in, 147, *148*
Spatial orientation, *see* Direction finding
Star-compass orientation, 187
Starlings, 182, 183
Strychnine, rhythmic sensitivity, 88
Suncompass orientation, *see* Direction finding
Tail-waggle dance, 175, *176*
  marathon dancing, 178, 179
Talitrus, *see* Sand hopper
Temperature
  effect on period, *17*, 23, 25, 29, *46*, 48, 54, 57, *71*, 73, 76, 83, 97, 213, 252, 253
  effect on phase, *11*, 15, 16, *23*, 48, *73*, *74*, 249, 250, 253, *286*, 287, 291, 312
  entrainment by, *11*, *15*, 16, *23*, *73*, *113*
  initiation of rhythms, 97, 98, 100
  pulse experiments, *74*, 98, *100*, 291, 312
Temperature phase setting rhythm, *11*, 15–17, 19, 23, *74*, 249, 250, 253, *286*, 287, 291
Temperature rhythm
  body, *128*, 131, *132*, *133*, *140*, *150*, 162, 241, 269
  entrainment by abnormal "days," 135, 152
    light-dark cycles, *164*
    10 Hz electric field, *159*
  heart beat, 145, 146
  as related to performance, *139*
  as related to time perception, *136*, *137*, 138
  during sleep deprivation, *134*
Theophylline, 354
Tidal rhythms, by organism
  fiddler crab, *2*, 3, 94, 108, *109*
    green shore crab, *97–99*, *100*, *112*, 113
    penultimate-hour crab, 95, *96*, *97*, 108
    sand hopper, 100, 101, *102–104*, 108, 115, *116*
  definition of, 3
  entrainment by, inundation cycles, 111, *112*
    light-dark cycles, 69, 108–*110*, 245, 292, *293*, *294*
    mechanical agitation cycles, 115, *116*
    pressure cycles, 114
  innate nature of, 98, *100*
  phase-setting of, *117*
Time perception rhythm, as a function of body temperature, *136*–138
Time sense
  bees, 75–*77*, 238, 239
    effect of translocation, 76
    temperature insensitivity, 76
    subterranean aspects of, 76
  man, *136*–138
Tobacco, photoperiodic response, 194
Transients, 73, 74, 241, *286*, *290*

Translocation experiments
  bee, 76
  fiddler crab, 85, 216
  oyster, 215, *216*
  sand hopper, *189*
Transmeridian travel, *see* Longitudinal travel
Transplant
  heart, 146, *147*
  kidney, 152
Uca, see Fiddler crab
Valinomycin, 34, 49, *335, 336*, 337
Vascular plants, rhythmic cell division, 49, 51, 143
Vertical migration rhythms, 98, 101, *105, 106*
Vole, photoperiodic response, 207
Volunteering, monthly rhythm in, 170
Wasp, parasitic, *202*
Water uptake rhythm in bean
  annual rhythm, 270, *272*
  monthly rhythm, 266, *268*, 269
  role of magnetic field, 223
White-crowned sparrow, activity rhythm, *67*, 68, *70*
Woodchuck, annual feeding rhythm, 270, 272
x-Irradiation, rhythmic sensitivity, 88
Yeast, DPN reduction rhythm, 305, 324, *325*, 327, *340*
Zeitedachtnis, 75
Zeitsinn, 75
Zugunruhe, 66, *67*, 181

A 6
B 7
C 8
D 9
E 0
F 1
G 2
H 3
I 4
J 5